Mostly Cloudy

Cloud Physics for Meteorologists

Sonia Lasher-Trapp

DEPARTMENT OF
CLIMATE, METEOROLOGY & ATMOSPHERIC SCIENCES
UNIVERSITY OF ILLINOIS URBANA-CHAMPAIGN

Sundog Publishing LLC
Madison, Wisconsin

Library of Congress Control Number: 2024946253

Ordering information and online resources: Instructors and students are encouraged to visit sundogpublishing.com to take advantage of substantial savings off the retail price for direct orders and to view errata and other resources. Examination copies may be requested by qualifying instructors at sundogpublishing.com/pages/instructors-examination-copy.

For prospective authors: Founded by a university professor, Sundog Publishing offers a non-traditional, author- and student-centered model for publishing quality academic textbooks. Visit sundogpublishing.com/blogs/news/for-authors for more information.

ISBN (print): 978-1-944441-00-5
ISBN (electronic): 978-1-944441-03-6

Sundog Publishing LLC
2911 Robin Ct
Madison, Wisconsin 53711
United States

Printed in the United States of America.

10 9 8 7 6 5 4 3 2 1

Cover photo and design by Grant W. Petty

This book was typeset in LaTeX.

Dedication

To all my former instructors, as well as past and current mentors, colleagues, and students, as an expression of my gratitude for the knowledge you have shared with me, and for giving me opportunities to work with you to gain more.

Preface

At present, many undergraduate atmospheric science programs across the U.S. do not require a full semester course in cloud physics, nor do many graduate programs. Instead, they may include cloud physics within a single-semester general physical meteorology course that must also cover atmospheric radiation and/or atmospheric thermodynamics. Unfortunately, some programs may contain very little education in cloud physics.

Why is a working population of atmospheric scientists knowledgeable in cloud physics needed? I would argue, for example, that:

- Our current daily weather forecasts based on numerical weather prediction models exhibit significant skill in forecasting the likelihood of precipitation falling on a given day, but enhanced understanding of precipitation processes is needed to improve their poorer performance in predicting precipitation accumulation at the ground;
- Our ability to correctly interpret radar and satellite data to understand weather phenomena depends upon an understanding of the microphysical processes within the clouds;
- Our ability to understand the current climate and to predict climate change depends upon the radiative effects of clouds and their influence on the hydrologic cycle, which are in turn controlled by cloud types, their persistence, and their precipitation.

This book aims to serve as a resource for acquiring foundational knowledge in cloud physics and physical meteorology. It is suitable for independent study, as part of a physical meteorology course, or even for a full semester course in cloud physics. Designed as a concise, affordable, and easy-to-follow guide, this book is intended for advanced undergraduate students but should also be valuable for graduate students and scientists new to meteorology and cloud physics.

Although the field of cloud physics is primarily dominated by physicists, some researchers, including myself, were educated and trained in meteorology programs and also teach in such programs. This book strives to fill a unique niche by serving as a "bridge" between these fields, placing the microscale topics of cloud physics within the context of the larger-scale courses typically taught in a meteorology curriculum. Many meteorology students are delighted to discover an entirely new aspect of clouds that they had not previously considered, and some have even decided to pursue careers in this direction. I hope to encourage more students to do the same.

This book addresses theoretical concepts, utilizes observations collected in nature or in laboratories, and frequently discusses numerical modeling at both detailed and larger scales. The field of cloud physics necessarily depends on all three approaches, and it is essential that students new to cloud physics are introduced to them all. Relying on multiple methods of scientific inquiry facilitates understanding and learning, as well as advances the frontiers of our current knowledge.

I regret that I could not include a chapter on the importance of electric fields within clouds and the influence of electrically charged hydrometeors on the processes leading to precipitation. Much is still to be learned from a concerted and coordinated effort in merging the fields of cloud physics and atmospheric electricity. The absence of this topic is not intended to diminish its importance or discourage future study, which is greatly needed. Rather, it reflects the substantial effort—and thus, the considerable amount of time—that would have been required to produce an up-to-date summary from the current scientific literature that is both thorough and comprehensible to an undergraduate audience.

I am indebted to many people who helped bring this book to fruition. First and foremost is the contribution and consistently enthusiastic support of its publisher, Dr. Grant Petty. He carefully edited this book through multiple iterations, sometimes suggesting additional material to make it more complete, and even contributing figures and ideas of his own. He was essential in making the dream of publishing a quality, low-cost textbook for students worldwide a reality. We have both been aware of the lack of an affordable, comprehensive undergraduate textbook on cloud physics for our students; we hope we have remedied this situation.

Invaluable feedback was provided by multiple faculty colleagues who "test drove" earlier versions of this book in their classes. These individuals include Professors Angela Rowe, Sandra Yuter, and Nicole Riemer. I cannot thank them enough for their suggestions and encouragement.

Of course, I also used earlier versions of this book in my own classes, and I am grateful to those students who endured the abundant typos, some errors, and some awkwardness in the text of earlier versions. Often, by addressing their questions, I recalled things I had forgotten or realized I had gaps in my own knowledge that needed to be remedied.

Other colleagues agreed to review single chapters of this book. These individuals include Dr. Jeff Frame, Dr. Steven Krueger, Dr. Angela Rowe, Dr. Jeff French, Dr. Alexandria Johnson, Dr. Jennifer Griswold, Dr. Nicole Riemer, Dr. David Mechem, Dr. Adele Igel, Dr. Sandra Yuter, Dr. Alicia Klees, and Dr. Eric Nielsen. Their feedback was invaluable in improving the text, content, and figures.

While working on this book during my sabbatical at the University of Bern, I also had the opportunity to receive feedback from colleagues for whom English is a second language: Lena Wilhelm, Dr. Monika Feldmann, Dr. Severin Kaderli, Dr. Alice Portal, Dr. Edgar Dolores-Tesillos, and Duncan Pappert. Their assistance helped make this book more understandable to readers worldwide.

The culmination of all this effort lies before you. I hope you find this textbook enjoyable to read and useful for your future scientific pursuits.

Sonia Lasher-Trapp
University of Illinois Urbana-Champaign, December 2024

Contents

CHAPTER 1

Cloud Classification and Characteristics

Clouds are ubiquitous throughout the Earth's atmosphere. At any given time, various types of clouds can be observed around the globe, as shown in Fig. 1.1. They range from thin clouds at either low or high altitudes to deep clouds spanning nearly the entire vertical depth of the troposphere. These clouds may contain liquid water, ice, or both. Some produce rain at the Earth's surface, while others produce snow or even hail. Most do not precipitate at all.

Clouds are indicative of various atmospheric conditions and weather patterns at specific times and locations. Synoptic-scale lifting of air, due to the passage of low-pressure systems and their associated fronts, often creates a variety of cloud types, depending on the local stability of the atmosphere. The buoyancy of air due to solar heating frequently leads to cloud formation, as does orographic lifting caused by winds blowing against terrain. Owing in part to the movement and evolution of weather systems and the atmospheric conditions they induce, clouds and the precipitation they produce have finite lifetimes.

Clouds are not just passive products of current atmospheric conditions; they actively alter the state of the atmosphere by transporting and processing heat, moisture, atmospheric particulates, and gases. They also affect the Earth's radiation balance: some clouds provide effective "shade," cooling the Earth, while others act like a thermal "blanket," warming it.

We begin our study of cloud physics at larger scales, ranging from a few to as many as hundreds of kilometers. The goal is to begin building a bridge from your prior meteorological knowledge at these scales to

Fig. 1.1: The Blue Marble. *Courtesy of NASA Visible Earth.*

the much smaller, detailed microscale processes that are the main subject of this book. These smaller scales range from centimeters and millimeters down to even micrometers. Though tiny by meteorological standards, they are central to the formation of droplets and ice within clouds, and thus for the larger-scale properties of clouds, including their ability to produce precipitation.

1.1 International Cloud Classification Scheme

As scientists, it is essential to communicate clearly and effectively about the phenomena we observe or study. Therefore, we develop classification schemes that are universally used within a specific field, similar to those in biology for categorizing different plants or animals.

In this regard, atmospheric science is no different, and the need for classifying cloud types was recognized long ago. Luke Howard first pub-

Fig. 1.2: *Left:* Shallow cumuli (*cumulus humilis*) over a beach in Miami, Florida. *Right:* A much deeper and more energetic *cumulus congestus*, as seen from a commercial aircraft.

lished a cloud classification scheme in 1803. Although others existed at the time, Howard's scheme was adopted by a global organization in 1896, with some modifications. Today, the World Meteorological Organization (WMO) maintains the **International Cloud Atlas** and has a panel that can approve new cloud types. The current cloud classification scheme is available online, with color pictures and descriptions, at cloudatlas.wmo.int.

1.1.1 Roots of the classification system

The International Cloud Classification Scheme uses a "tree-like" structure beginning with *genera*, which are further subdivided into *species* and *varieties*. Clouds may be categorized solely by their genera, or with more detail using the name of their genera followed by the name of their species or variety, or both. Categories of *supplemental features* and *accessory clouds* are also included. The former descriptor *étages* is also sometimes still used. The upper level of the classification scheme, the genera, consists of names with five Latin roots: *cumulus, stratus, cirrus, nimbus,* and *altus*. We begin by addressing the meanings of these roots.

i) Cumuliform clouds

The root **cumulus** refers to an "accumulation," in the sense of a heaping up of something. Clouds with a cumulus-like character are called **cumuliform**. Many cumuliform clouds exhibit an appearance (Fig. 1.2) often described as "fluffy," "piled up," or "cauliflower-like." Individual cumuliform clouds range from having depths of a few tens of meters, with similar or greater widths, to having both depths and widths of 10 kilometers

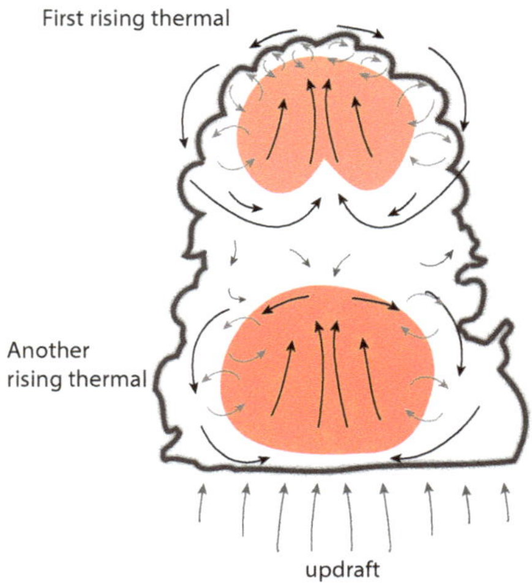

Fig. 1.3: Simplified depiction of two thermal circulations within an actively growing cumulus cloud. Red shading shows each rising thermal with its larger overturning circulations shown as black arrows. Circulations of smaller turbulent eddies within the cloud are shown as gray arrows. Air within the updraft supplies the cloud with heat and moisture.

or more. Some may also have much greater depths than widths, being several kilometers tall but only a kilometer or less wide.

Cumuliform clouds result from localized rising plumes of buoyant air—often referred to as **thermals**. As buoyant air rises, ascending motion occurs at its center, comprising the cloud *updraft*. Diverging winds near its top force air outward. Mass continuity necessitates some compensating descending air motions along the sides of the thermal. Converging winds bring air back toward the base of the thermal. All actively growing cumuliform clouds consist of one or more of these overturning *thermal circulations*. Consequently, these clouds are often referred to as **convective clouds**, highlighting their vertical overturning motions due to the buoyant air they encompass.

Two examples of thermal circulations are shown in Fig. 1.3 for a hypothetical, moderately deep cumuliform cloud. The first (older) thermal circulation is collocated with the ascending top of the cloud. A second (newer) thermal circulation can follow it from beneath and push the cloud top even higher. Each primary thermal circulation breaks down in time into smaller and smaller overturning eddies, resulting in the cauliflower-like appearance of cumulus clouds (Fig. 1.2, right).

Cumuliform clouds exhibit the largest vertical accelerations of any clouds due to the buoyant air they contain. Additionally, they contain more *turbulence*, i.e., random fluctuations in wind velocity, compared to other cloud types.

Cumuliform clouds can have very short lifetimes, sometimes lasting twenty minutes or less, or much longer, with some extreme cases enduring for several hours. Certain cloud complexes containing cumulus elements can persist for six hours or more. The precipitation produced by cumuliform clouds tends to take the form of *showers*, i.e., relatively short-lived precipitation lasting from minutes to an hour. Showers are often characterized by their sudden onset and cessation, which can result from the movement of localized precipitating cumuli over a stationary observer on the ground, or a precipitating cloud with a very short lifetime. This showery precipitation can vary greatly in intensity from low to very high, sometimes greatly exceeding 75 mm (approximately three inches) of accumulation in less than an hour.[1] The most intense precipitation rates found on Earth are produced by cumuliform clouds for reasons we will see in the remainder of this book.

Cumuliform clouds can be excellent "vertical mixers" of the atmosphere. Due to the significant depths they can attain and the substantial vertical motions within these depths, they can transport large volumes of air within their updrafts several kilometers or more above Earth's surface, sometimes even throughout the entire troposphere and into the lower stratosphere. The downdrafts of the deepest cumuliform clouds can also carry air from the mid-to-upper troposphere back down to Earth's surface. Measurements and numerical simulations have demonstrated that these clouds play a critical role in the vertical transport of heat, momentum, moisture, trace gases, and atmospheric particles within the troposphere and sometimes even into the lower stratosphere.

ii) Stratiform clouds

The root **stratus** means to "spread" or "flatten out," or "cover with a layer." Clouds having this general character are called **stratiform**. Stratiform clouds are thus layered clouds (Fig. 1.4) that can cover tens or hundreds of kilometers in horizontal extent. They typically have much smaller vertical depths compared to cumuliform clouds, ranging from less than a kilometer to one or two kilometers, but occasionally far deeper (see *nimbostratus*, p. 11).

Stratiform clouds usually result from gradual cooling and/or moistening of stable air; that is, air having temperature and humidity profiles that are unfavorable for convective motions. They exhibit much slower

---

[1]The world record for the greatest amount of rainfall in one hour is 304.8 mm (12 inches), which fell in Holt, Missouri on 22 June 1947 over a period of 42 minutes.

Fig. 1.4: *Left:* Two stratiform cloud layers over Southwest England, as seen from the cockpit of a research aircraft flying between the layers. The top layer is an *altostratus*, while the bottom layer is a *stratocumulus*. *Right:* A *stratocumulus* over London, England, as viewed from the ground.

vertical air motions compared to cumuliform clouds, resulting in reduced turbulence. These clouds can persist for many hours or even days.

Precipitation from stratiform clouds tends to be less intense than from cumuliform clouds and to generally be more spatially uniform. The large horizontal extent of these clouds, along with their potentially long lifetimes, can produce precipitation of long duration, lasting sometimes several hours or longer at any given location.

Stratiform clouds, characterized by their shallowness and weaker vertical motions, typically mix atmospheric constituents over much smaller vertical distances than cumuliform clouds. However, they can facilitate the transport of air over extensive horizontal distances through slow, slantwise ascent, especially in association with frontal systems. This horizontal transport can span much greater distances than the vertical transport facilitated by cumuliform clouds.

iii) Cirriform clouds

The Latin word **cirrus** means "a curl, lock, or tuft of hair", and can also refer to the shape of a horse's tail. It thus aptly describes the delicate, wispy appearance of **cirriform** clouds (Fig. 1.5) found in the upper troposphere. These clouds form at such high altitudes and low temperatures that they usually consist entirely of ice crystals. An elongated or stretched-out appearance may result from the strong vertical wind shear commonly found at these high altitudes.

Cirriform clouds can result from local or broad areas of dynamically lifted and cooled air, such as along the upper levels of a front or within

Fig. 1.5: *Top: Cirrocumulus* (left) and *cirrus uncinus* (right) clouds during the day. *Bottom: Cirrostratus* at sunset near St. Louis, MO.

particular quadrants of a jet streak, or result from local instability in the upper troposphere. They can also manifest as the remnants of deep thunderstorms, or even from aircraft exhaust (referred to as *contrails*). Cirriform clouds tend to contain vertical motions with strengths lying between the extremes of stratiform and cumuliform clouds. Their lifetimes can vary significantly; for example, some are composed of tiny individual elements that may only last thirty minutes, while other more widespread forms can persist for several hours or longer.

Cirriform clouds can weakly precipitate ice crystals that fall from their bases as *fall streaks*, but these ice crystals sublimate (convert back to water vapor) in the drier air below the cloud long before reaching the distant ground. Precipitation that evaporates or sublimates before reaching the Earth's surface is referred to as **virga**.

iv) Additional roots

Two more roots are used to complete the cloud classification scheme. The root **altum**—which is Latin for "height" and which is closely related to the word "altitude"—is used for clouds having bases in the middle of the troposphere. The root **nimbus** specifies that a particular cloud is producing precipitation falling toward the surface of the Earth. These two roots are never used as cloud types alone but instead modify one of the other three cloud roots discussed earlier.

1.1.2 Genera

From combinations of the five Latin roots described in the previous section (cumulus, stratus, cirrus, alto, nimbus), ten different cloud **genera** comprise the cloud classification scheme. These ten genera are depicted in Fig. 1.6.

We classify clouds partly by the height of their bases. Many cloud characteristics are influenced by the tropospheric air properties near their bases, a topic we will discuss in future chapters. The ten cloud genera fall into three different height ranges, sometimes referred to by weather observers as **étages**. Cloud base heights are conventionally reported in terms of their distance above ground level (AGL).[2]

Low clouds have bases ranging from ground level to about 2 km above it. Four of the ten cloud genera have the low bases characterizing this étage: cumulus, cumulonimbus, stratus, and stratocumulus (Fig. 1.6, lower section). *Mid-level clouds* have bases between approximately 2 and 7 km AGL, and consist of the two genera containing the prefix "alto": altocumulus and altostratus, as well as nimbostratus (middle section). Finally, *high clouds* have bases ranging from approximately 7 to 13 km AGL. This étage only includes the three genera of cirriform clouds: cirrus, cirrocumulus, and cirrostratus (upper section).

The ranges of cloud base heights just described for each étage are appropriate for midlatitude clouds. Outside of the midlatitudes, the ranges of the two higher étages differ in polar and tropical regions, due to the respective differences in typical troposphere depths with latitude. For both mid-level and high clouds, the upper limit of the range of cloud base

[2] In mountainous terrain where there is no unambiguous ground level, the heights might more appropriately refer to values above sea level.

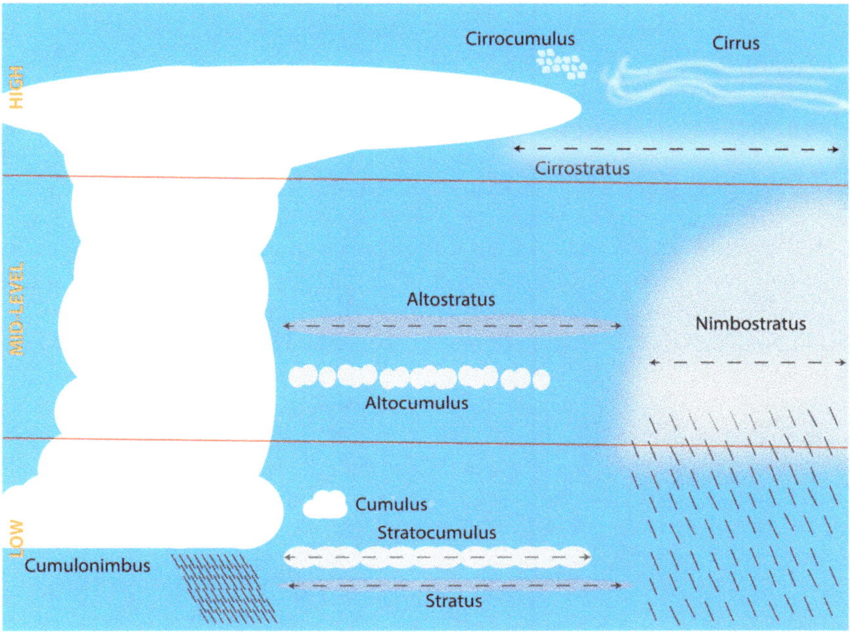

Fig. 1.6: Simple depiction of the ten genera of clouds, where the heights of their bases determine whether they are considered low, mid-level, or high clouds, as labeled on the far left in yellow.Opacity of the pictured clouds denotes their relative maximum magnitudes of liquid and/or ice mass content. Horizontal dashed arrows in the stratiform clouds denote that these clouds should be pictured with much greater horizontal extent than shown here.

heights can be one to three kilometers lower in polar regions than in midlatitude regions, and in tropical regions, it can be one to three kilometers higher.

Knowledge of these ten cloud genera is often sufficient for communicating the most important scientific information. It is recommended that students use the online International Cloud Atlas and practice identifying these clouds. To assist in this effort, we now briefly examine the general characteristics of the ten cloud genera. The following descriptions are informal and cover the most common cases.[3]

---

[3]For the formal criteria used to distinguish cloud genera, visit cloudatlas.wmo.int/en/clouds-genera.html.

i) Low clouds

Stratus (St) typically has a thickness ranging from around 100 to 500 meters. It is generally shallow compared to other cloud types, often forming in uniform, featureless layers that can cover large areas of the sky. Its base may be much closer to the ground than is common for other cloud types; indeed, fog is essentially a stratus cloud having a base touching the Earth's surface. Stratus clouds contain very weak updrafts, typically ranging from 1 to 10 cm s^{-1}. They also contain significantly lower **liquid water content** or **ice water content**—defined as the mass of liquid water or ice per unit volume—than many other clouds, generally around 0.1 g m^{-3}. Stratus clouds are typically not considered turbulent. They often lack precipitation; any drizzle or snow they may produce is very light.

Cumulus (Cu) is characterized by distinct separations between individual elements and appears as rising mounds with flat bases. The height of its base can range from a few hundred meters AGL in humid air masses to at least two kilometers in dry, desert air masses. Its dimensions can range from less than one kilometer to several kilometers in width and/or depth. The updrafts of cumulus clouds generally exceed 1 m s^{-1} but are less than 10 m s^{-1}. These clouds often contain liquid and/or ice water content exceeding 1 g m^{-3}. Over land, they very rarely precipitate, but cumulus over the ocean may produce small rain or snow showers.

Stratocumulus (Sc) forms in a shallow layer like stratus, but contains readily visible structural elements, which may be rolls, ripples, or cumulus-like puffs. These structures usually result from overturning within the cloud layer due to shallow buoyant convection and/or mechanical mixing associated with boundary layer turbulence. Stratocumulus can also form from the horizontal spreading of remnants of dying cumulus clouds. Stratocumulus clouds tend to have updraft speeds and liquid water or ice contents that are comparable to those for shallow cumulus clouds. They can also sometimes produce light precipitation (drizzle or light snow) at the Earth's surface.

Cumulonimbus (Cb) produces precipitation at the Earth's surface, often copiously[4] and at rates greater than any other cloud genus. These clouds can extend over the entire depth of the troposphere and may even penetrate through the tropopause into the lowermost stratosphere. A key distinguishing feature relative to deep cumulus clouds is that the tops of cumulonimbus clouds tend to be *glaciated*, i.e., consisting entirely of ice, conferring a cirriform appearance there (often referred to as an *anvil*) that is quite different from the "hard" cauliflower appearance of shallower cumulus.

---

[4]Occasionally, cumulonimbus clouds in dry desert air masses may have very high bases and produce only virga, with rain fully evaporating before reaching the surface.

Typical cumulonimbus cloud widths range from a few kilometers to ~10 km. The deepest cumulonimbus clouds contain the strongest updrafts of all cloud types on Earth, ranging from 10 to 50 m s^{-1}. They also contain the greatest liquid water and ice contents, often exceeding 5 g m^{-3}, and the most turbulence, making them hazardous for aircraft to intercept. Cumulonimbus clouds are responsible for virtually all lightning, hail, and significant tornadoes.

This genus encompasses all thunderstorms, from the most mundane to the most severe. *Supercell thunderstorms*, although not a formal cloud type within the International Cloud Classification Scheme, are notable because the air ascending within the storm updraft also rotates around a vertical axis, either cyclonically or anticyclonically. The rotating updraft produces unique dynamical motions within the storm that increase the storm's lifetime and sometimes generates large tornadoes, damaging winds, and/or extremely large hailstones.

Cumulonimbus clouds can also be components of larger-scale groups of clouds that produce severe weather. These groups of clouds are not named in the International Cloud Classification scheme, as they usually contain some stratiform and cirriform cloud elements. Examples of these larger-scale phenomena are *mesoscale convective complexes*, *mesoscale convective systems*, and *hurricanes*.

ii) Mid-level clouds

Altostratus (As) has its base in the middle étage and is otherwise very similar to stratus, being mostly featureless and containing similarly weak vertical velocities and small liquid water or ice water content. Thin altostratus is sometimes somewhat translucent so that a blurry sun can be vaguely seen through it. Altostratus clouds often result from large-scale lifting of stable, humid air at mid-levels, typically in connection with an advancing warm front. Altostratus is a common precursor to nimbostratus.

Nimbostratus (Ns) is darker and more diffuse in appearance than other stratiform genera, as it produces mostly continuous precipitation (rain or snow) and completely blocks out the sun when observed from the ground. The cloud base may be difficult to distinguish due to falling precipitation. Because they often originate as altostratus clouds that deepen over time and begin to precipitate, their bases frequently grow downward into the low étage, sometimes approaching the ground. It follows that nimbostratus clouds are typically much deeper than either stratus or altostratus clouds, often several kilometers deep or more. Like other stratiform clouds, they contain weak vertical velocities, but their greater depths allow them to have larger liquid water or ice water contents than other stratiform clouds, but still less in magnitude compared to cumuliform clouds. Compared to cumulonimbus clouds, their precipitation is of lower intensity and tends to be widespread and long-lived, with smaller raindrops than those produced by cumulonimbus clouds.

Altocumulus (Ac) combines a layer-like quality with structural elements that vary depending on atmospheric conditions. These elements may appear cumuliform when instability is present, or as rolls and ripples caused by vertical wind shear acting on stable air, analogous to wind-driven ripples on a pond's surface. In fact, of all the genera, altocumulus arguably encompasses the widest range of visual cloud forms. Altocumulus is usually more akin to stratocumulus than to cumulus, the most important difference being its higher base, but with similar vertical velocities and liquid water or ice water contents. A rule of thumb used by weather observers is that altocumulus elements are between about 1° (about a finger's width at arm's length) and 5° (about a hand's width) when viewed overhead, while stratocumulus elements are larger than 5°.

iii) High clouds

Cirrocumulus (Cc) contains cumuliform elements or ripples that indicate a shallow layer of unstable air and/or wind shear at that altitude. As such, it may be regarded as the cirriform analog of altocumulus, except cirrocumulus will contain only ice. As seen from the ground, elements have an angular diameter smaller than 1°. They thus appear small to the eye, but these elements have diameters around 100 m, entirely comparable to the elements seen in stratocumulus and altocumulus. Cirrocumulus may undergo a life cycle in which it begins as a field of separated cells or ripples that eventually weaken and become more diffuse, eventually transforming into ordinary cirrus. Cirrocumulus tends to have vertical velocities on the order of 0.1 m s^{-1} and can be mildly turbulent due to the overturning air that produces it. Its ice water content is on the order of 0.1 g m^{-3}. Due to its great height, it is incapable of producing precipitation at the ground but may produce virga.

Cirrus (Ci) consists of separate fibrous elements containing only ice. If vertical wind shear is present over the cloud depth, part of the cloud can be advected great distances downstream from its source region. As a result, it can extend hundreds of kilometers or more in width. Cirrus is typically, but not always, less than 1 kilometer deep. Vertical motions within cirrus would tend to be on the order of 0.01 m s^{-1}. Because of the very limited amount of water vapor available in the cold upper troposphere where cirrus resides, the water content of cirrus (in the form of ice) is the smallest of all cloud types, in the range of ~0.01 to 0.1 g m^{-3}. Cirrus can be initated by wave motions propagating through the atmosphere at high altitudes or be the remnants of decaying cirrocumulus. It is not as turbulent as cumulus and typically does not produce virga.

Cirrostratus (Cs) is a thin, sheet- or veil-like ice cloud occurring at the typical heights of cirrus, usually without much if any visible structure. It is never thick enough to strongly dim the sun, distinguishing it both from darker, grayer altostratus and from the dense anvil of a dying cumulonimbus. It would tend to have vertical velocities on the order of 0.01 m s^{-1}, and ice water content comparable to cirrocumulus. Cirrostratus is not turbulent, and does not produce virga. A unique hallmark of cirrostratus is the common appearance of a 22° halo (white luminous ring of light) centered on the sun. Cirrostratus stereotypically arises far in advance of an approaching warm front where high-level humid air is gliding up and over a colder air mass. In that situation, it may slowly extend downward to produce altostratus and ultimately nimbostratus. It can also result from a remnant cumulonimbus anvil that has thinned and expanded over time.

1.1.3 Species, varieties, supplementary features, and accessories

Further classification of clouds beyond their genera provides additional details. These details may be important especially for aviation purposes or for communicating cloud evolution over the course of a day or night.

Species are generally based on the cloud shape or internal structure. Some are suggestive of the clouds building, filling in, or decaying, and thus are useful to forecasters and aircraft pilots. Examples include the appearance of *altocumulus castellanus*, which might signify that deeper cumulonimbus clouds will develop soon, or the sudden transformation of *stratocumulus castellanus*, which has deeper cumuliform elements, to *stratocumulus fractus*, resulting from the cloud breaking up into smaller elements and soon dissipating. A few examples of species are given in the top section of Table 1.1, but many more exist.

Varieties can help identify visual features of clouds, particularly (but not exclusively) those related to transparency. For example, *stratus opacus* clouds are thicker and block the sun, while *stratus translucidus* clouds are thinner and not only allow more sunlight to reach the ground but also sometimes allow other clouds to be seen above them. These distinctions have been historically important in aviation weather observations for assessing the potential impact of cloud layers on flight safety. While they can potentially provide useful information for studies of radiation and climate or plant growth or for predicting daily maximum temperatures, other cloud characteristics—such as genus, altitude, and sky coverage—are more commonly referenced for these broader applications. Some examples of varieties are given in the bottom section of Table 1.1.

Finally, contained within the International Cloud Classification Scheme are **supplementary features** that clouds can have, as well as **accessory clouds**, which are smaller clouds that may accompany or merge with a larger cloud. These features and clouds are typically identified by their names without specifying the cloud they inhabit or accompany. A few common examples are provided in Table 1.2, but more extensive lists can be explored further by consulting the online WMO Cloud Atlas.

Table 1.1: Common species and varieties of clouds.

Species	Meaning	Example
uncinus	hooked	cirrus uncinus
castellanus	castle	altocumulus castellanus
floccus	tuft of wool	cirrus floccus
lenticularis	lentil	altocumulus lenticularis
fractus	to fracture	stratus fractus
humilis	near ground, small	cumulus humilis
mediocris	moderate vertical depth	cumulus mediocris
congestus	to heap up	cumulus congestus

Variety	Meaning	Example
undulatus	having waves	stratus undulatus
vertebratus	having vertebrae	cirrus vertebratus
translucidus	translucent	altostratus translucidus
opacus	shadowy, thick	altostratus opacus
spissatus	dense	cirrus spissatus

Table 1.2: Common cloud supplementary features and accessory clouds.

Supplementary Features	Meaning
incus	anvil
mammatus	udder
virga	stick
arcus	bow, arch

Accessory Clouds	Meaning
pileus	cap
pannus	shred of cloth

Fig. 1.7: *Left:* A field of *cumulus mediocris* in Indiana. *Right:* A *cumulonimbus* in Illinois, including an anvil at its top.

Although the information provided by the visual classification of all these cloud types is useful to weather forecasters, aviation, and climate change research, the increasing automation of surface weather stations around the world is eliminating much of this detail in hourly observations. Researchers are currently using artificial intelligence algorithms based on live camera imaging to try to reproduce this information.

1.1.4 Insights from cloud classification

The International Cloud Classification Scheme is based on the visual appearance of the cloud, but it also communicates other information important for weather and climate issues, briefly discussed here.

i) Large-scale environmental conditions

The cloud type communicates information about the state of the atmosphere at that time. For example, shown in Fig. 1.7 (left) is a field of fairly shallow cumulus (*cumulus mediocris*). This cloud type and its scattered pattern across the sky suggest that the lower troposphere is mildly unstable due to warm thermals released from the land surface, likely caused by solar heating.

If these clouds were instead occurring along a single line, we might suspect an approaching cold front or another surface boundary where air is converging and forcing upward motion into the unstable air above it. On the other hand, if the cumulus develops more deeply into *cumulus congestus* and then *cumulonimbus* (Fig. 1.7, right), then we know that the troposphere has become unstable over a much deeper layer. The cumuliform element popping up above the anvil in the same photo is referred to as an "overshooting top," indicative of an extraordinarily strong updraft within

Fig. 1.8: New contrails, older contrails (more spread out) and other very thin cirrus over Indiana.

Fig. 1.9: *Stratocumulus* over the Grand Canyon.

the storm and thus extreme amounts of instability over the troposphere at that location.

In the upper troposphere, the relative humidity of that air can be qualitatively indicated by the longevity of cirrus clouds formed from contrails, as seen in Fig. 1.8. Higher humidity limits the evaporation that would otherwise cause their immediate disappearance after being formed by the aircraft exhaust, allowing them to mix with the surrounding air and spread out over time.

In the lower troposphere, the stratocumulus layer shown in Fig. 1.9 is indicative of a low-level *temperature inversion*, a layer over which temperature increases with height, located immediately above its top (not visible in this image). This situation is common after the passage of a cold front.

Students can gain valuable insights into the atmosphere by classifying a cloud and then exploring thermodynamic and synoptic-scale weather data to answer the question, *"Why is that cloud there right now?"* Chapter 2 takes a closer look at the various sources of clouds, offering deeper understanding of the unseen atmospheric processes that shape the clouds we observe.

ii) Implications for Earth's energy budget

Clouds are not only products of local weather processes; they also contribute to local, regional, and global weather and climate by influencing the energy balance of both the atmosphere and the Earth. Thus, the presence of a cloud at a given location and time should also motivate students to consider these effects.

Latent heating/cooling

First consider a non-precipitating cloud. When a cloud forms, water vapor condenses into liquid water, and latent heat is released into the air wherever that condensation occurs, warming the air. If the same cloud then later dissipates (evaporates), that heat energy is extracted from the air, and there is no net change in the internal energy of the atmosphere.

Now consider when a cloud produces precipitation that reaches the ground. The precipitation that owes its origin to condensed water vapor clearly has not all evaporated, so some of the associated latent heat energy previously released to the air during condensation has not been extracted by an equivalent amount of evaporation, but remains within the atmosphere. Precipitation thus results in a significant source of net heating for the atmosphere at the cloud level. On the other hand, precipitation evaporating as it falls beneath the cloud extracts latent heat over a depth of the atmosphere where no prior condensation or latent heat release has occurred. So precipitation is also a source of net cooling of the atmosphere beneath a cloud.

Radiative heating/cooling

A characteristic of almost all clouds is that they block some or most of the solar (shortwave) radiation that would otherwise reach and heat the Earth's surface, thus acting as "shade." This blocking is manifested mainly in the reflection of solar radiation back into space, where it is lost forever. Thus, the reflective properties of clouds, by themselves, have a cooling effect on the climate. Considering only the solar radiative effects, an increase in the cloud horizontal extent and/or cloud thickness (the latter sometimes referred to as the *brightness*) would be expected to lead to a reduction in global average temperature.

However, clouds are also efficient emitters of longwave (thermal infrared) radiation, with the intensity of the radiation they emit to space being proportional to the fourth power of the absolute temperature at their tops. Clouds with low, warm tops therefore emit much more longwave radiation to space than clouds with high, cold tops. The Earth's surface, which also emits longwave radiation to space proportional to the fourth power of its absolute temperature, is usually the warmest level in the atmospheric column, and thus it emits the most longwave radiation of all.

The presence of clouds, typically with colder tops relative to the Earth's surface, reduces the amount of longwave radiation escaping into space, thus acting like a "blanket." The magnitude of this reduction depends on the temperature difference between the Earth's surface and the cloud top. The colder the cloud top, the smaller the amount of Earth's longwave radiation that it will allow to be emitted to space. Any reduction in outgoing longwave radiation acts as a warming factor to the global climate, with the effect being most pronounced for clouds with the coldest tops.

Considering the impact of clouds on both shortwave and longwave radiation fluxes at the top of the atmosphere, it has been shown that clouds can exert either a net warming or net cooling effect on the global climate. This depends on whether the reduction in outgoing longwave radiation outweighs the reduction in absorbed solar radiation. At the current time, observations and numerical modeling suggest that clouds are producing a net cooling effect upon the Earth, but this may change as cloud frequencies or other characteristics change in a changing climate.

Thick, extensive, and highly reflective cloud layers —such as stratocumulus, broad fields of shallow cumulus, as well as the extensive lines of cumulonimbus in equatorial regions—generally have a net cooling effect. Their greater liquid/ice water content, greater depth, and greater horizontal coverage of the Earth increase their reflection of solar radiation compared to other cloud types. They are especially effective at reducing incoming solar radiation when they are present over an ocean surface, which alone is not very reflective.

In contrast, clouds with high, cold tops—especially cirrus, cirrostratus, and cumulonimbus—can have a net warming effect. The potential warming effect of cirriform clouds on the global climate is a result of having the coldest temperatures of all clouds, maximizing their reduction of outgoing longwave radiation emitted by the Earth's surface. Any increase in the pervasiveness of cirriform clouds, for example due to higher relative humidity in the upper troposphere, or an increase in contrails from high-altitude air traffic, would add to the factors known to be warming the Earth's climate.

On a more local scale, clouds also significantly influence the energy balance at the Earth's surface. During the daytime, clouds block varying amounts of solar radiation, reducing daytime maximum temperatures. At night, clouds trap longwave radiation, leading to significantly warmer nighttime minimum temperatures and making dew or frost less likely on the ground.

1.2 Other Cloud Categories

To understand and communicate differences in cloud characteristics that cannot be detected visually by a human observer, we also make use of a few other categories of cloud classification. These classifications are not part of the International Cloud Classification Scheme, but their labels often serve as useful indicators of particular cloud characteristics that are important for microphysical processes within them. These labels are typically used to convey their temperature, phase (liquid/ice), and/or the number of cloud droplets they contain.

1.2.1 Based on temperature

Because the temperature within the troposphere generally decreases with altitude, the cloud depth influences the internal cloud temperature. If the temperature within any part of the cloud is less than 0 °C, it may be referred to as a **cold cloud**, and microphysical processes that involve ice may need to be considered. However, this term is not universally applied; some scientists refer to cold clouds only as those having temperatures entirely below 0 °C. Conversely, clouds whose depths only span temperatures warmer than 0 °C are referred to as **warm clouds**, where microphysical processes involving liquid drops are primarily considered.

1.2.2 Based on phase

The temperature range spanned by a cloud throughout its depth influences which phases of hydrometeors can exist, even temporarily, within it. Cloud physicists may often use the shorthand descriptors **glaciated** or **ice clouds** to denote those clouds that contain only ice. It is no coincidence that these include the cold clouds described in the previous section, but these terms can also refer to only a particular region within a cloud that contains liquid elsewhere.

In contrast, **liquid** clouds contain only liquid water and generally overlap with the warm cloud category. However, many clouds contain **supercooled liquid water**, i.e., liquid droplets at temperatures below 0 °C, for reasons we will explore in later chapters. **Mixed-phase clouds** contain both water drops and ice. These clouds may have temperatures at their bases greater than or near 0 °C but temperatures at their tops much less than 0 °C. In some regions of a mixed-phase cloud, the liquid and frozen hydrometeors may be intermixed, while in other areas, there might be regions containing solely liquid or solely ice.

1.2.3 *Based on air mass origin*

Five typical categories often used to describe clouds are based upon the origin of the air mass in which they reside: *maritime, continental, transitional/coastal, urban,* and *Arctic.* The type of air mass implies general ranges of water vapor available for hydrometeor growth and, especially, general ranges of particle concentrations in the air that in turn can influence the number of droplets the clouds contain. Both are major factors in determining whether shallow clouds will be able to produce precipitation.

As we will discuss in Chapter 7, cloud droplets owe their existence to a subset of the natural and anthropogenic particles in the air that we call cloud condensation nuclei, discussed more thoroughly in Chapter 8.

The prevalence of these particles in a given air mass helps determine the number of small droplets within clouds forming in that same air mass. We quantify the number of droplets in the cloud using the *droplet number concentration,* typically expressed as the number of droplets per cubic centimeter of air. For example, a droplet number concentration of 100 cm^{-3} implies that 100 droplets would exist within any cubic centimeter of air within a cloud, whereas 1000 droplets would be expected within that same volume if the droplet number concentration is 1000 cm^{-3}. We often characterize clouds by the maximum droplet number concentration they contain.

i) Maritime clouds

Air masses over the world's oceans, far from land or other pollution sources, generally contain far fewer particles in the air than air masses over land. Therefore, typical maximum droplet number concentrations of clouds occupying maritime air masses can be quite low, between 50 and 200 cm^{-3}. Maritime air masses also normally contain abundant water vapor, especially in tropical regions. Describing clouds as **maritime** thus implies that they have low droplet number concentrations and exist in very humid environments.

ii) Continental clouds

Over land, air masses contain significantly more particles, thereby increasing maximum cloud droplet number concentrations there to values greater than 500 cm^{-3}. Air masses influenced by large cities contain even more particles, driving droplet number concentrations even higher within clouds in those air masses, often exceeding 1,000 cm^{-3}. Both continental and urban air masses can sometimes contain large amounts of water vapor, but might contain far less, depending on the season and their geographical location, including their proximity to bodies of water. Considering all these factors, using the term **continental clouds** communicates the expectation of high droplet number concentrations, while the term **urban clouds** would connote extremely high droplet number concentrations. However, owing to the large ranges possible, neither of these classifications is able to communicate a particular amount of water vapor.

iii) Coastal clouds

Along coastlines, clouds can be influenced by either maritime or continental air masses, depending on the wind direction. When affected by both air masses daily, as in the case of sea breeze circulations, these clouds typically exhibit intermediate droplet number concentrations, approximately between 200 and 500 cm^{-3}, reflecting a mixture of these two distinct air mass types. If the prevailing winds are off the ocean or off the land for several days, however, droplet number concentrations will be more characteristic of maritime or continental clouds, respectively. Coastal regions can also experience amounts of water vapor typical of maritime or continental air masses, depending on the direction of the wind. The category **coastal clouds** then implies that they have moderate droplet number concentrations, possibly extending to the smaller or larger values characteristic of maritime and continental clouds, and wide ranges of water vapor possible in their surrounding environment.

iv) Arctic clouds

Usually far removed from anthropogenic influences, **Arctic clouds** tend to have lower droplet number concentrations, possibly ranging from the low values typical of maritime clouds to the moderate values associated with coastal clouds. They would rarely have high droplet number concentrations like continental clouds, and especially not have values anywhere similar to those of urban clouds. Arctic air masses would generally contain less water vapor due to their lower temperatures, but Arctic clouds are still capable of producing precipitation.

Now that you are aware of the typical properties of different cloud categories, in the next chapter we discuss the thermodynamic principles and quantities needed to understand the conditions giving rise to these different types of clouds.

Questions for Review

1. What are the five Latin roots used in the International Cloud Classification Scheme, and what characteristics are associated with each root?
2. Name the ten cloud genera and the level (low, mid-level, high) of their bases.
3. What do the species and variety categories generally communicate about the clouds? Give some examples.
4. What are some examples of supplementary features and accessory clouds?
5. What can the presence of different cloud types indicate about the current atmospheric conditions at your location?
6. How do clouds actively alter the state of the atmosphere beyond being just a product of current conditions?
7. How do clouds affect Earth's energy budget?
8. What other cloud categories might cloud physicists use, and what characteristics does each of these categories imply?

Homework problems and other resources for students and instructors can be found at bit.ly/4dYBwB5, or scan this code:

Further Reading

The WMO International Cloud Atlas can be found at cloudatlas.wmo.int.

CHAPTER 2

Thermodynamics Review

Thermodynamics governs some of the larger-scale dynamical features characteristic of the cloud types discussed in Chapter 1 and plays a critical role in determining whether a cloud can exist at all. Here, we first review essential concepts in atmospheric thermodynamics that are key to understanding cloud formation and then examine cloud formation under different dynamic and thermodynamic conditions.

2.1 Adiabatic Parcels

A *parcel* of air is loosely defined as a mass of air with a small enough volume that its characteristics are uniform throughout, but large enough to contain a very large number of air molecules. While its volume is not precisely prescribed, a useful visualization might be based on the assumption of one million (100^3) cubic meters of air. The air outside this parcel is considered its *environment*. The thermodynamic properties of the parcel (e.g., pressure, temperature, water vapor content) are initially identical to those of the surrounding environment from which it originated, but the parcel properties can change as it moves horizontally or vertically.

The concept of an **adiabatic parcel** is commonly used in atmospheric thermodynamics. An adiabatic parcel is assumed to have no exchange of mass or energy with its environment. In the context of the following discussion, we will consider what happens to an adiabatic air parcel as it moves vertically in the troposphere.

2.2 Dry Adiabatic Lapse Rate

If an adiabatic air parcel is *unsaturated*, i.e., not starting within an existing cloud, then its thermodynamic state as it ascends or descends in the atmosphere can be well approximated by the First Law of Thermodynamics applied to dry air, in the form:

$$dq = 0 = C_p \, dT - \alpha \, dp, \tag{2.1}$$

where dq is the heating/cooling experienced by the air parcel in J kg^{-1}, C_p=1004 J kg^{-1} K^{-1} is the specific heat at constant pressure, dT is the temperature change of the parcel in kelvin, α is the specific volume of the parcel in m^3 kg^{-1} (equivalent to the inverse of the air density ρ), and dp is the change in the parcel pressure in pascals, where it is assumed that the pressure within the parcel instantly equilibrates to that of the environment when the parcel moves upward or downward.

For this unsaturated, adiabatic parcel, $dq = 0$, because no heat is supplied to or removed from the parcel as it moves, consistent with the definition of an adiabatic process. Any ascent of the parcel leads to a reduction in its pressure ($dp < 0$) as it equilibrates with the decreasing environmental pressure with height in the atmosphere. According to (2.1), this decrease in pressure is accompanied by a decrease in the parcel temperature ($dT < 0$).

We can substitute a different expression for dp by making use of the hydrostatic balance equation

$$dp = -\rho g \, dz, \tag{2.2}$$

where g is the acceleration due to gravity (9.81 m s^{-2}) and z is the height in meters, and we can then rearrange terms to find the rate of change of temperature with height as an unsaturated air parcel moves upward or downward in the atmosphere:

$$\Gamma_d = -\frac{dT}{dz} = \frac{g}{C_p}. \tag{2.3}$$

Γ_d is called the **dry adiabatic lapse rate** and has a constant value equal to 9.8 °C km^{-1}.

2.3 Moisture Variables

At the most basic level, the formation of clouds in Earth's atmosphere is determined by thermodynamic principles, particularly those governing water vapor **saturation**. Saturation is the condition where the water vapor and liquid water in a given volume of air are in exact equilibrium.

Clouds form when the air attains and exceeds saturation, causing excess water vapor to change phase, creating a mixture of air, water vapor, and water droplets and/or ice. Here, we review the various ways of characterizing both the absolute water vapor content of the air and its proximity to saturation.

2.3.1 *Saturation vapor pressures e_s and e_{si}*

The **saturation vapor pressure** e_s is the partial pressure of water vapor in air overlying a flat surface of pure liquid water that is required for both phases of water to be in equilibrium. Importantly, it is a function of temperature only, for reasons that are discussed in Sec. 6.1.2. It is given by the Clausius-Clapeyron equation, which can be approximated as

$$e_s = 611 \exp\left[\frac{L_v}{R_v}\left(\frac{1}{273.15} - \frac{1}{T}\right)\right] \tag{2.4}$$

where e_s is given in pascals, L_v is the latent heat of vaporization (= 2.5 x 10^6 J kg^{-1} at a temperature of 273.15 K), R_v is the gas constant for water vapor (= 461.5 J kg^{-1} K^{-1}), and T is the air temperature in kelvin.[1]

A similar equation can be derived for the saturation vapor pressure with respect to a flat surface of pure ice:

$$e_{si} = 611 \exp\left[\frac{L_s}{R_v}\left(\frac{1}{273.15} - \frac{1}{T}\right)\right] \tag{2.5}$$

where e_{si} is again given in pascals, L_s is the latent heat of sublimation (= 2.83 x 10^6 J kg^{-1} at a temperature of 273.15 K), and all other symbols are the same as in (2.4). The difference in the saturation vapor pressures with respect to liquid water and ice is revisited in Sec. 9.4.

2.3.2 *Vapor pressure e*

While the saturation vapor pressure is a theoretical value that depends only on the temperature, the **vapor pressure** e is a measure of the *actual* water vapor content of the air at a given time and location. It represents the pressure exerted exclusively by water molecules in the vapor state within a volume of air. It is given by the Ideal Gas Law for water vapor,

$$e = \rho_v R_v T \tag{2.6}$$

---

[1] The latent heats of vaporization, L_v, and of sublimation, L_s, are functions of temperature, but this temperature dependence is neglected in the integration used to produce (2.4) and (2.5). For research-level accuracy, (6.1) or (6.2) should be used instead.

where e must have units of pascals, ρ_v is the vapor density, i.e., the mass of water vapor per unit volume of air in kg m^{-3}, and all other variables are the same as in (2.4).

2.3.3 Relative humidity RH

The **relative humidity** of a parcel of air is defined as

$$RH = \frac{e}{e_s} \times 100\%. \tag{2.7}$$

This value is unitless, simply expressed as a percentage, since e and e_s share the same units, typically pascals.

The relative humidity with respect to ice, RH_i, can be expressed with a similar equation that substitutes e_{si} for e_s in (2.7), for temperatures less than 0 °C:

$$RH_i = \frac{e}{e_{si}} \times 100\%. \tag{2.8}$$

The relative humidity has the practical function of describing how close an air parcel is to saturation; by definition, *air at saturation has a relative humidity of exactly 100%*. Importantly, it depends *both* on the actual vapor content of the parcel (via e) *and* on the temperature of the parcel (via e_s). It is therefore not a good measure of the absolute vapor content of the air, because it varies strongly with changing temperature even if the vapor content remains fixed.

2.3.4 Dewpoint temperature T_d

Although vapor pressure, e, is a perfectly good measure of absolute vapor content, it is not measured or reported directly in standard weather observations. Even if known, vapor pressure alone does not intuitively indicate how close a parcel is to saturation at a given temperature. Furthermore, its value can change as the temperature and/or volume of the air parcel changes, even for constant absolute vapor content.

The closely related and commonly reported alternative is therefore the **dewpoint temperature**, T_d, which is the temperature to which air must be cooled isobarically, and without the addition of any more water vapor, to be brought to saturation. If the vapor pressure e is known, it can be calculated using (2.4), substituting e for e_s in that equation, and then solving for T. By performing such a calculation, we are effectively asking "At what temperature would the actual vapor pressure e equal the saturation vapor pressure e_s?" The answer is the dewpoint temperature, T_d.

The difference between the actual air temperature and the dewpoint temperature, $\Delta T_d \equiv T - T_d$, is called the *dewpoint depression*. Like the relative humidity, it conveys information about how close an air parcel is to saturation. When air is saturated, $T = T_d$, and $\Delta T_d = 0$. Otherwise, $T > T_d$ and $\Delta T_d > 0$.

2.3.5 *Water vapor mixing ratio* q_v

In atmospheric thermodynamics and cloud physics, yet another moisture variable proves very important: the **water vapor mixing ratio** q_v, which is defined as the ratio of the mass (or density) of water vapor in a parcel to the mass (or density) of dry air in the parcel.[2] This variable is particularly useful because it is conserved even when the air pressure and/or air temperature changes, provided no water vapor is added or removed from the air.

The mixing ratio is related to the vapor pressure e and atmospheric pressure p (both in the same units, such as pascals) by

$$q_v = \frac{0.622\, e}{p - e}. \tag{2.9}$$

As written, q_v is unitless. However, for convenience, the mixing ratio is often expressed in units of grams of water vapor per kilogram of dry air. These units are obtained by multiplying the result of (2.9) by 1000. For reference, when the dew point temperature near sea level is 21 °C (70 °F)—a typical summertime value in the central U.S.—then $q_v \approx 16\ \mathrm{g\ kg^{-1}}$.

2.3.6 *Virtual temperature* T_v

The buoyancy of an air parcel relative to its environment is often of great importance in cloud physics—for example, when considering the strength of cloud downdrafts (Sec. 12.2.1) as well as some processes of cloud formation discussed later in this chapter. Air that is less dense than its environment is *positively buoyant* and will ascend relative to it, while air that is more dense than its environment is *negatively buoyant* and will descend relative to it.

The density of air is a function not only of its temperature and pressure, but also of the amount of water vapor it contains. For the same temperature and pressure, air that contains more water vapor has a lower density relative to air that contains less water vapor. It is thus useful to define a

---

[2]The water vapor mixing ratio is closely related to a similar quantity, the *specific humidity*, which is the mass ratio of water vapor to *total* air, including the vapor component. Under normal atmospheric conditions, the mixing ratio and specific humidity are numerically almost identical.

variable that takes into account both temperature and water vapor combined when considering the buoyancy of different kinds of air at the same pressure.

We call this variable the **virtual temperature**, T_v, defined as the temperature that dry air (lacking water vapor) would need to have in order to have the same density as a given sample of moist air (containing water vapor) at the same pressure. It is calculated using

$$T_v = (1 + 0.61 q_v) T, \tag{2.10}$$

where q_v is the (unitless) water vapor mixing ratio and T is the temperature in kelvin. At any given air temperature T, a greater amount of water vapor increases its virtual temperature, increasing its buoyancy.

2.3.7 Saturation mixing ratio q_{vs}

For a given atmospheric temperature and pressure, we may obtain the **saturation mixing ratio** by simply replacing e in (2.9) with e_s:

$$q_{vs} = \frac{0.622\, e_s}{p - e_s}. \tag{2.11}$$

The saturation mixing ratio is the theoretical value of the water vapor mixing ratio required for the air to be just saturated. Like the mixing ratio q_v, q_{vs} is also commonly expressed in units of grams per kilogram.

Curves of constant q_{vs} are plotted on standard thermodynamic diagrams like skew-Ts, permitting numerical values of q_{vs} to be estimated at a glance for any combination of temperature and pressure. The same set of curves allows the determination of the actual mixing ratio q_v from dewpoint temperature and pressure.

It should be noted that the relative humidity (2.7) can also be approximated using the water vapor mixing ratio and the saturation mixing ratio:

$$RH \approx \frac{q_v}{q_{vs}} \times 100\%. \tag{2.12}$$

2.3.8 Saturation ratio S

We will also find useful a variable called the **saturation ratio**, S, defined as the ratio of the vapor pressure to the saturation vapor pressure of the air:

$$S = \frac{e}{e_s}. \tag{2.13}$$

This ratio is again unitless. It is identical to the equation for relative humidity (2.7) except that it is not expressed as a percentage. For unsaturated air, $S < 1$, and for saturated air, $S = 1$. Air in Earth's clouds can become supersaturated, i.e., $S > 1$, although typically by only a few hundredths.

A saturation ratio with respect to ice can also be expressed as

$$S_i = \frac{e}{e_{si}}. \tag{2.14}$$

Saturation ratios with respect to ice may be as large as 1.5 in Earth's clouds.

2.3.9 *Supersaturation s*

Supersaturation is the extent to which the air exceeds the condition of saturation (100% relative humidity). In this book, we will denote the supersaturation by s (lowercase, rather than the uppercase S used for the saturation ratio). It can thus be expressed as:

$$s = \left(\frac{e}{e_s} - 1\right) \times 100\%, \tag{2.15}$$

and for supersaturation with respect to ice:

$$s_i = \left(\frac{e}{e_{si}} - 1\right) \times 100\%. \tag{2.16}$$

In Earth's clouds, the value of s rarely exceeds 1–2%, while s_i can be as large as 50%.

2.3.10 *Cloud water mixing ratio q_c and cloud ice mixing ratio q_i*

Up to this point, we have focused only on variables that quantify water vapor in the air or its proximity to saturation. But a cloud is also composed of liquid droplets and/or ice—collectively referred to as *hydrometeors*—with the space between them occupied by air that still contains some water vapor. Despite how they appear to the human eye, clouds are not continuous liquid water or ice. They are mixtures of hydrometeors and air (Fig. 2.1), and that air also contains molecules of water still in the vapor state.

Due to certain particles in the air that we will discuss in Chapters 7 and 8, it is impossible for the air to become saturated and not produce at least some cloud droplets or ice particles. As a result, clouds contain at least two, and possibly all three, phases of water in coexistence. Thus, when the air is supersaturated—that is, when the relative humidity inside a cloud exceeds 100% with respect to liquid water or ice—water vapor in excess of the saturation value either **condenses** to form cloud droplets or **deposits** to form ice crystals. These processes are addressed in detail in Chapter 9.

Fig. 2.1: Idealized illustration of a very magnified volume within a cloud containing saturated air (gray) and liquid and/or frozen hydrometeors (circles).

We therefore sometimes need to quantify the mass of liquid water or ice within a cloud. Similar to the water vapor mixing ratio q_v, we can define mixing ratios for the amount of liquid water or ice in saturated (cloudy) air. For air that contains liquid drops, we define a **cloud water mixing ratio** q_c as the total mass of small droplets per unit mass of dry air. For air containing ice, a **cloud ice mixing ratio** q_i quantifies the total mass of small ice crystals per unit mass of dry air. In many clouds, neither q_c nor q_i exceeds 1 g kg^{-1}, though they can reach much larger values in some cumuliform clouds.

2.4 Saturated adiabatic lapse rate

When a saturated (cloudy) air parcel ascends adiabatically, it cools, and the excess water vapor condenses into liquid water. During this process, latent heat is released,[3] which slows the rate of cooling. Conversely, if a saturated parcel containing liquid water descends and warms due to compression, some of the liquid water evaporates back into water vapor. The required latent heat is extracted from the air inside the parcel, thereby slowing the rate of warming.[4]

[3] Latent heating and cooling are discussed in greater detail in Sec. 6.1.5.

[4] As a saturated air parcel descends, just enough liquid water evaporates to maintain saturation. However, if the parcel continues to descend and warm after all the liquid water has evaporated, it then warms at the dry adiabatic lapse rate during its descent.

In summary, phase changes either add heat to or extract heat from the air parcel, causing dq in (2.3) to be nonzero for a saturated air parcel. However, this can still be considered an adiabatic process because the heating or cooling arises solely from internal energy transformations within the parcel, without any exchange of heat with its surroundings.

In this context, the First Law of Thermodynamics for a saturated adiabatic parcel can be written as:

$$dq = -L_v \, dq_{vs} = C_p \, dT - \alpha \, dp, \tag{2.17}$$

where heat is either added to or removed from the air parcel (dq) depending on the phase change of the water mass. For a liquid water–water vapor system, this phase change in water mass is represented by the change in the water vapor saturation mixing ratio q_{vs}, as given by (2.11), multiplied by the latent heat of vaporization L_v.[5]

For example, if the parcel ascends and cools, q_{vs} also decreases (as a result of e_s decreasing, according to 2.11). The amount of the decrease in q_{vs} corresponds to the amount of water vapor that will condense to grow droplets. Condensation of water vapor in turn releases latent heat within the parcel, so that the $-L_v \, dq_{vs}$ term in (2.17) makes $dq > 0$, i.e., heating of the parcel, because dq_{vs} is negative in this case.

Conversely, the descent of the cloudy parcel decreases its volume and leads to an increase in temperature of the parcel due to compression. The saturation mixing ratio q_{vs} increases as a result, which then decreases the relative humidity in the parcel below saturation, causing the droplets to partially—or perhaps completely—evaporate. Evaporation of the droplets requires energy to be extracted from the surrounding air within the parcel, producing a net cooling ($dq < 0$), because dq_{vs} is positive in this case.

We can derive a different lapse rate for the ascent or descent of saturated parcels in the atmosphere. Starting from (2.17) and again making use of the hydrostatic equation (2.2) to eliminate dp, we have

$$-L_v \, dq_{vs} = C_p \, dT + g \, dz. \tag{2.18}$$

Dividing both sides of this equation by $C_p \, dz$, and then recognizing that the last term in (2.18) is now Γ_d, we can solve for $-\, dT / \, dz$:

$$-\frac{dT}{dz} = \frac{L_v}{C_p}\frac{dq_{vs}}{dz} + \Gamma_d. \tag{2.19}$$

[5]Values of L_v at different temperatures are listed in Table 6.1.

Using the chain rule, we can rewrite dq_{vs}/dz as

$$\frac{dq_{vs}}{dz} = \frac{dq_{vs}}{dT}\frac{dT}{dz}, \tag{2.20}$$

and substituting this expression back into (2.19) and simplifying yields the **saturated adiabatic lapse rate** Γ_s:

$$\Gamma_s = -\frac{dT}{dz} = \frac{\Gamma_d}{1 + \frac{L_v}{C_p}\frac{dq_{vs}}{dT}}. \tag{2.21}$$

As a result of latent heat release (or uptake), the saturated adiabatic lapse rate is always less than the dry adiabatic lapse rate. That is, because q_{vs} increases as T increases, the denominator of (2.4) always exceeds 1.

Unlike the dry adiabatic lapse rate, the saturated adiabatic lapse rate is *not* constant due to the dependence of the saturation mixing ratio q_{vs} on temperature T. The saturated adiabatic lapse rate generally ranges from 6–7 °C km^{-1} in the middle troposphere. At lower, warmer altitudes, Γ_s is smaller, and at higher, colder altitudes, Γ_s is larger. In extremely cold air, around −50 °C, the saturated adiabatic lapse rate approaches the dry adiabatic lapse rate as condensation becomes a negligible source of latent heating.

In Fig. 2.2, air parcels in both (a) and (b) start at the same altitude z_0 and temperature T_0, but the air parcel in (a) is unsaturated, while that in (b) is saturated. If both parcels are lifted to a height of z_1, the release of latent heat from condensation within the saturated parcel results in its temperature decreasing more slowly with height. Thus, it only reaches a temperature of T_1 for this increase in altitude, while the unsaturated air parcel cools more quickly for the same amount of ascent, reaching a lower temperature T_2. For the saturated parcel, water vapor is still present in the air surrounding the droplets, but the amount decreases as the droplets grow during the parcel's ascent.

Now suppose that at the height of z_1 in Fig. 2.2, the environmental air temperature is between T_2 and T_1. The unsaturated air parcel in (a), now being cooler than its environment, will descend and warm until it reaches the same temperature as the surrounding air. In contrast, the saturated air parcel in (b) will be warmer than the environmental temperature at z_1 and will continue to ascend until it cools to match the environmental temperature at some higher altitude. This atmospheric layer from z_0 to z_1 is termed *conditionally unstable*: unsaturated air parcels will not be buoyant within this layer, while saturated air parcels will be buoyant. Such conditions are common in the troposphere and can lead to significant vertical cloud development if air parcels are initially forced upward and cool sufficiently to become saturated. We explore this process further, as well as other cloud formation processes, in the next section.

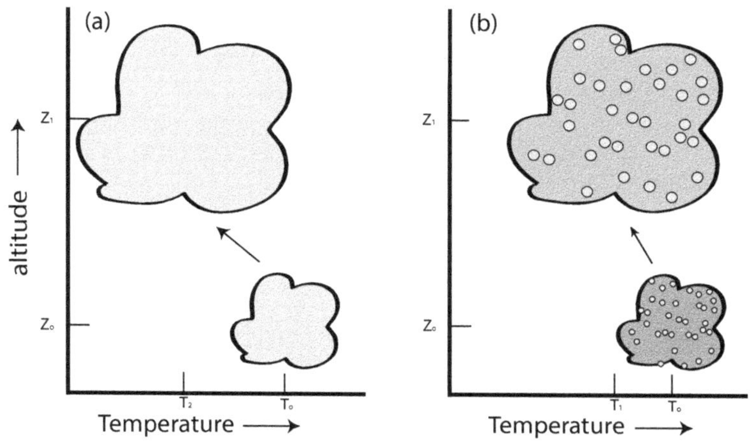

Fig. 2.2: Illustration depicting the temperature decrease during adiabatic ascent from an initial height of z_0 to a greater height z_1 for (a) an unsaturated air parcel and (b) a saturated (cloudy) air parcel. Darker shading indicates a greater water vapor mixing ratio; cloud droplet sizes are indicated by the size of small circles in the saturated case. Note that $T_2 < T_1 < T_0$. If the environmental temperature outside of the parcels at a height of z_1 is between T_1 and T_2, the layer between z_0 and z_1 would be called conditionally unstable.

2.5 Methods of bringing air to saturation

We have discussed how to quantify the conditions under which air will be saturated, but now we address the means by which it actually reaches saturation. We examine three different processes: cooling the air, adding water vapor to the air, or both cooling the air and adding water vapor.

Mathematically, we can see from (2.7) that to bring the relative humidity to 100% (saturation) or higher (supersaturation), the numerator can be increased, the denominator can be decreased, or both the numerator can be increased while the denominator is decreased. Increasing the numerator requires increasing the mass of water vapor in the air. Decreasing the denominator requires lowering the temperature of the air, as e_s depends solely on temperature. Cooling and moistening the air concurrently will both increase the numerator and decrease the denominator simultaneously. Some typical processes in the atmosphere that bring the air to saturation by one of these methods are summarized in this section.

i) Adiabatic ascent

As an air parcel ascends through the atmosphere, its pressure always closely matches the decreasing pressure of its surroundings. It therefore expands. If the ascent is adiabatic, then internal energy must be expended to do the work of expansion, and the parcel cools according to the dry or saturated adiabatic lapse rate, as appropriate. To determine if an unsaturated parcel might ascend to the altitude at which it becomes saturated, we make use of traditional adiabatic parcel theory, applying the graphical analysis methods commonly performed on skew-T plots of atmospheric sounding data.

The height to which unsaturated air must be lifted and thereby cooled to first reach saturation is called the **lifting condensation level** (LCL). At this level, the air will have cooled sufficiently to make its relative humidity equal to 100%. The height of the LCL is dependent on the air parcel temperature and the amount of water vapor it contains.

An air parcel ascends because it is either forced upward by some mechanism or has positive buoyancy—meaning it is less dense than the surrounding air and thus experiences a positive (upward) buoyancy force. The distinction is intrinsically tied to the static stability of its environment, which depends on the vertical lapse rates of temperature and moisture. Air residing within a stable layer must be forcibly lifted to ascend.

Forced adiabatic ascent: A parcel of air can become saturated when forced to ascend, such as along a frontal boundary or when winds push air upward against high terrain. When air parcels ascend under these conditions, they generally do so slowly. Consequently, the upward motion is typically on the order of a few centimeters per second, resulting in more laminar (i.e., smooth in appearance and lacking turbulent structure) clouds. Most stratiform clouds originate from air at higher altitudes, not near the ground, that has been forced upward. This sequence of events is summarized in the blue boxes on the left side of Fig. 2.3.

Buoyant adiabatic ascent: The "free" ascent of buoyant air that cools adiabatically to saturation is also a common source of clouds. During daytime hours, solar heating may be sufficient to warm air parcels near the ground and make them positively buoyant. These parcels then ascend freely as unsaturated thermals, cooling at the dry adiabatic lapse rate as they rise. Often, a succession of unsaturated thermals acting over one to several hours is required to modify the stability of the environment so that the thermals can eventually ascend to their point of saturation, the LCL.[6] Beyond this point, they can ascend and cool at the saturated adiabatic lapse rate.

The vertical acceleration of an air parcel is proportional to its buoyancy—that is, the difference in densities, which in turn is propor-

[6]Occasionally, strong solar heating during summer months can warm an air parcel near the ground to its so-called *convective temperature*. At that point, a single thermal has suffi-

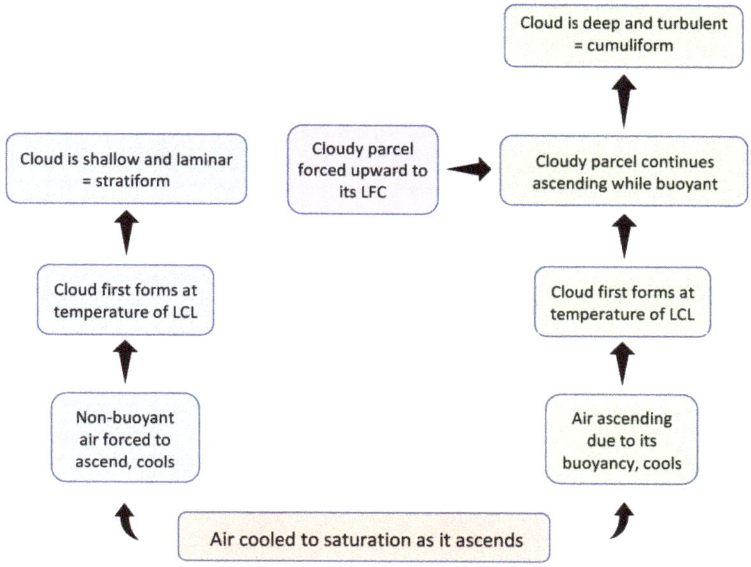

Fig. 2.3: The production of clouds from air ascending adiabatically. *Left column* (blue boxes): forced adiabatic ascent. *Right column* (green boxes): buoyant adiabatic ascent. Forced ascent followed by buoyant ascent isrepresented by the pathway starting at lower left (blue box), proceeding along arrows through the center (purple box) to the upper right (green box).

tional to the difference in virtual temperatures between the air parcel and its environment. Consequently, this acceleration can vary significantly. Typically, the upward motion ranges from one to ten meters per second, with higher speeds leading to the formation of turbulent clouds. Cumuliform clouds often form through this process, usually from air that originated near the ground. This sequence of events is summarized in the green boxes on the right side of Fig. 2.3.

Forced ascent followed by buoyant ascent: A cloud can also form from a combination of the two previously described processes when the atmosphere is conditionally unstable, as briefly mentioned at the end of Sec. 2.4. Initially non-buoyant air may be forced to ascend and cool past the LCL to the **level of free convection** (*LFC*), where it becomes buoyant. Above the LFC, the cloudy air ascends due to its positive buoyancy, which results in the formation of a deep and turbulent cumulus cloud. This path in Fig. 2.3

cient buoyancy to cool to its point of saturation, called the *convective condensation level* (CCL). However, this is far less common.

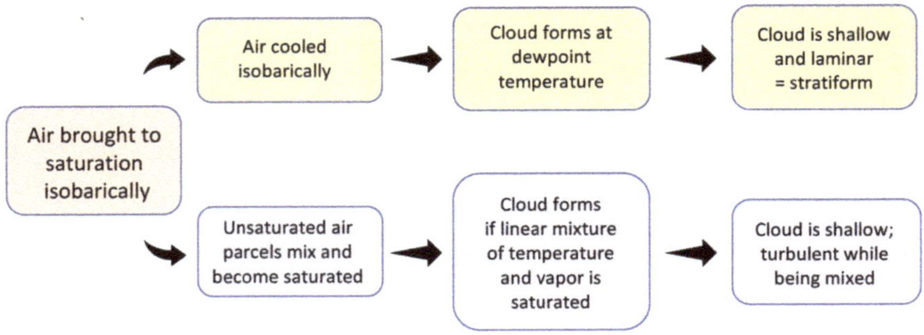

Fig. 2.4: The production of clouds by isobaric processes. *Top row*: isobaric cooling. *Bottom row*: adiabatic, isobaric mixing of two kinds of air.

starts at the lower left set of blue boxes for forced ascent, passes through the center (purple) box, and ends at the top green box in the upper right. This is a very common process for the formation of deep cumulus clouds.

ii) Isobaric cooling

In this process of cloud formation, without any ascent of the air, and thus no change in pressure, the air cools to the point of saturation. For example, the air very near the ground can cool radiatively overnight to the point of saturation, which can produce fog. If we can predict the overnight low temperature accurately, we can then determine whether air parcels might isobarically cool to saturation, assuming the temperature decreases to the dewpoint temperature. The clouds formed by this mechanism are laminar, have almost no vertical velocity, and tend to be shallow. This mechanism typically creates stratiform clouds, especially fog (Fig. 2.4, top row).

iii) Adiabatic, isobaric mixing

The adiabatic, isobaric mixing of two different types of unsaturated air in close proximity—for example, along a front, or when jet aircraft exhaust is injected into cold upper-tropospheric air—can result in saturation. This process is summarized in the bottom row of Fig. 2.4. Clouds formed by isobaric mixing are initially turbulent due to the mixing of the two types of air that initiate them, but eventually the mixing ends and vertical velocities will cease. Continued dilution by drier environmental air will inevitably bring the relative humidity below 100%. Beyond that point, these clouds will persist only for the duration needed to evaporate the droplets or sublimate the ice crystals.

To determine whether air parcels might reach saturation by this means, we need to ascertain whether, after mixing, both the new saturation vapor

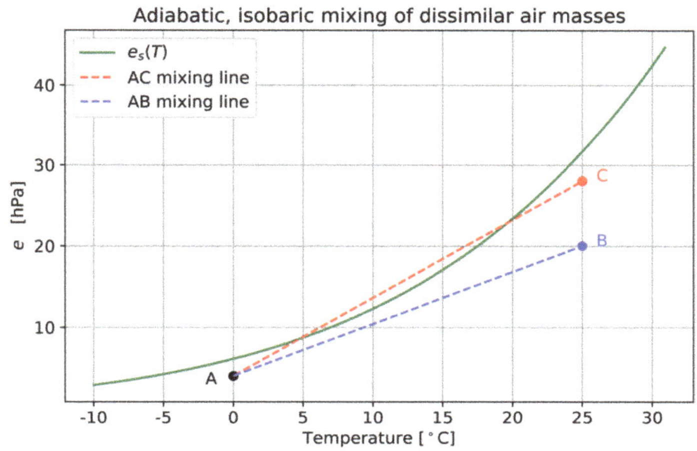

Fig. 2.5: Illustration of the conditions that can lead to saturation and cloud formation when two air masses are mixed adiabatically and isobarically, as a function of their combined vapor pressure and temperature, as discussed in the main text. Mixtures of air mass 'A' (black dot) with air mass 'B' (blue dot) lie along the blue dashed curve. Mixtures of air mass 'A' (black dot) with air mass 'C' (red dot) lie along the red dashed curve. The green curve denotes the saturation vapor pressure at each temperature.

pressure and the new vapor pressure produce a relative humidity that exceeds 100%. Both the temperature and the water vapor mass can be assumed to mix linearly, in proportion to the amount of their respective air masses in the mixture.

For example, the results of the linear mixing of two air masses with different properties are shown in Fig. 2.5. Air mass 'A' (black dot) has a temperature of 0 °C and a relative humidity of 60%. If it is mixed isobarically with air mass 'B' (blue dot), which has a temperature of 25 °C and a relative humidity of 63%, then all possible mixtures will consist of combinations of temperature and vapor pressure that fall on the dashed blue mixing line. In this case, the mixing line never lies above the saturation vapor pressure curve, so e is always less than e_s, and saturation would never be achieved.

However, if we now consider air mass 'A' mixing with air mass 'C' (red dot), a different result occurs. Air mass 'C' has the same temperature as 'B' but a higher relative humidity of 88%. Mixtures of 'A' and 'C' fall on the dashed red mixing curve, much of which lies *at or above* the saturation vapor pressure curve, where e is equal to or greater than e_s. Therefore, a range of different proportions of 'A' and 'C' would lead to the mixture being saturated (or even supersaturated), resulting in visible cloud or fog formation.

An easily observed example is the formation of a cloud when you exhale on a very cold morning. The exhaled air is much moister and warmer

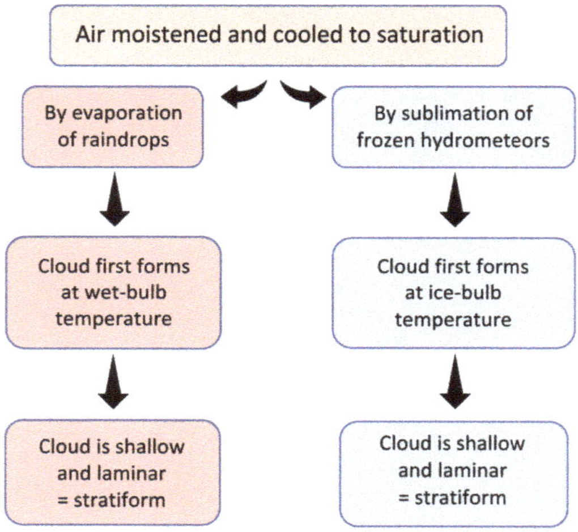

Fig. 2.6: The production of clouds by latent cooling of evaporating (left column) or sublimating (right column) hydrometeors.

than the ambient air outside your mouth, and initially, neither the exhaled air nor the ambient air contains visible cloud particles. (If your friend looks inside your mouth, they would not see a cloud inside!) However, when these two types of air mix through turbulent processes, the resulting mixture reaches saturation, and a visible "cloud" forms as you exhale. This cloud does not persist for long in the cold, unsaturated environment, and the droplets quickly evaporate, causing the cloud to dissipate shortly after it is formed.

The formation of contrails works similarly. The warm, moist jet exhaust is ejected into the surrounding colder and drier air, and when these two types of air mix, they create saturated air, i.e., cloud. The longevity of the contrail is determined by the relative humidity of the environment. If the environment has a low relative humidity, the contrail disappears (evaporates) almost instantly.

iv) Evaporative cooling and moistening

The *evaporation* of liquid precipitation, or *sublimation*[7] of frozen precipitation, as it falls into unsaturated air increases the mass of water vapor in the unsaturated air and cools it through latent heat loss. Either process may be

[7]Sublimation is the conversion of ice to water vapor when $RH_i < 100\%$.

strong enough to saturate the air, as summarized in the separate columns of Fig. 2.6, and depends upon the unsaturated air reaching its **wet-bulb temperature**, T_w, or **ice-bulb temperature**, T_{wi}, for liquid or frozen precipitation, respectively. You are likely already familiar with these concepts. They indicate the amount of cooling by either evaporation or sublimation that will result in the air just becoming saturated. Higher wet-bulb or ice-bulb temperatures are favored for saturating the air, requiring less cooling and less additional water vapor.

When the wet-bulb or ice-bulb temperature is reached by this process, it typically occurs over depths of less than a few kilometers above the ground, and is therefore more associated with the production of fog. For example, within the heaviest precipitation of a thunderstorm, raindrops may evaporate when they fall into initially unsaturated air below the cloud. If sufficient evaporation occurs, the air can become saturated, further decreasing visibility for the unfortunate automobile drivers passing through it!

Questions for Review

1. How is an air parcel defined?
2. How is an adiabatic air parcel defined?
3. Which variables describe the amount of water vapor in the air?
4. Which variables describe how close the air is to saturation or by how much it exceeds saturation?
5. At what rates do unsaturated and saturated air parcels cool or warm as they ascend or descend, and why do these rates differ?
6. What does it mean for an air parcel to be positively or negatively buoyant?
7. Why does increasing the amount of water vapor in an air parcel increase its buoyancy, and which meteorological variable conveniently accounts for this effect?
8. What processes bring air to saturation in the atmosphere, and which cloud types or characteristics are commonly associated with each of these processes?
9. How do particular cloud formation processes depend upon the static stability of the atmosphere?
10. What thermodynamic "level" marks the height of the cloud base when a cloud forms by either forced or buoyant adiabatic ascent?

Homework problems and other resources for students and instructors can be found at bit.ly/4dYBwB5, or scan this code:

Further Reading

For a review of atmospheric thermodynamics:

Petty, G. W., 2008: *A First Course in Atmospheric Thermodynamics*. Sundog Publishing

For more on meteorological processes producing clouds [advanced]:

Houze, R. A., Jr., 2014: *Cloud Dynamics*. 2nd ed., Academic Press

CHAPTER 3

Cumulus Entrainment

The previous chapter examined various ways by which air can be brought to saturation to form clouds, noting that processes involving the ascent of buoyant air typically produce turbulent cumuliform clouds. In Chapter 1, we observed that cumuliform clouds generally consist of thermals (Fig. 1.3). In this chapter, we briefly examine how the air motions associated with thermals influence the amount of liquid water and/or ice in cumuliform clouds, which in turn affects their potential to produce precipitation.

We begin by introducing a thermodynamic quantity, likely new to you, that builds on the concept of an adiabatic parcel. It helps estimate how much liquid water could potentially exist in a cloud and offers a broad perspective on its precipitation potential. We will then compare this theoretical value to actual measurements in real clouds to see how thermally driven motions influence the amount of liquid water or ice in a cumuliform cloud. These influences usually cause the saturated parcels within clouds to deviate from being adiabatic, so we conclude by discussing how often one might encounter truly adiabatic parcels within cumuliform clouds.

3.1 Adiabatic Liquid Water Content

3.1.1 Definition and trend with height inside a cloud

In discussing the International Cloud Classification Scheme in Chapter 1, we noted the tendency of various clouds to have bases that were "low" in the atmosphere, at "mid-levels," or "high" in the atmosphere. Besides issues related to aviation, why should we be concerned about the height of cloud bases?

Higher cloud bases are typically colder, because temperature generally decreases with height in the troposphere. For clouds formed by air ascending (and cooling) adiabatically, *the temperature and pressure at the cloud base uniquely define the characteristics of the source air*, and thus determine the theoretical maximum mass of **condensate** (liquid and ice combined) per unit mass of dry air that can be produced as the cloudy air parcels ascend.[1]

The temperature and pressure at which the cloud base occurs uniquely determine the vertical profile of the **adiabatic liquid water content** (L_A) of the cloud: the mass of liquid water per unit volume of air within a cloudy adiabatic parcel that is ascending from the cloud base. It represents a *theoretical upper limit* for how much liquid water mass can be present in a cloud at a given altitude.[2] As air cools and water vapor condenses onto droplets, and provided that none of the droplets fall out of the parcel (following the rules of adiabatic ascent defined in Chapter 2), the higher the parcel ascends from the cloud base, the more liquid water mass per unit volume of air it can contain.

We typically express L_A in units of g m^{-3}, and we can compare it to the observed liquid water content, L, which is calculated as the total mass contributed by all the droplets per unit volume of cloudy air. By comparing L_A to L, we can determine how close the cloudy parcel is to the adiabatic limit. Adiabatic, or nearly adiabatic, saturated air parcels will have the highest temperatures (and thus the greatest buoyancy) compared to other

[1] From the cloud base (LCL) on a skew-T chart, the pressure and temperature there define the saturation mixing ratio q_{vs} at cloud base. If the cloudy parcel is lifted along a saturation adiabat, notice how q_{vs} decreases with decreasing temperature and pressure. Generally speaking, if we assume that the relative humidity stays near 100% ($q_v \sim q_{vs}$), vapor in excess of q_{vs} must convert to liquid droplets or ice.

[2] This is true for any type of cloud that is formed by lifting the air, whether it be buoyant or forcibly lifted. However, the adiabatic liquid water content is not useful for clouds produced by isobaric processes.

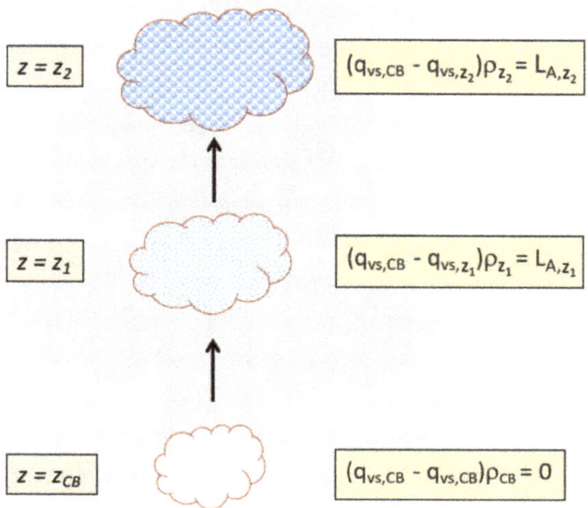

Fig. 3.1: Illustration of the increase in adiabatic liquid water content L_A with height z for an adiabatic parcel based on (3.1). "CB" denotes values at cloud base. For heights $z_2 > z_1 > z_{CB}$, $L_{A,z_2} > L_{A,z_1} > L_{A,CB}$.

cloud parcels and will contain the most liquid water content to potentially convert to precipitation.

The formula for computing the adiabatic liquid water content can be expressed as:

$$L_{A,z} = (q_{vs,CB} - q_{vs,z})\, \rho_z \tag{3.1}$$

where q_{vs} denotes the saturation water vapor mixing ratio at a given height in g kg^{-1}, the subscript CB denotes the value at the height of cloud base, and the subscript z denotes some height above cloud base. The difference in saturation mixing ratio is then converted to units of g m^{-3} by multiplying it by ρ_z, the local dry air density at the altitude z expressed in kg m^{-3}.

Figure 3.1 illustrates how the adiabatic liquid water content of a saturated parcel increases as it ascends from the cloud base within a cloud updraft. At the cloud base, when the air parcel first reaches saturation, $q_{vs,z}$ equals $q_{vs,CB}$, so by (3.1), $L_A = 0$. As the parcel ascends above the cloud base to a height of z_1, it cools, and q_{vs,z_1} becomes less than $q_{vs,CB}$. Consequently, the excess water vapor greater than q_{vs,z_1} condenses into liquid, resulting in L_{A,z_1} attaining a positive value given by (3.1). As the parcel continues to ascend to a greater height z_2, q_{vs} decreases further, causing

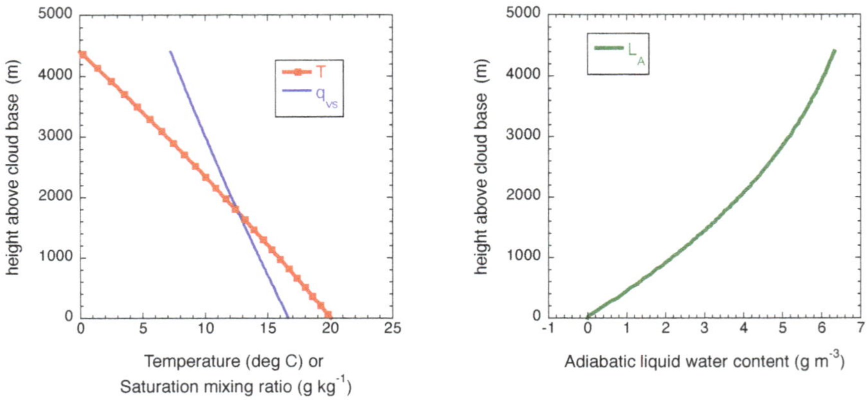

Fig. 3.2: Temperature and saturation vapor mixing ratio (left) and adiabatic liquid water content (right) versus height above the cloud base for an ascending, saturated adiabatic air parcel. Cloud base temperature is 20 °C and cloud base pressure is 900 hPa.

additional water vapor to condense into liquid. Therefore, L_{A,z_2} at height z_2 is greater than L_{A,z_1} at z_1.

Equation (3.1) can be evaluated numerically (i.e., with a computer program) to compute L_A as a function of height above the cloud base for a particular cloud base temperature and pressure. Results of such calculations are shown in Fig. 3.2. The left plot shows that the cloudy air parcel cools as it ascends, but at a rate less than the dry adiabatic lapse rate of 9.8 °C km^{-1}—in this example closer to 4.5 °C km^{-1}, because of the release of latent heat as condensation occurs. The saturation mixing ratio (q_{vs}) also decreases with altitude due to the temperature decrease with altitude. Assuming that any amount of water vapor (q_v) that exceeds q_{vs} results in condensation, the adiabatic liquid water content must increase as the parcel ascends (Fig. 3.2, right).

These calculations demonstrate that the cloud will contain more liquid water at its top compared to its base, assuming the adiabatic condition holds. For instance, if the cloud depicted in Fig. 3.2 extends to a height of 1,000 m above its base, it could potentially have a liquid water content as high as 2.3 g m^{-3}. If it continues to grow to a height of 4,000 m above its base, the adiabatic upper limit suggests a theoretical maximum liquid water content approaching 6 g m^{-3}. The calculations also suggest that deeper warm clouds (with temperatures above 0 °C throughout their depths) can theoretically contain more liquid water near their tops than shallower clouds, provided that they have the same cloud base conditions and that the adiabatic assumption remains valid. Thus, deeper clouds, such as many cumuliform types, would tend to have the largest L_A val-

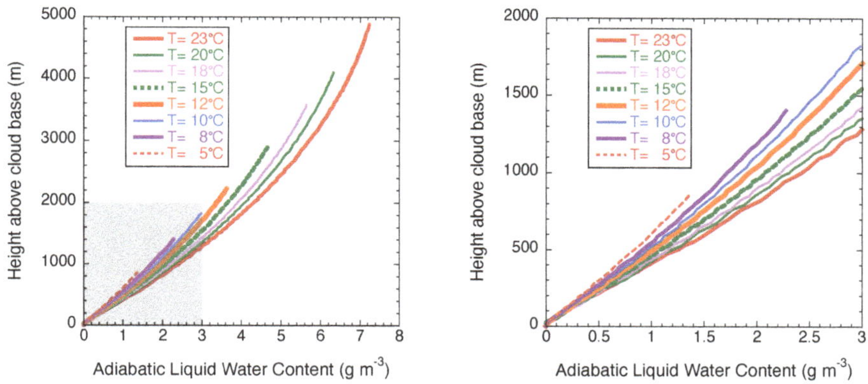

Fig. 3.3: *Left*: Adiabatic liquid water content versus height above cloud base, for cloud bases at a pressure of 900 hPa and temperatures as shown in the legend. *Right*: Zoomed-in region within the gray square shown on the left, revealing detail at the lower altitudes and smaller values. For both plots, calculations are only shown until the parcel temperature decreases to 0 °C.

ues and therefore theoretically contain the most condensed liquid water available for conversion to precipitation.

3.1.2 Dependence on cloud base temperature

The adiabatic liquid water content at any given height within a cloud is strongly dependent on the cloud base temperature. For the same height above the cloud base, clouds with colder bases have smaller adiabatic liquid water content compared to those with warmer bases. As shown in Fig. 3.3, the adiabatic liquid water content increases more rapidly with altitude for warmer cloud bases than for colder ones due to the nonlinear dependence of saturation vapor pressure on temperature.

Figure 3.3 also indicates that clouds with lower altitude (higher temperature) bases tend to have greater liquid water content than those with higher altitude (lower temperature) bases. Consequently, deep clouds with low-level bases potentially contain more liquid water available for precipitation than those with mid-tropospheric bases. This suggests that clouds in warmer, tropical environments (with warmer bases) should have more condensed liquid water available for precipitation than those in cooler, higher-latitude environments (with colder bases). For instance, cumulonimbus clouds forming over the Intertropical Convergence Zone (ITCZ) in tropical regions might yield greater precipitation rates than those over land at higher latitudes. However, other factors, such as cloud updraft speed, are equally important and will be discussed in future chapters.

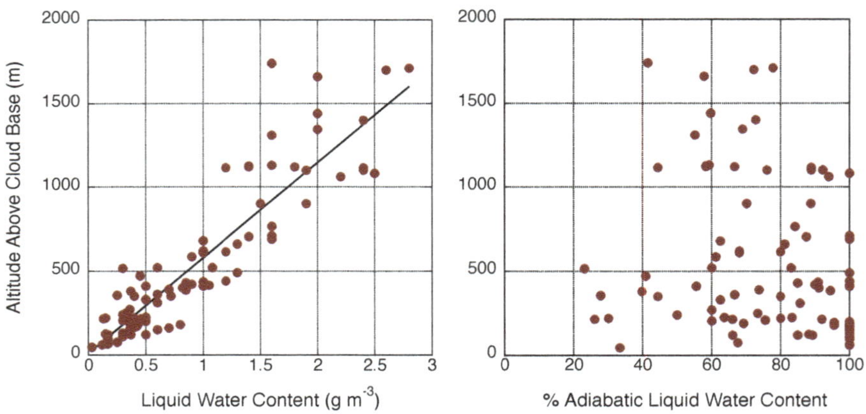

Fig. 3.4: *Left:* Measurements taken within Florida cumuli showing liquid water content versus altitude above cloud base, with a curve fit to the data. *Right:* The same data, now plotted as the percent of the adiabatic liquid water content for each observation height.

You have likely already considered this aspect in synoptic meteorology courses, where the amount of water vapor in a given environment is often characterized by a quantity called the *precipitable water*, the integrated water vapor within a column of air from the surface to the top of the atmosphere, often expressed as the condensed liquid water equivalent in units of millimeters. Precipitable water observations commonly show much higher values in the Tropics (often 50–70 mm) compared to midlatitudes (15–45 mm). This usually means that more water vapor is available for tropical clouds to potentially convert into precipitation, provided their updrafts are strong enough.

Despite the usefulness of the adiabatic parcel concept, clouds in reality often contain parcels that do not fully adhere to adiabatic assumptions. One major reason is the formation and fallout of precipitation, whereby the parcel loses some of its condensed water to its environment. Such a loss constitutes a direct violation of adiabatic principles, which assume that the total mass of the parcel remains constant.

However, even clouds that do not produce precipitation often contain few or no adiabatic parcels. Fig. 3.4 shows examples from aircraft measurements of the liquid water mass within non-precipitating cumulus clouds near the coast of Florida. The measured liquid water content values do increase with altitude (as indicated by the line fit to the data on the left), but they often contain less liquid water than the adiabatic values at those altitudes (right plot). *No adiabatic parcels were observed at higher altitudes.* Some process is acting to make many of the cloudy parcels fail the adiabatic assumptions, and thus they contain less liquid water that could potentially be converted to precipitation. We will examine the reason for this behavior in the next section.

3.2 Entrainment: Departure from Adiabaticity

Up to this point, we have considered the hypothetical situation in which a cloudy air parcel is a *closed* system, i.e., it does not exchange heat or mass with its surroundings. Under these conditions, the measured liquid water content L in the cloud at a given altitude would be equivalent to the adiabatic liquid water content, L_A. Air parcels ascending into the base of the cloud updraft are generally assumed to be adiabatic.

However, real clouds are *open* systems that exchange mass and heat with their environment. This behavior directly violates adiabatic principles, which state that neither heat nor mass can be exchanged with the parcel's environment. Cumulus clouds draw in drier and cooler air from outside their horizontal and top boundaries. This process is referred to as **cumulus entrainment**, *eddy entrainment*, *lateral entrainment*, or simply *entrainment*. It results from the overturning motions of turbulent eddies that often give cumulus clouds their cauliflower-like appearance (shown in Fig. 3.5; also discussed in Chapter 1).

Some of the earliest aircraft measurements within cumulus clouds in the 1940s noted that their internal temperatures and liquid water contents were much less than the adiabatic values. Stratocumulus clouds undergo a life cycle that depends significantly on eddy entrainment, which primarily occurs at their tops. Cirrocumulus clouds may also experience this type of entrainment to some extent.

Numerical models, when run at high spatial resolution, can capture some of the eddy motions that produce entrainment. Figure 3.6 shows the results of such a model simulation, with color shading indicating the liquid water content at each location in the cloud. Note how the different colors, representing various values of the liquid water content, are arranged into nearly horizontal stripes just above the cloud base. This pattern indicates that this part of the cloud is adiabatic, as the liquid water content depends solely on height (refer back to Fig. 3.1 and Eq. 3.1), causing the colored liquid water content values to change predictably with height.

However, as the turbulent eddies develop at higher altitudes within the cloud, the striped pattern in the colors disappears, and higher and lower values become mixed throughout the cloud. The liquid water content is influenced by several thermals within the cloud, as discussed in Chapter 1. Some precipitation is also beginning to fall within the cloud, increasing liquid water content values at certain locations in the middle and lower levels (green shading).

Once unsaturated air is entrained into the cloud from outside its boundaries by turbulent eddies, it mixes with and dilutes the cloudy air. We have already discussed the mixing of two different kinds of unsaturated air in Chapter 2 and how this mixing can potentially lead to cloud

Fig. 3.5: The view from an aircraft of a developing thunderstorm (foreground), where turbulent eddies that laterally mix environmental air into the cloud are clearly visible.

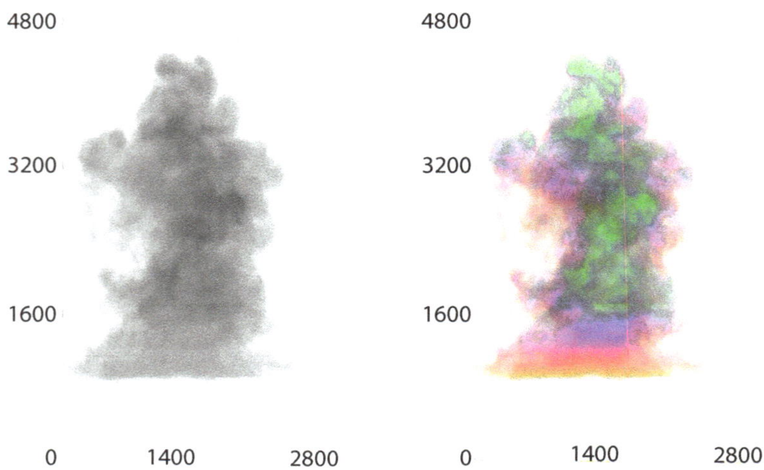

Fig. 3.6: Simulated cumulus cloud growing in an environment with no wind; scale in meters shown on vertical and horizontal axes. *Left:* Photorealistic cloud shown in grayscale with darker (wetter) areas blocking out more light. *Right:* Colors represent ranges of cloud water mixing ratio: yellow 0–1 g kg^{-1}, orange 1–2 g kg^{-1}, pink 2–3 g kg^{-1}, purple 3–4 g kg^{-1}, and green >4 g kg^{-1}. *Adapted from Cooper et al. (2013). © American Meteorological Society. Used with permission.*

formation. Here, *mixing* refers to the process that combines saturated and unsaturated air, or that combines different parcels of saturated air within the cloud. Mixing results primarily from turbulent motions and acts to make the cloud properties more uniform—although never completely uniform.

The environmental air entrained into a cumulus cloud is generally cooler and drier than the buoyant parcels inside an actively growing cumulus cloud. Thus, the consequences of entrainment typically include:

- a decrease in L, the cloud liquid water content[3] due to evaporation of the droplets within the cloud, possibly leaving less to be converted into precipitation; and
- a decrease in cloud temperature that occurs not only from mixing with cooler air but also from the latent heat needed to evaporate the droplets. This temperature reduction decreases cloud buoyancy, thereby limiting the maximum height the cloudy parcels can attain and ultimately restricting the cloud depth. This is a primary reason some cloud tops do not reach the equilibrium level diagnosed from a thermodynamic sounding.

3.3 The Shedding Thermal Model

Entrainment can be envisioned in a cumulus cloud using a conceptual model called the **shedding thermal model**, which has been supported by observations worldwide. As discussed in Chapter 1 and shown in Fig. 3.7, a thermal circulation occurs as a buoyant air parcel (shaded turquoise region) rises, with ascending motion in its center (black arrows). Diverging winds near its top (red arrows) carry some cloudy air outward into the environment (referred to as *detrainment*), strengthening the descending motion at the thermal sides (dashed green arrows) through evaporative cooling of the ejected cloudy air. Finally, converging winds near the thermal base (blue arrows) close the circulation and entrain dry air into the cloud. The thermal circulation is located near the top of a buoyant cloud turret.

The (detrained) air pushed outward over the entire top of the thermal is also evident in the right-hand panel of Fig. 3.7, where a high-resolution simulation of a cumulus cloud is colored red where air is moving outward relative to the cloud boundary. Regions colored blue indicate where air is being entrained, i.e., where air is moving inward across the cloud boundary, and this is clearly seen at the base of the highest thermal.

---

[3]If ice is present in the cloud, it too can decrease (sublimate) with the addition of entrained environmental air.

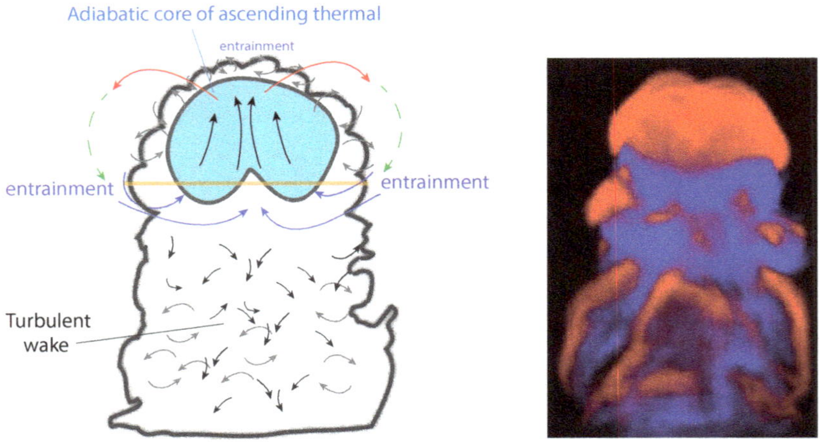

Fig. 3.7: *Left*: The shedding thermal conceptual model of cumulus entrainment, with arrows colored blue denoting major entrainment of environmental air and arrows colored red denoting major detrainment of cloudy air. Horizontal gold line corresponds hypothetically to the aircraft data shown in Fig. 3.8. *Adapted from Blyth et al. (1988). © American Meteorological Society. Used with permission.*
Right: A simulated cumulus cloud with regions of entrainment (shaded blue) and detrainment (shaded red). The simulation lacked the resolution to reveal the smaller eddies that can entrain smaller amounts of environmental air but the results illustrate the important role of the thermal circulation in entrainment.

The left panel in Fig. 3.7 has been simplified to depict only a single thermal in the cloud. In reality, multiple thermals often coexist within a single cumulus cloud, as indicated by the red/blue velocity patterns shown in the right panel. Over time, the primary thermal circulation also breaks down into smaller overturning eddies, complicating the simple picture shown here.

Although some entrainment can occur near the top and sides of the thermal via small overturning eddies, more significant entrainment occurs through the converging winds at the bottom of the thermal, as indicated in Fig. 3.7. Once the air is entrained and mixes with the cloud, buoyant mixed parcels may continue to ascend, while negatively buoyant parcels will descend and be "shed" into the "trailing wake" of the ascending thermal. In the center of the thermal, the "core" (shaded turquoise in the left panel) retains its adiabatic properties. However, in the trailing wake region, the temperature and liquid water content are reduced compared to adiabatic values, as this cloudy air contains mixtures of cloudy and dry environmental air.

If the shedding thermal model is correct, one might wonder whether clouds have a hole in their center where dry air has been entrained. If a research aircraft flies through the center of a cumulus cloud—specifically,

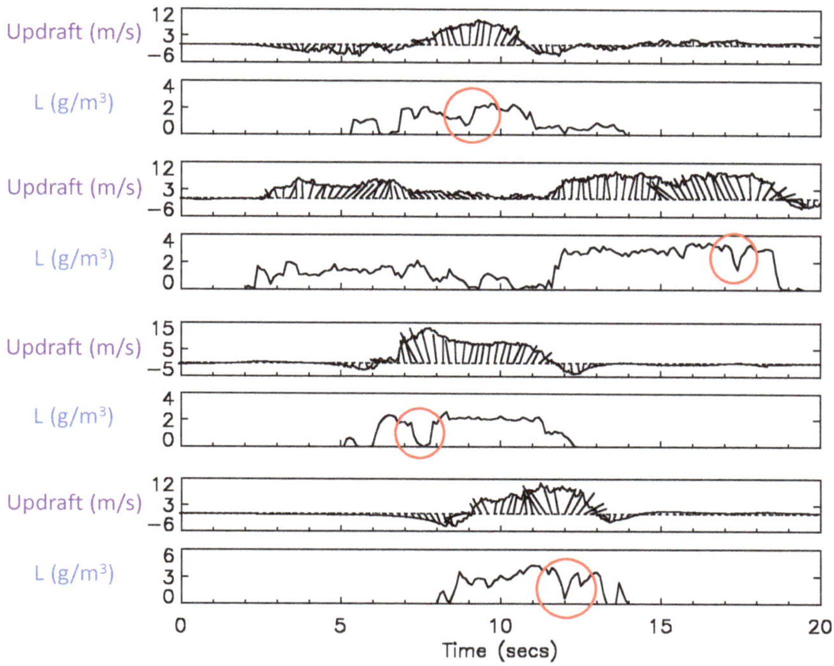

Fig. 3.8: Four examples of aircraft transects through growing cumulus clouds where reduced, or even zero, liquid water content (L) was measured within a strong cloud updraft (circled regions). Each transect has two panels of data: the upper panel depicts vertical velocity and wind vectors across the cloud, and the lower panel depicts measured liquid water content at the same locations. The aircraft would have been passing through the cloud near the region shown by the gold horizontal line in the left panel of Fig. 3.7. *Adapted from Blyth et al. (2005). ©1974 John Wiley and Sons, used with permission.*

at the height of the horizontal gold line shown in the left panel of Fig. 3.7—and measures the liquid water content there, would one observe a decrease at the center of the cloud? The answer is "yes," but the research aircraft must be at the right location at the right time to measure this hole. This is because the entrained air will quickly mix with the cloud and either produce a mixture of cloudy air that is still buoyant and ascends, or that is no longer buoyant and descends into the trailing wake. Thus, the hole is a transient feature. Figure 3.8 shows an example of some aircraft measurements where the decrease, or sometimes a complete "hole," in the center of the cloud was observed.

Cloud droplets or ice within the cloud, when pushed outside its boundaries by detrainment, will evaporate in the clear air, thus cooling and humidifying it. This process can create a "shell" of humid air around

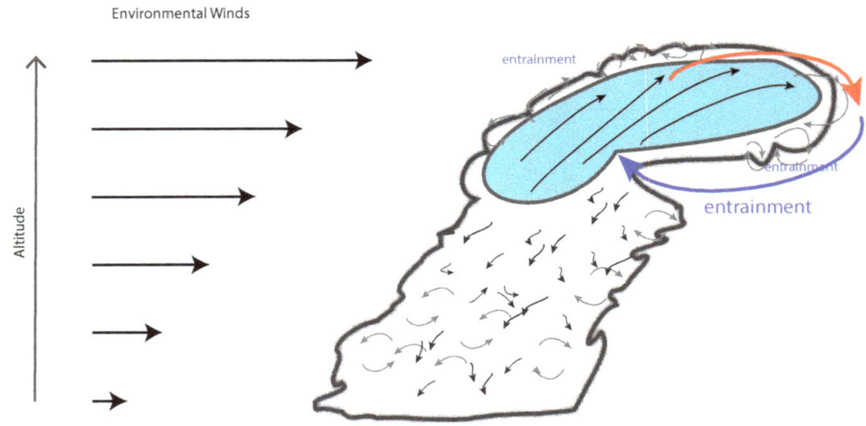

Fig. 3.9: Schematic of the alteration of the primary thermal circulation and the resulting entrainment pattern when a cumulus cloud leans with height due to vertical wind shear, as shown by the black arrows at the far left.

clouds and is one of the ways in which clouds impact their surroundings. In the presence of a stable atmospheric layer, such as an inversion, cloud detrainment over an extended period (e.g., an hour or more) can cool the air sufficiently to destabilize the layer, promoting cloud development to greater heights than would otherwise be possible. This impact of clouds on their environment operates on a much larger scale than the localized humid air shell around a single cloud.

Finally, it should be noted that up to this point, we have considered entrainment occurring under conditions where the wind profile in the cloud environment—as measured by a rawinsonde—is weak or nearly constant with altitude. However, *vertical wind shear*, a situation in which the winds in the cloud environment change speed or direction with altitude, can enhance entrainment. In such cases, the cloud may lean in one direction with altitude, altering the symmetry of the primary thermal circulation depicted in Fig. 3.7. The upshear branch of the thermal circulation is inhibited by the environmental winds, whereas the downshear branch is enhanced, as shown in Fig. 3.9. Entrainment through other mechanisms in deep cumulonimbus clouds that develop in environments with very strong wind shear remains an active area of research.

3.4 Do Adiabatic Parcels Exist?

At this point, you may feel that some of your instructors in meteorology have misled you! After all, we use the concept of adiabatic parcels and the "parcel method" to study atmospheric thermodynamics, to forecast the heights of cloud bases through the lifting condensation level (LCL) and the convective condensation level (CCL), as well as the maximum limits of cloud and storm updraft speeds and tops using convective available potential energy (CAPE) and the equilibrium level (EL) on a thermodynamic diagram. We do not typically adjust these values for the effects of entrainment.

At the time you were learning these concepts, you may not have fully appreciated that they represent upper limits on cloud or storm characteristics. Particularly for deep convective storms, we are always interested in knowing "how bad it could be," i.e., the upper limits, as weather forecasters prepare to warn the public if needed. The adiabatic calculations provide the theoretical upper limits.

While the majority of cloudy air parcels observed in shallow clouds may not be adiabatic, a few can be adiabatic. Figure 3.10 shows an aircraft transect through the lower level of a shallow (about three-kilometer deep) Florida cumulus cloud, with the shaded area indicating a region that is likely adiabatic. These "adiabatic cores" found in some clouds are characterized by nearly constant liquid water content and low turbulence.

Limited measurements from weather balloons released into the updrafts of deeper cumuli or thunderstorms have documented the presence of adiabatic parcels throughout the depth of these clouds or storms, situated in the very interior of their updrafts. Rare measurements from research aircraft have also confirmed the existence of adiabatic parcels in thunderstorm updrafts.[4] The fact that deeper clouds and storms often extend up to their equilibrium levels, and have maximum updraft speeds consistent with the buoyancy suggested by the CAPE within their environments, supports the theoretical standpoint that adiabatic parcels do exist in these cases.

Entrainment can be limited in clouds or storms with wider updrafts, perhaps explaining why deeper and larger clouds and storms tend to contain adiabatic parcels within the centers of their updrafts. In these cases, the overturning circulations of the primary thermal circulation, and consequently the smaller turbulent eddies, cannot "reach" the very center of the wide updraft, allowing adiabatic parcels to exist there. Numerical simulations of supercell thunderstorms have found that the large width of their rotating updrafts limits the reach of entrainment into the cores of their

[4]Scientists on board the aircraft noted that the turbulence outside the adiabatic core of the storm was very intense, but the ride within the adiabatic core was so smooth that they would never have known the aircraft was inside the storm!

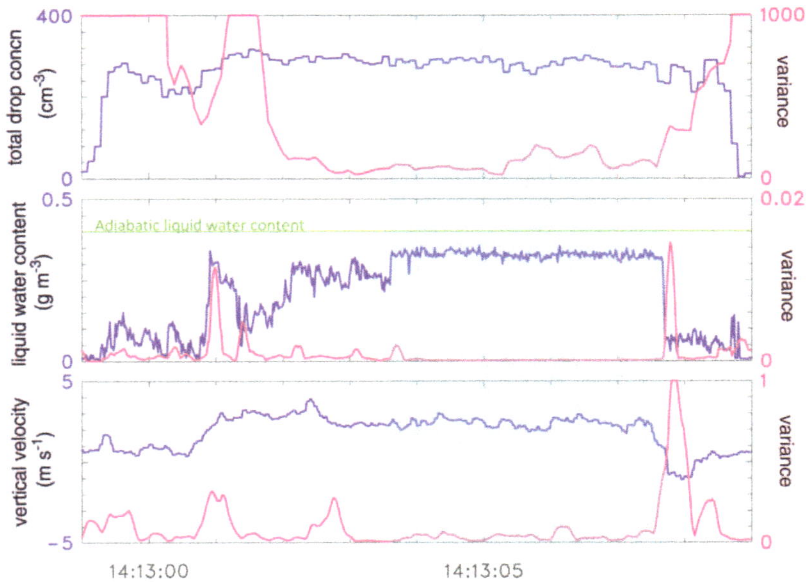

Fig. 3.10: An aircraft transect through a Florida cumulus cloud represented by purple lines depicting cloud drop number concentration, liquid water content, and vertical velocity as functions of time across the cloud. Variance in each quantity is plotted as a pink line. Shaded area denotes the likely location of an adiabatic region within the cloud. Adiabatic liquid water content is denoted as a green line in the second panel; here it is likely that the measured L was slightly less than the adiabatic value due to measurement uncertainties.

updrafts. The special dynamical motions of these rotating updrafts may also produce overturning "ribbons" of air (Fig. 3.11), as well as turbulent eddies, both capable of entraining some environmental air, but in limited quantities.

In summary, aside from some cloudy air parcels within very strong updrafts, the majority of air parcels in clouds and storms will have temperatures and liquid water content less than their respective adiabatic values. Due to the entrainment of environmental air into the cloud, sounding-based estimates of cloud characteristics usually overestimate the actual values. When analyzing cloud observations, it can be challenging to determine how much of the deviation in cloud liquid water content from adiabatic values is due to entrainment compared to losses from precipitation fallout. Consequently, numerical simulation is currently the primary method for studying these respective effects.

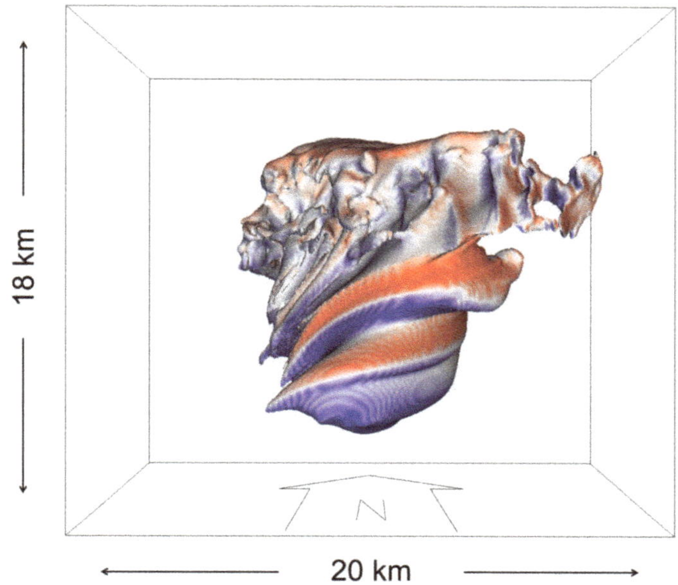

Fig. 3.11: The core of the rotating updraft of a simulated supercell thunderstorm, highlighting locations of entrainment (blue) and detrainment (red) as a result of the overturning, elongated circulations ("ribbons") produced by the rotating updraft. *Adapted from Lasher-Trapp et al. (2021). © American Meteorological Society. Used with permission.*

Questions for Review

1. What is the adiabatic liquid water content, and why is it important?
2. Describe the typical vertical profile of the adiabatic liquid water content within a cloud, and explain how the amount at any given level differs for clouds with different cloud base temperatures.
3. What is cumulus entrainment, and how are the cloud characteristics altered as a result?
4. What is the primary air circulation pattern associated with the shedding thermal conceptual model, and how does it differ for a cloud that tilts with height in an environment with strong vertical wind shear?
5. How common are adiabatic air parcels within clouds, and what processes cause cloudy parcels to deviate from adiabatic behavior?

Homework problems and other resources for students and instructors can be found at bit.ly/4dYBwB5, or scan this code:

Further Reading

For adiabatic concepts [advanced], see Chapter 2 of:

Curry, J. A., and P. J. Webster, 1998: *Thermodynamics of Atmospheres and Oceans*. Elsevier

For more on entrainment [advanced], see Chapter 7 of:

Houze, R. A., Jr., 2014: *Cloud Dynamics*. 2nd ed., Academic Press

CHAPTER 4

Hydrometeor Types and Characteristics

We now transition from the "macroscale" perspective, which is typically the focus of operational meteorologists, to the "microscale" realm. On this smaller scale, we will explore the characterization and behavior of water droplets and ice crystals within clouds, along with their potential evolution into precipitation.

A **hydrometeor** is any liquid droplet or mass of ice in the atmosphere that has formed from water vapor, including those produced by collections of multiple droplets and/or ice particles originally formed from water vapor. It is essential to understand the names and characteristics of different hydrometeor types to communicate effectively with other scientists. Some of the terms used by a cloud physicist might differ slightly from those used by a weather forecaster and therefore the general public.

4.1 Frequently Used Units

Let us first briefly review the tiny scales and units common in cloud microphysics before starting our discussion of hydrometeor types and sizes. These scales are much smaller than those traditionally used in meteorology and are also smaller than the ones we discussed when examining the macroscale characteristics of clouds in Chapters 1 through 3.

The smallest hydrometeors are often measured in micrometers (μm), a unit of length defined as 10^{-6} m. Larger hydrometeors are measured in units of millimeters (10^{-3} m) or centimeters (10^{-2} m), with the following equivalents:

$$1{,}000\ \mu\text{m} = 1\ \text{mm} = 0.1\ \text{cm} = 0.001\ \text{m}. \tag{4.1}$$

To slightly confuse matters, different research communities have varying preferences for describing the sizes of hydrometeors. Observations of hydrometeor sizes are usually reported in terms of their *diameters, D*, because this is how the instruments measure them. When hydrometeors are not spherical, their sizes are reported as their maximum dimension, which is the largest length of each hydrometeor, sometimes as projected onto the 2-D sensor. Conversely, numerical models tend to use calculations in terms of hydrometeor *radius, r*, (or equivalent spherical radius if aspherical hydrometeors are considered), because the equations for hydrometeor growth that models use are most conveniently expressed in terms of radius.

Thus, care must be taken when comparing and interpreting the results of various studies that might be reporting hydrometeor sizes in different ways. This attention is most important when comparing quantities involving powers of D and r. For example, the volume of a sphere is given by $(4/3)\pi r^3 = (\pi/6)D^3$.

We will often use a quantity called the **number concentration** (sometimes referred to as the *number density*) of hydrometeors, which is the number of hydrometeors per unit volume of air. When hydrometeors are abundant, their concentration is usually reported as the number of hydrometeors per cubic centimeter of air, denoted as N cm^{-3}, where N is typically expressed as an integer.

Some hydrometeors are commonly present in much smaller number concentrations, so they are expressed as the number of hydrometeors per liter of air; that is, as N L^{-1}. For the largest hydrometeors, which are commonly the fewest in number, their number concentrations are more conveniently reported as the number of hydrometeors per cubic meter of air, denoted as N m^{-3}.

For reference, the same number concentration of hydrometeors can be expressed as:

$$1\ \text{cm}^{-3} = 10^3\ \text{L}^{-1} = 10^6\ \text{m}^{-3}. \tag{4.2}$$

There is thus a factor of 1,000 difference between each common unit of volume.

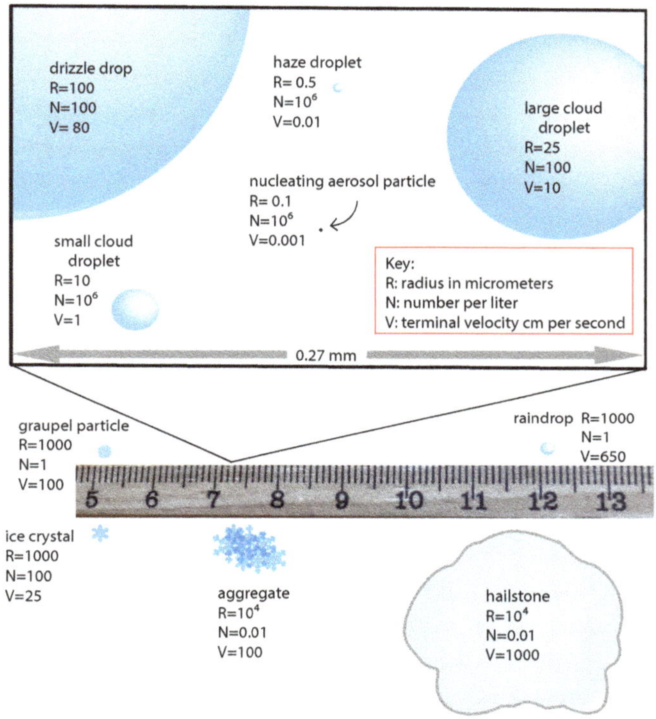

Fig. 4.1: Schematic of sizes (in terms of approximate radius), number concentrations, and fall speeds of various liquid and frozen hydrometeors typical of Earth's atmosphere, compared to a ruler labeled in centimeters. Top panel is magnified relative to the ruler, for visibility. The number concentrations shown for aerosol particles and hydrometeors can often be several orders of magnitude lower, especially at larger sizes. *After McDonald (1958).*

4.2 Aerosol Particles

Small particulates in the atmosphere that do not consist of water in one of its phases are referred to as **aerosol particles** (AP). Some can facilitate a change of water phase, that is, they help *nucleate* cloud droplets or ice. The AP of greatest abundance, which are most important for creating water drops in clouds, are present in concentrations from tens up to a few thousand per cubic centimeter of air (the higher value is equivalent to 10^6 L^{-1} as shown in Fig. 4.1), with radii on the order of 0.1 μm. Chapter 7 will take a closer look at the characteristics necessary for these particles to nucleate droplets or ice in clouds, and Chapter 8 will further discuss their size ranges and number concentrations, as well as their sources and sinks in the atmosphere.

4.3 Types of Liquid Hydrometeors

Liquid hydrometeors in clouds can span five orders of magnitude in size. Their relative scales are shown in the top two-thirds of Fig. 4.1. Under especially favorable conditions, growth from the smallest to the largest liquid hydrometeors can occur in as little as 15 minutes, a remarkable feat of nature.

4.3.1 *Haze droplets*

Haze droplets are the smallest liquid hydrometeors typically found in the atmosphere. These droplets are often 0.5 μm in radius or smaller, and can be as tiny as a tenth of a micrometer in radius. In extreme cases, they can be up to 1 μm in radius. Haze droplets can occur in concentrations reaching several thousand per cubic centimeter of air (10^6 L^{-1}, as shown in Fig. 4.1). When present in sufficient numbers, these minute droplets contribute to the atmospheric phenomenon known as **haze**, in which the air is not yet cloudy (saturated) but is significantly less transparent than on a clear, dry day. Haze is commonly observed during the summertime when the local relative humidity is high, as well as in urban or industrial environments impacted by pollution. It can also be found in maritime air masses that contain abundant sea salt particles generated by breaking waves and sea spray. The formation of haze droplets is discussed in Chapter 7.

4.3.2 *Cloud droplets*

Cloud droplets are small water droplets with radii ranging from 1 to 100 μm, with typical sizes lying in the range of 5 to 30 μm. In polluted environments, small cloud droplets may be present in numbers as high as several thousand per cubic centimeter of air ($\sim 10^6$ L^{-1}), whereas in the cleanest air over oceans, the number of larger cloud droplets may be as low as one per ten cubic centimeters (10^2 L^{-1}). The characterization of cloud types as *continental* or *maritime* in Chapter 1 is usually intended to indicate the relative number of cloud droplets in the clouds. The nucleation of cloud droplets will be thoroughly examined in Chapter 7 and their growth from water vapor in Chapter 9.

In clouds with temperatures warmer than 0 °C, cloud droplets exist exclusively in the liquid state. In colder clouds, cloud droplets can often be *supercooled*, meaning they can remain in the liquid state despite temperatures being significantly below 0 °C, even if ice particles are present. However, in clouds colder than approximately −40 °C, essentially all cloud droplets freeze, as discussed in Chapter 6.

4.3.3 Drizzle drops

Drizzle drops are liquid drops larger than cloud droplets but smaller than raindrops. Definitions can differ slightly even among cloud physicists: drizzle drops are often defined as those having radii between 50 μm and 1,000 μm (1 mm), but some studies use this term for the large cloud drops depicted in Fig. 4.1 with a radius as small as 25 μm.[1] They can range in number concentration from one to several hundred droplets per liter of air. Drizzle drops can also be supercooled in clouds at temperatures far below 0 °C, theoretically down to −40 °C, but as a general rule, the larger the drop, the less likely it is to remain supercooled at very low temperatures.

As we will see in Chapters 9 and 10, not all clouds are capable of producing drizzle drops, as additional microphysical processes must occur beyond their growth from water vapor alone. They can also be produced by the melting of moderately sized frozen hydrometeors, as discussed in Chapter 12.

4.3.4 Raindrops

Raindrops are the largest liquid hydrometeors in clouds, typically defined as those having radii exceeding 1 mm (1000 μm).[2] Their number concentrations are the smallest of all liquid hydrometeors, with typical values ranging from several up to thousands per cubic meter of air (1 L^{-1}). Most shallow clouds are incapable of producing raindrops in their lifetime, for reasons we will examine in Chapter 10. Like drizzle drops, raindrops can also be produced by the melting of large frozen hydrometeors, discussed in Chapter 12.

Raindrops can also exist in a supercooled liquid state at temperatures well below 0 °C, sometimes reaching the ground as *freezing rain*; however, due to their larger size, they are even less likely than drizzle to persist in this state at very low temperatures.

---

[1] In the context of WMO weather reporting practices, the term drizzle is used slightly differently still, referring to fairly uniform precipitation composed exclusively of closely spaced, fine drops with radii of 100 to 250 μm. Thus, this definition requires an assessment of the drop frequency (close spacing, fairly uniform precipitation) and considers drizzle to be within a smaller subset of sizes than the definition used by the cloud physicist.

[2] Since the operational meteorology definition of drizzle differs from that of the cloud physicist, the definitions for raindrops must also differ. According to U.S. reporting practices, raindrops are liquid hydrometeors with radii greater than 0.25 mm.

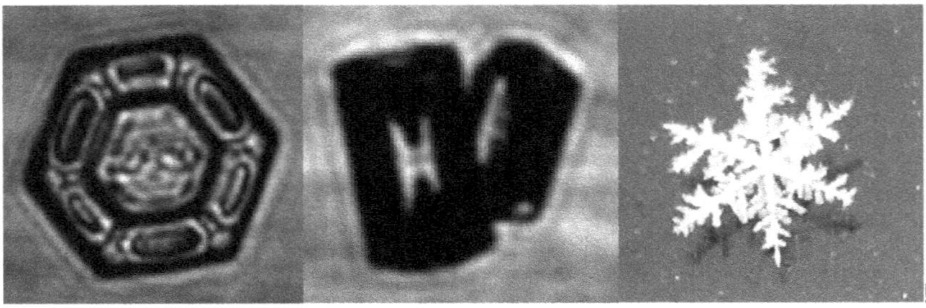

Fig. 4.2: Three general types of ice crystal habits: a plate (left), columns (center; two are shown in this image), and a dendrite (right). *Left and center images from the Cloud Particle Imager; photo on the right by S. Lasher-Trapp.*

4.4 Types of Frozen Hydrometeors

Frozen hydrometeors in clouds can span five or more orders of magnitude in size, as shown in the lower part of Fig. 4.1. They usually exist in lower number concentrations than liquid drops of comparable sizes. This is because the special aerosol particles that nucleate ice in Earth's atmosphere (not pictured in Fig. 4.1) are typically present in much lower number concentrations, 100 L^{-1} or less.

Because frozen hydrometeors are often non-spherical, measuring and modeling their growth is more challenging. To easily compare the scales of liquid and frozen hydrometeors in our discussion here and in Fig. 4.1, we conceptualize a hypothetical sphere around each type of frozen hydrometeor and use its radius as a representative scale.

4.4.1 Ice crystals

We will use the term **ice crystals** to refer to individual particles of ice with a clearly crystalline form. There is no set limit on the sizes defining an ice crystal, but individual crystals tend to range from a few micrometers up to about a millimeter in radius. Their number concentrations can also span orders of magnitude but are often 100 L^{-1} or less, except for some special cases of enhanced ice production discussed in Chapter 7.

Ice crystals can be further classified based on their specific shapes or **habits**. At a general level, there are three basic habits: *columns*, *plates*, and *dendrites*, as shown in Fig. 4.2. However, there are many variations in ice crystal habits as shown by the examples in Fig. 4.3, requiring their own classification scheme. This classification scheme, along with the conditions required for these different habits to grow, is addressed in Chapter 9.

Many ice crystals defy easy classification because they are combinations of different habits. They may also lack the classic symmetry that is typically expected. Combinations of habits that grow directly from water

Fig. 4.3: Images of ice crystals and a possible supercooled droplet (bottom left) collected by an aircraft in cold stratiform clouds. *Courtesy of J. Stachnik.*

vapor alone, without the collection of other ice crystals, are referred to as "polycrystals." The fourth image in the top row of Fig. 4.3 is an example of a polycrystal, but several other images are candidates as well. Polycrystals tend to form at low temperatures (less than −20 °C).

4.4.2 Aggregates

Aggregates are clumps or clusters of two or more ice crystals that are stuck together. Aggregates can have a maximum dimension ranging from 1 mm to tens of millimeters (radii roughly from 0.5 mm to 1 cm, represented as $10^4 \mu$m in Fig. 4.1), which greatly exceeds the maximum size of raindrops. Aggregates of dendrites are commonly seen at ground level, as are aggregates of needles (extremely long and thin columns with jagged edges) or even combinations of several crystal types, as shown in Fig. 4.4. Like ice crystals, their number concentrations can span orders of magnitude, but measurements at the ground are not likely to exceed 10 m^{-3} (0.01 L^{-1}). The process of aggregation is addressed in Chapter 11.

When cloud physicists use the term "snow," they are generally referring to aggregates. Weather forecasters and observers, however, do not typically differentiate between individual ice crystals and aggregates, describing both as "snow," since both contribute to surface accumulations

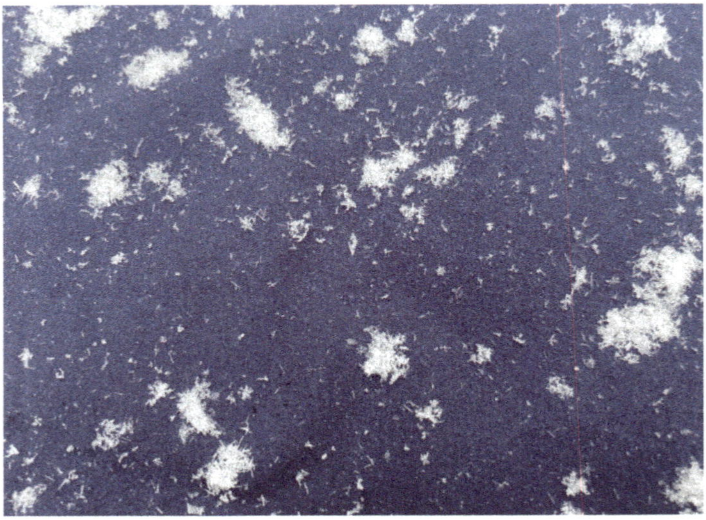

Fig. 4.4: Aggregates composed of dendrites and/or needles, with some individual crystals also shown.

that are indistinguishable to commuters and skiers alike. However, understanding the processes of precipitation formation requires recognizing such distinctions.

4.4.3 Graupel

When frozen hydrometeors collide with supercooled droplets, the droplets freeze upon impact, forming a bumpy or granular layer of ice on the frozen hydrometeors, called **rime**. The top set of images in Fig. 4.5 shows examples of ice crystals rimed to various degrees. If the rime covers the surface of ice crystals to the extent that their original habits cannot be distinguished, the frozen hydrometeors are collectively referred to as **graupel** (far right). Graupel particles do not have a defined minimum size, but they do have an upper size limit of 2.5 mm (2,500 μm) in radius, a limit shared by both cloud physicists and weather observers. Larger graupel particles can be comparable in size to raindrops. The number concentrations of graupel vary widely depending on cloud type and other factors, but can reach values on the order of 1 L^{-1}.

Graupel particles often occur in two basic shapes: spherical or conical, as shown at the bottom of Fig. 4.5. The conical shape arises when the particle maintains a fixed orientation (base side down) while growing. Graupel particles have lower densities than pure ice due to tiny air pockets trapped between the collected droplets as they freeze. Rime also

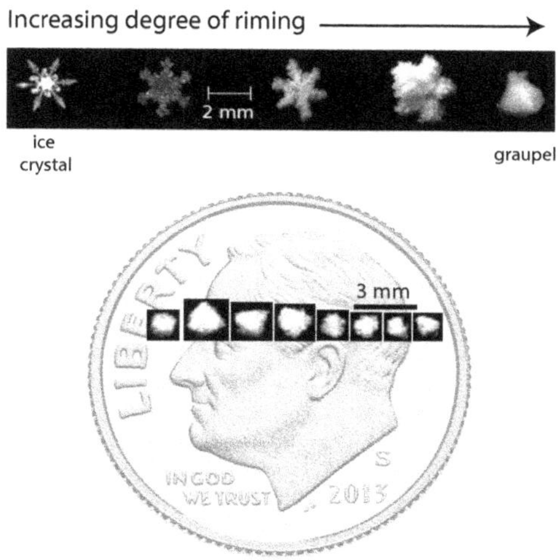

Fig. 4.5: *Top row:* Examples of ice crystals with increasing amounts of rime, eventually becoming a graupel particle if the original habit is indistinguishable. *Bottom row:* Examples of conical graupel and spherical graupel with scale relative to a U.S. dime (about 18 mm in diameter). *Top image from Praz et al. (2017) (CC-BY). Bottom image courtesy of S. Yuter.*

imparts an opaque (milky-white) color to the particles, as opposed to clear ice. Riming and the formation of graupel are discussed in Chapter 11.

Weather forecasters and observers often use the term "graupel" when these hydrometeors reach the ground from convective storms in warm seasons. This occurs especially at higher elevations, where they may not have time to melt first. However, in winter, they may instead refer to these hydrometeors as "snow pellets."

4.4.4 Hailstones

In extreme cases, when there are abundant supercooled drops available for a frozen hydrometeor to collect, sufficient rime can accumulate and increase the hydrometeor size to greater than 2.5 mm in radius. These hydrometeors are then called **hailstones**, or collectively, **hail** (Fig. 4.6). They tend to have a higher density than graupel particles, as less air is trapped within the rime due to its rapid accumulation. In some cases, layers within the hailstones may contain only clear ice with no air.

The number concentration of hailstones estimated from measurements can vary widely. Smaller hailstones may number up to 10 per cubic meter of air, whereas hailstones larger than 2.5 cm in radius generally have much lower number concentrations, on the order of 10^{-3} m^{-3} or less.

Hailstones are commonly defined as having a minimum radius exceeding 2.5 mm; a larger size is shown in Fig. 4.1 but is typical of commonly

Table 4.1: Size categories of hailstones: severity labels as currently used by the U.S. National Weather Service, and larger categories as proposed by Kumjian et al. (2020).

Size Category	Max. dimension (cm)
Small/sub-severe	< 2.5
Large/severe	≥ 2.5
Significantly severe	≥ 5
Giant	≥ 10
Gargantuan	≥ 15

Fig. 4.6: *Top:* A large hailstone with a golf ball (nearly ∼4.5 cm diameter) and a U.S. quarter (∼2.5 cm diameter) shown for reference. *Bottom:* A record-setting hailstone that fell in Vivian, South Dakota, on 23 July 2010. The hailstone broke the United States record for largest hailstone by diameter (20.1 cm/7.9 inches) and weight (0.88 kg/1 pound, 15 ounces). *Top photo by S. Lasher-Trapp; bottom photo courtesy of the Aberdeen SD National Weather Service office.*

observed hailstones. As they grow to larger sizes, they tend not to be spherical and are more commonly sized by their maximum dimension, as shown in Table 4.1. Meteorologists in the U.S. classify hailstones as "large" when their maximum dimensions equal or exceed 2.5 cm (1 inch). Due to their great velocities and masses, large hailstones can sometimes damage crops, buildings, and vehicles, as well as cause injury to livestock and humans. Thunderstorms producing (or forecast to produce) large hailstones are accordingly labeled as "severe," emphasizing the potential for such damage.

4.5 Other Hydrometeor Terms

Weather forecasters often need to warn the public about hazards caused by various types of frozen or freezing precipitation. These terms are briefly mentioned here to emphasize that they do *not* result from unique microphysical processes within the cloud. Instead, their formation is influenced by the temperature profile between the cloud base and the ground. This topic will be covered in detail in Chapter 12.

Sleet, sometimes referred to as "ice pellets," is composed of transparent or translucent, round or irregular frozen hydrometeors with radii less than 2.5 mm. It essentially consists of frozen raindrops. For sleet to form, snow or graupel must first fall through a layer of warm air that melts it partially or completely into raindrops or at least watery slush. These particles subsequently refreeze as they pass through a deep lower layer with a wet-bulb temperature below 0 °C. Sleet often bounces when it hits hard surfaces.

Snow grains are similar to graupel, having formed through riming, which gives them an opaque white appearance. However, they are much smaller, typically less than 0.5 mm in radius.

Freezing rain or **freezing drizzle** consists of supercooled raindrops or drizzle drops that freeze immediately upon contact with a surface. Similar to sleet, they occur when frozen hydrometeors fall through an elevated layer of the atmosphere warm enough to completely melt them into rain or drizzle. They subsequently descend into a colder, shallow layer of air above the surface that cools the drops to well below 0 °C again. However, unlike sleet, there is insufficient time for the drops to refreeze before they reach the surface. Freezing precipitation can lead to significant accumulations of clear ice or glaze on roads, sidewalks, trees, and utility poles, causing damage and hazardous conditions for pedestrians and drivers.

4.6 Hydrometeor Size Distributions

Beyond knowing the total number concentration for a given kind of hydrometeor, it is often of interest to know their number concentration in relation to their size. As discussed in previous sections of this chapter, the size ranges and *total* number concentrations of different types of hydrometeors can often span multiple orders of magnitude. These details of hydrometeor (and aerosol particle) populations are typically presented in the form of a **size distribution**. The size distribution indicates how many hydrometeors of given sizes are present in a volume of air. Knowledge of the

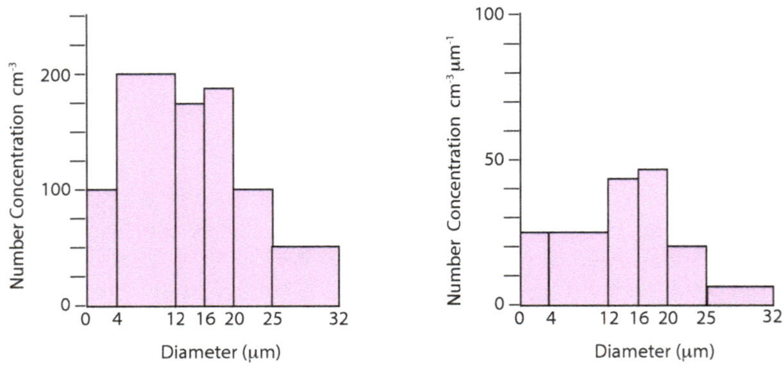

Fig. 4.7: *Left*: Example of a size distribution of droplets as produced by the raw data from an instrument that measures hydrometeor sizes in uneven bin widths. *Right*: The same data, but values on the vertical axis have been normalized by the bin widths to scale the numbers similarly. The total number concentration N is identical for both plots and equal to 813 cm^{-3}.

size distribution is crucial for predicting microphysical processes within numerical models and when computing radar backscatter and other physical phenomena that depend on hydrometeor size.

4.6.1 Discrete size distributions

The size distribution can be represented in a number of ways. The simplest method places the hydrometeors into "bins" according to their size (usually diameter, but sometimes radius). Shown in the left panel of Fig. 4.7 is a size distribution of cloud droplets measured by a commonly used instrument. The **number concentration** (n_D) is the number of droplets with diameters between the lower and upper bounds of each bin. To find the total number of droplets (**total number concentration**, N) in the population per unit volume of air, the individual number concentrations in all the bins are added together, i.e.,

$$N = \sum_i n_{D,i} \quad , \tag{4.3}$$

where the subscript i ranges from 1 to the maximum number of size bins for that instrument. Aerosol particles and cloud droplets are typically present in larger number concentrations, so N is expressed as the number per cubic centimeter of air (cm^{-3}). Frozen hydrometeors are typically far fewer in number, so the number concentrations are instead quantified in units of number per liter of air (L^{-1}). For the largest hydrometeors like raindrops, aggregates, and hailstones, the number per cubic meter of air (m^{-3}) is used.

To determine the liquid water content L (as discussed in Chapter 2) for a given cloud droplet size distribution, one sums the mass within all its bins. Specifically,

$$L = \rho_l \sum_i \frac{\pi}{6} D_i^3 n_{D,i}, \tag{4.4}$$

where ρ_l is the density of pure liquid water (1 g cm^{-3}), D_i is the midpoint of the diameter range in a given bin, and it is assumed that the cloud droplets are small enough to be treated as perfect spheres with volume $\pi D_i^3/6$. This computed liquid water content can be compared with measurements from other instruments that assess total water mass (discussed in Chapter 5) and/or the adiabatic values calculated in Chapter 3.

Microphysical probes that count and size hydrometeors and store these data in bins will be discussed in detail in Chapter 5. For some probes, the bins are uniformly spaced, e.g., all having widths of 10 or 100 μm, but for some other probes, the bins are unevenly spaced to retain the greatest accuracy, taking into account how the probe counts and sizes the hydrometeors.

These uneven bin widths can be misleading. For example, the left plot in Fig. 4.7 suggests that the most frequent droplet sizes in this population are those between 4 and 12 μm in diameter. But upon further inspection, one sees that this bin is much wider, so that it contains a larger size range of droplets compared to other bins. So perhaps that bin just has more droplets because it contains a greater variety of sizes?

Thus, when dealing with uneven bin widths, it is useful to *normalize* the values on the vertical axis by the width of each bin to enable "fair" comparisons between bins. This normalization is applied in the plot on the right, using the same data as shown on the left, but now plotting the **normalized number concentration** (n) on the vertical axis.

Note how the units on the vertical axis have changed between the two plots. This is an easy way to tell if the number concentrations have been normalized. The raw data have number concentrations expressed in units of number per unit volume of air. The normalized number concentrations have units of number per unit volume of air, *per unit size* of hydrometeor (diameter or radius). We now see that the true peak (the "mode") in the distribution is at larger sizes, 16–20 μm in diameter, for this particular example. In the plot on the left, it only appeared that the peak was at smaller sizes because the wider bin included many more droplets.

When the number concentrations are normalized, one can then find the total number concentration of hydrometeors in the population by

$$N = \sum_i n_i \Delta D_i, \tag{4.5}$$

where it is now necessary to multiply the number in each bin by its respective bin width, ΔD. Convince yourself, using (4.3) and (4.5), that the total number of hydrometeors in each plot in Fig. 4.7 is the same.

4.6.2 Surface area and volume contributions

Sometimes, a population of hydrometeors is quantified based on the contribution of each size interval to the *surface area* or *volume* rather than to number concentration. The hydrometeor surface area is useful in certain atmospheric chemistry problems and for calculations to quantify how reflective of solar radiation a cloud may be. The hydrometeor volume can approximate its water mass, particularly for populations of water droplets with known and constant density.[3] Figure 4.8 depicts a population of cloud droplets in all three ways. The red bins show the droplet size distribution, which reveals that there are far more smaller droplets than larger ones in this population. The peak number of droplets in this case occurs at a radius of 10 μm.

When the same data are instead represented as a surface area distribution (blue bins), the peak in the distribution shifts toward larger sizes (now 18 μm). This shift occurs because the surface area is proportional to the square of the radius of the droplets. Therefore, even though the smaller droplets are more numerous, their collective surface area is not as large as that at somewhat larger sizes.

The last plot (green bins) presents the equivalent volume distribution for this sample of droplets. Since the volume is proportional to the cube of the droplet radius, the peak in the distribution has shifted even further toward larger radii (20–22 μm). The volume distribution also indicates that these droplet sizes dominate the overall mass of the population, as all droplets have the same density.[4] An essential point to remember is that while the smallest droplets are often more numerous, it is typically the larger hydrometeors, albeit fewer in number, that contribute the most to the overall water mass.

4.6.3 Gamma distributions

Sometimes, idealized analytical functions are fitted to hydrometeor or aerosol particle size distributions to facilitate their representation in numerical models. Although many analytical functions exist, the most com-

[3] A more customary distribution to show in this case is the mass distribution, which, for aerosol particles or hydrometeors of the same composition, would be just the volume distribution multiplied by the hydrometeor density.

[4] If the distribution contained ice with different densities, the volume distribution would not be proportional to the mass distribution.

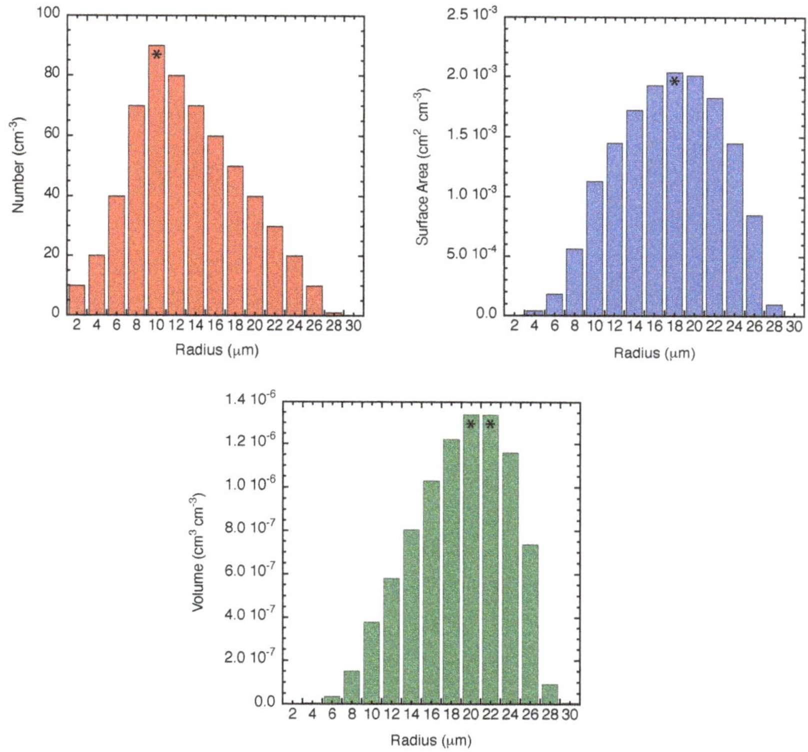

Fig. 4.8: Sample data for droplets showing the droplet size distribution (red bins; top left), the surface area distribution for the same droplets (blue bins; top right), and the volume distribution of the same droplets (green bins; bottom). Each quantity is expressed in its respective units per cubic centimeter of air. An asterisk marks the size where the peak in each distribution occurs.

monly used is the three-parameter *gamma distribution*, given by

$$n = N_0 D^{\mu} e^{-\Lambda D}, \tag{4.6}$$

where n represents the number of hydrometeors per unit volume (e.g., m^{-3}) per unit size interval (e.g., mm^{-1} diameter). The parameter N_0 controls the overall amplitude, while the parameters μ and Λ determine the shape of the distribution, which can vary significantly, as shown in Fig. 4.9.

For smaller aerosol particles, cloud droplets, and ice crystals, the full three-parameter gamma distribution often fits the measured size distributions very well. As we shall see in later chapters, measured size distributions for these smaller particles and hydrometeors tend to exhibit small-to-moderate number concentrations at the smaller sizes, a peak in number

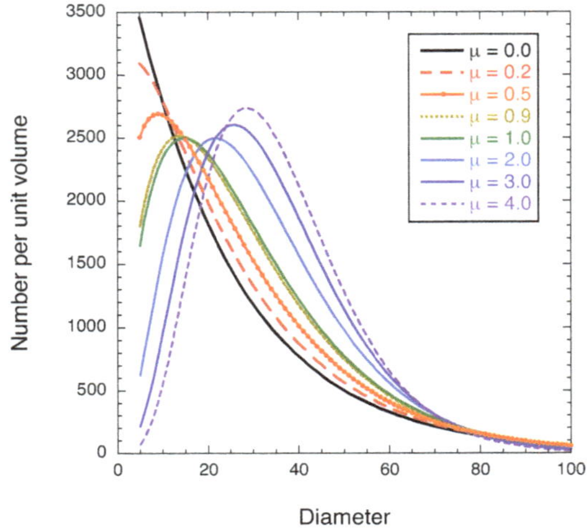

Fig. 4.9: Examples of various shapes of hydrometeor size distributions as represented by gamma distributions using (4.6) with different values of the parameter μ (shown in legend) while holding N_0 and Λ constant. In this plot, volume and diameter units are not specified, as any pertinent units can be used for the particular hydrometeors of interest.

concentration at moderate sizes, and then a long tail of decreasing number concentration for the largest sizes, similar to the colored lines in Fig. 4.9.

4.6.4 Exponential distributions

For the special case that $\mu = 0$, the gamma distribution simplifies to a negative *exponential distribution*, given by

$$n = N_0 e^{-\Lambda D} \tag{4.7}$$

and is depicted as the black curve in Fig. 4.9. By integrating over (4.7), one can find that the total number concentration of hydrometeors in this distribution is

$$N = \frac{N_0}{\Lambda}. \tag{4.8}$$

This negative exponential distribution is useful for representing the larger hydrometeors produced by clouds, as discussed further in the next sections.

i) Marshall-Palmer distribution for raindrops

Raindrop size distributions measured at the ground are important for a variety of reasons, including obtaining accurate rainfall amounts and understanding their ability to scavenge pollution (e.g., aerosol particles) as

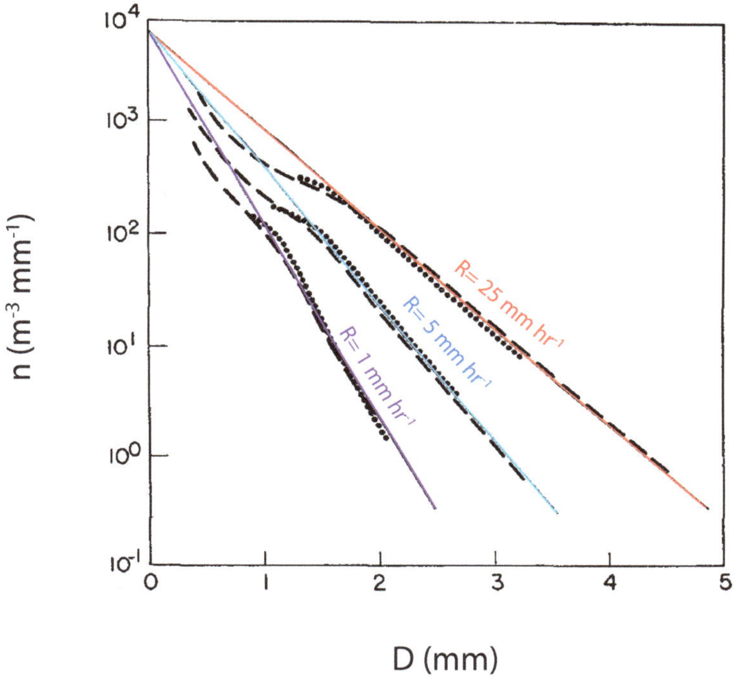

Fig. 4.10: Examples of different observed raindrop size distributions for the labeled rain rate R (dots are one set of observations; dashed lines are another). The corresponding Marshall-Palmer distributions are depicted as solid lines following (4.7), (4.9), and (4.10) for the indicated rain rates. *Adapted from Marshall and Palmer (1948).*

they fall. As we will see later when discussing radar estimation of precipitation, much work has been done in the past to understand and quantify raindrop populations at ground level.

In fairly steady rain—i.e., neither at the very start nor the very end of a rain shower—the raindrop size distribution can often be described by an exponential distribution, as expressed in (4.7). This means that the number of raindrops per unit volume decreases exponentially with increasing size. If the vertical axis of normalized raindrop number concentration is on a logarithmic scale, the decrease in number concentration with increasing raindrop size appears linear, as indicated by the straight lines in Fig. 4.10.

Actual raindrop size distributions can vary widely from storm to storm depending on the meteorological environment. However, the model commonly assumed when no other specific information is available is the **Marshall-Palmer raindrop size distribution,** named for the scientists who first published a study of their raindrop measurements on the ground from thunderstorms in 1948. It is an exponential distribution of the form given by (4.7), with the parameters N_0 and Λ defined as follows:

$$N_0 = 8000\,\mathrm{m}^{-3}\,\mathrm{mm}^{-1} \tag{4.9}$$

and

$$\Lambda = 4.1R^{-0.21}\,\mathrm{mm}^{-1}, \tag{4.10}$$

where R is the rain rate in mm hr^{-1}. Examples for specific rain rates are shown as solid lines in Fig. 4.10.

An interesting feature of the Marshall-Palmer distribution is that the intercept parameter N_0 is independent of rain rate; only the slope parameter Λ varies. As rain rate increases, the slope flattens, and larger drops become much more numerous while the number of very small drops remains nearly unchanged.

Raindrop observations since those of Marshall and Palmer have indicated that N_0 can differ substantially across storms in the same geographical area, and even during the rainfall from a single storm. Thus, most current models and algorithms do not assume a constant N_0 for the raindrop size distribution.

ii) Exponential distributions for graupel, snow, and hail

In addition to raindrops, the size distributions of snow, graupel, and hail are often represented as negative exponential functions. The added complexity of fitting gamma distributions may not justify much improvement in the representation of these measured size distributions (Fig. 4.11). The exponential distributions used for these hydrometeors have different values for the parameters N_0 and Λ compared to those for raindrop size distributions. Although all types of precipitating hydrometeors typically include many more small hydrometeors and far fewer large ones, significant deviations from an idealized exponential distribution can and do exist.

4.7 Hydrometeor Bulk Density

In our investigation of the growth of various hydrometeors, we will see that their density (mass per unit volume) greatly influences their fall speed. Additionally, higher density hydrometeors transport more water mass to the ground as precipitation.

We will assume in this textbook that all liquid hydrometeors have the standard density of water, ρ_l =1 g cm^{-3}. Departures from this value due to temperature are small and usually ignored.

The densities of frozen hydrometeors, on the other hand, are much more variable. At the molecular level, the density of *pure* ice, ρ_i, is approximately 0.917 g cm^{-3}, with minor variations due to temperature. However,

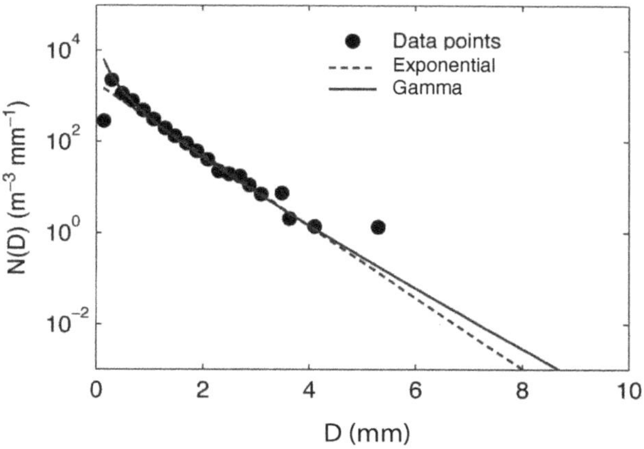

Fig. 4.11: An aggregate size distribution observed at ground level during a snowfall event in Colorado based upon 2D video disdrometer data (black circles). The low data point at the smallest size is presumed to be due to instrument bias in measuring the smallest aggregates. Corresponding best fits of an exponential distribution following (4.7) and of a gamma distribution (4.9) are also shown. *Adapted from Brandes et al. (2007). © American Meteorological Society. Used with permission.*

as mentioned earlier, air can make up a significant fraction of the volume of some ice crystals, rimed particles, and aggregates, potentially reducing the average density of the particles well below that of pure ice.

As a result, we often have the need to describe the effective or **bulk density** of a frozen hydrometeor, which can vary due to its shape, its size, and any added mass due to rime. The bulk density ρ_x for any given hydrometeor of type x is defined as

$$\rho_x = \frac{\text{Mass}}{\text{Volume}}. \tag{4.11}$$

Here, "Mass" is the total mass of the hydrometeor, and "Volume" refers to the volume of a simple, known geometric shape that encloses the entire particle. Calculating the true volume of certain crystal habits, rimed crystals, and aggregates, whether rimed or not, can be very challenging due to their asymmetry.

Thus, for mathematical simplicity, we approximate the volume of the frozen hydrometeor using a shape for which the volume calculation is straightforward. For example, a plate or dendritic crystal can be approximated by a flattened disk with radius r and height h, such that its volume is given by $V = \pi r^2 h$. A similar formula applies to a columnar ice crystal, which would have significantly smaller r and much larger h.

More complex treatments might use the volume of a right hexagonal prism, as shown on the left side of Fig. 4.12. For more complicated or irreg-

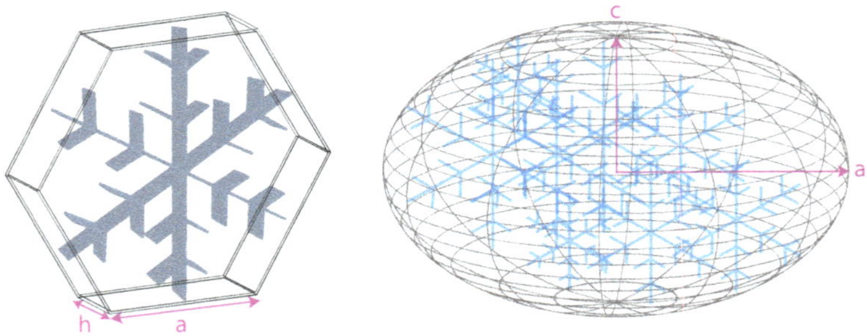

Fig. 4.12: Illustrations showing how the volumes of an ice crystal (left) or aggregate (right) might be represented by simple shapes for calculating the bulk density as given by (4.11). The volume of a right hexagonal prism (left) is $V = (3\sqrt{3}/2)a^2h$ and of an oblate ellipsoid (right), $V = (4/3)\pi a^2 c$, using dimensions as labeled.

ular shapes, such as polycrystals or aggregates, the volume of an ellipsoid[5] of a comparable maximum dimension is often used (right panel).

Using the dendrite shown as an example, we can estimate its bulk density if certain other quantities are known. Suppose that the dendrite has a mass of 0.04 mg, a width of 2 mm, and a depth of 50 μm. Using straightforward geometry, we find that the values of a and h shown in Fig. 4.12 correspond to 0.1 cm and 0.005 cm, respectively, yielding a total volume of 1.3×10^{-4} cm^3. Dividing the crystal mass by this volume gives a bulk density of approximately 0.3 g cm^{-3}, which is much less than the standard bulk density of ice (0.917 g cm^{-3}).

Dendrites tend to have the lowest bulk densities, 0.2 g cm^{-3} or even less, while more "solid" ice crystals like columns and plates have a value equivalent to the standard density of ice (Table 4.2). Between these extremes, some plates, and definitely hollow columns, needles, and sheaths, contain "pockets" where only air resides, decreasing their bulk densities. Sector plates have large spaces of air between their sectors, so their bulk densities are less than those of plates, but they have more ice mass than dendrites of similar dimensions and thus greater bulk densities.

Aggregates often exhibit various unusual shapes and can be large in any single dimension, but they contain a significant amount of empty space (air). Some aggregate bulk densities may be less than 0.1 g cm^{-3}. For this reason, when a meteorologist determines the liquid equivalent of

---

[5] An *ellipsoid* is an ellipse rotated around either its major or minor axis to form a three-dimensional object with set formulas for its volume and surface area. We will see in Chapter 9 how the use of such shapes simplifies calculations of the growth of ice crystals.

Table 4.2: Bulk densities of ice crystal habits, as summarized from Ono (1970) and Fukuta and Takahashi (1999).

Crystal Habit	Density (g cm^{-3})
Solid columns	0.9
Plates	0.4 to 0.9
Hollow columns	0.7 to 0.9
Needles	0.6 to 0.9
Sheaths	0.75 to 0.9
Sector plates	0.4 to 0.6
Dendrites	0.2 to 0.3

a particular snowfall depth—by melting a carefully collected sample in a calibrated container—the resulting liquid depth is usually much smaller than the actual snow depth, with ratios near 1:10 being common.

The bulk density has been observed to decrease with increasing crystal/aggregate dimension. For this reason, meteorologists working with numerical weather prediction models or with algorithms that estimate snowfall from satellite or radar measurements commonly employ a **mass-diameter relationship**. These relationships are expressed as

$$m = aD^b, \tag{4.12}$$

where D typically represents the longest dimension of the crystal or aggregate. The coefficients a and b are constants determined for each crystal habit or aggregate type (e.g., aggregates containing few or many individual ice crystals, as well as the crystal habits within the aggregate) based on numerous observations.

This relationship, while simple to implement in a model or algorithm, can also vary substantially depending on geographical region and the conditions under which the ice grew (especially temperature). Researchers are currently studying probabilistic methods to represent the uncertainty in mass-diameter relationships.

The bulk density of frozen hydrometeors can increase significantly with the addition of rime on their surfaces. The bulk densities of graupel particles can range from 0.1 to 0.85 g cm^{-3} or greater, depending on the amount of air trapped within the rime. Graupel particles consisting of single small ice crystals with minimal rime might lie at the low end of this density range, while rimed frozen raindrops would be at the higher end. Hailstones, which grow through the accumulation of much larger amounts of supercooled water, can sometimes contain layers of very dense ice up to 0.9 g cm^{-3}, although they often also have layers with densities as low as 0.5 g cm^{-3}. These variations in bulk density have implications for fall speeds, which are addressed in the next section.

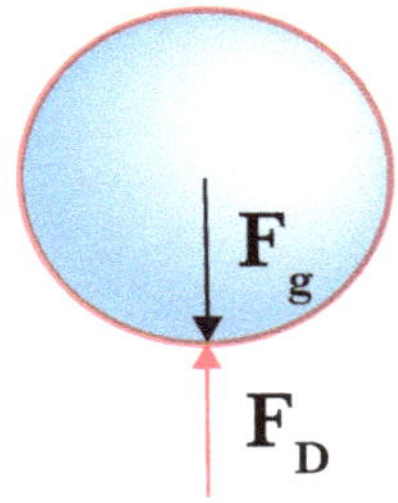

Fig. 4.13: The balance of forces acting upon a falling hydrometeor to achieve its terminal fall speed. The force due to gravity (F_g) acts on the center of mass of the hydrometeor. The drag force (F_D) acts over the entire surface of the hydrometeor (highlighted in red), but here we are only concerned with the net upward component. When the two forces are equal as shown, there is no net acceleration on the hydrometeor, so it falls at a constant speed, V_t.

4.8 Hydrometeor Fall Speeds

We now address how quickly a hydrometeor can fall in still air, i.e., its **terminal fall speed**, V_t. For any object, when the gravitational force, F_g, and drag force, F_D, acting on it are in exact balance (Fig. 4.13), the object falls at its terminal fall speed. This fall speed is constant because there is no net acceleration acting on the hydrometeor. The gravitational force is proportional to the hydrometeor mass. For all but the smallest cloud droplets and ice crystals, the drag force is usually proportional to the hydrometeor cross-sectional area perpendicular to the direction of fall.

It should be kept in mind that the actual speed at which a hydrometeor falls *relative to the ground* and the path it takes through a cloud depend not only on its fall speed but also on additional air currents acting on the hydrometeor, such as cloud updrafts, downdrafts, and small-scale turbulence.

4.8.1 Water drops

While the gravitational force acting on a water drop of a given size is easily calculated, the drag force is not, particularly for larger sizes. For smaller cloud drops, a simple idealized flow can be assumed, with air viscosity dominating the drag force. For larger drops, the generation of a turbulent wake in the air, as well as deformation of the largest drops from a perfect sphere, complicates the calculation of the total drag force acting on the drop. In these cases, both theoretical approximations and laboratory observations are used to estimate the drag force and, consequently, the drop fall speeds as a function of drop size.

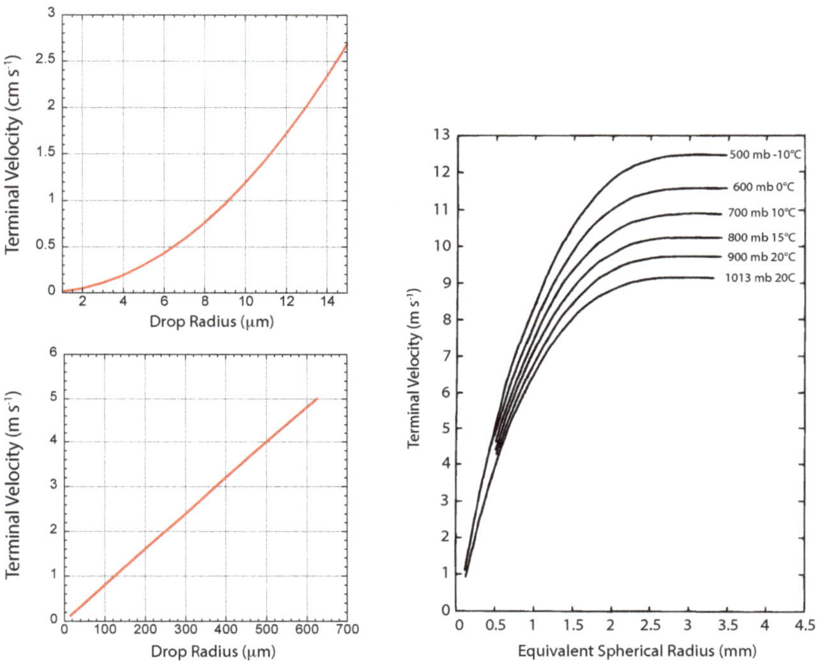

Fig. 4.14: Terminal fall speed versus drop radius for different size ranges. Red curves correspond to equations (4.13) or (4.14). Black curves show laboratory results from Beard (1976); equation (4.15) only represents the part of these curves from 0.625 to 1 mm in radius.

When high accuracy is not required, the following simplified expressions for the fall speeds of water drops can be used:

Size range	**Fall speed**		
$r \lesssim 15\ \mu$m	$V_t \approx C_1 r^2$	$C_1 = 1.19 \times 10^6\ \mathrm{cm^{-1}\ s^{-1}}$	(4.13)
$15 \lesssim r \lesssim 625\ \mu$m	$V_t \approx C_2 r$	$C_2 = 8000\ \mathrm{s^{-1}}$	(4.14)
$625 \lesssim r \lesssim 1000\ \mu$m	$V_t \approx C_3 \sqrt{r}$	$C_3 = 2010(\rho_0/\rho)^{1/2}\ \mathrm{cm^{1/2}\ s^{-1}}$	(4.15)

where ρ_0 is a reference density at sea level equal to 1.20 kg m^{-3} and ρ is the local air density around the drop, also in units of kg m^{-3}. All of these formulas give the fall speed in units of cm s^{-1} when r is expressed in centimeters.

Not surprisingly, Fig. 4.14 shows that, in general, the larger the drop, the faster it falls. However, the *rate* of increase in fall speed with size diminishes due to the rapidly increasing drag force as the drop size grows.

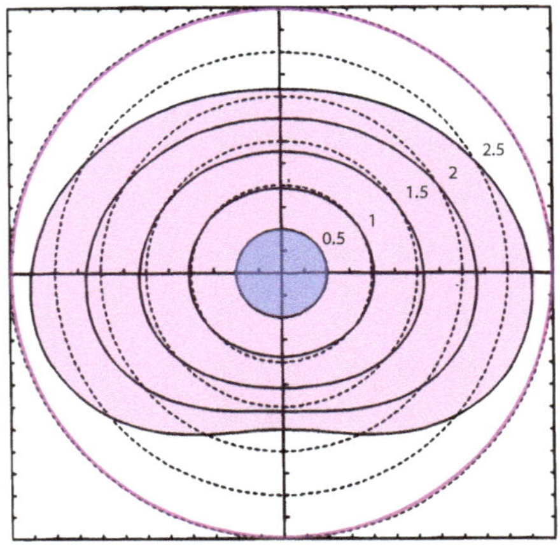

Fig. 4.15: True shape of falling raindrops (solid black curves) compared to the radius they would have if they were shaped as perfect spheres (dashed black curves), at the sizes labeled in millimeters. Raindrops having a radius of 0.5 mm (solid blue circle) do not deviate much from a spherical shape, but larger raindrops increasingly flatten. A raindrop 2.5 mm in radius (purple ring) would have the flattened shape shaded in purple. *Adapted from Beard et al. (1989). © American Meteorological Society. Used with permission.*

The smallest droplets have almost negligible fall speeds, but V_t increases quite quickly with size up to about 15 μm radius, reaching a maximum of a little over 2.5 cm s^{-1}. Over the size range of 15 to about 625 μm radius, which includes large cloud drops, drizzle, and the smallest raindrops, the terminal fall speeds increase up to nearly 500 cm s^{-1} (5 m s^{-1}). Larger raindrops can attain fall speeds ranging from 9 to over 12 m s^{-1}, depending on the local air density. A raindrop of a particular size falls more quickly at higher altitudes, where the local air density is less than at sea level.

For raindrops with radii larger than 1 mm, there is no simple formula for their fall speeds. Observations from laboratory studies show that the increase in fall speeds with increasing size begins to slow down when this size is reached, as shown in Fig. 4.14. For raindrops with radii greater than approximately 2.5 mm, the fall speed remains nearly constant with increasing size for a given local air density. This occurs because raindrops of this size begin to flatten as they fall, as shown by laboratory measurements (Fig. 4.15).[6] The flattened shape effectively acts as a "parachute," increasing the drag force on the raindrop and thus decreasing its fall speed.

[6]Despite how raindrops are often depicted, free-falling raindrops never have a classic teardrop shape! This common misconception likely stems from the appearance of raindrops

The distortion increases with the size of the raindrop, such that eventually the drag force increases at the same rate as the gravitational force, resulting in a nearly constant fall speed with increasing raindrop size. In Chapter 5, we will see how the flattening of larger raindrops can be used to improve radar-derived estimates of rainfall.

4.8.2 Ice crystals

Deriving the fall speeds of ice crystals from first principles is significantly more challenging than for water drops. The drag force acting on intricate crystal habits is difficult to model and is also influenced by any asymmetries in the crystal. The mass-diameter relationship from equation (4.12) can be used in a series of calculations to estimate the drag force, but a given m–D relationship is often not universally valid for a particular crystal habit, as noted earlier.

Even for observations of falling crystals, only approximate *mean* values of crystal fall speeds can be attained, and the actual fall speeds may have significant fluctuations for particles of similar size and habit. The relatively large area-to-volume ratio, crystal asymmetries, and low densities of ice crystals result in irregular motions, including components that are helical (falling along a path similar to a corkscrew), rotational (spinning about an axis), and/or sideways slipping (movement side-to-side without rotation). Some types of periodic motions can also be observed as ice crystals fall (Fig. 4.16).

Ice crystals generally fall with their largest dimension facing downward. In particular, flat, plate-like crystals tend to fall oriented horizontally, as seen in optical phenomena such as sun pillars and sundogs. But they can also exhibit some tilt from this orientation. This tilt can be caused by factors such as surface irregularities or inhomogeneities in the crystal density, leading to a lack of a single stable orientation. For instance, crystals may flutter downward like falling leaves, as depicted in the side views of falling plates in Fig. 4.16, or they may even tumble during their descent. The orientation of ice crystals relative to the ground changes over time due to these irregular motions. Consequently, their cross-sectional areas perpendicular to the direction of fall also vary over time. Unlike ice crystals, water droplets, which are usually spherical or only slightly flattened, maintain a mostly horizontal orientation. As a result, observed ice crystal

---

as they glide down a windowpane or when a water drop is about to break free from a dripping faucet.

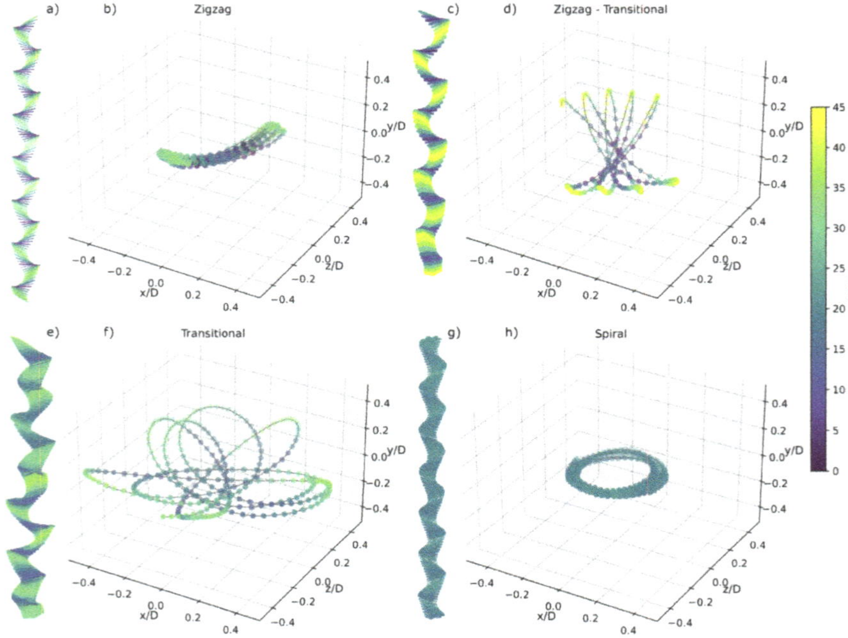

Fig. 4.16: *Top*: Some periodic fall patterns of hexagonal plates and broad-branched plates based on 3D printer models of the crystals. Vertical profiles in (a), (c), (e), and (g) show the side view of the plate as it falls, colored by its inclination angle θ, which is the angle from a perfectly flat orientation of 0° (scale shown right). Three-dimensional plots in (b), (d), (f), and (h) show ice crystal fall trajectories in space, again colored by inclination angle, where connected points represent the crystal center of mass and the distances along the horizontal and vertical axes are normalized by the plate diameter. *From Stout et al. (2024). (CC-BY)*

fall speeds, typically measured as time averages, inherently account for these orientation effects.

Plate-like ice crystals exhibit a wide range of fall speeds depending on their habit, as depicted in Fig. 4.17 (top). Generally, the larger the crystal of a given habit, the greater its fall speed. However, the fall speeds of the most branched crystals (dendrites, having very low mass-to-area ratios) do not significantly change with increasing size.

The fall speeds of plate-like crystals, including dendrites, range from 10 cm s^{-1} to approximately 70 cm s^{-1}. For comparison, a 1-mm-diameter raindrop falls at 300 cm s^{-1}, and a 2 mm raindrop falls at 600 cm s^{-1}. The effects of the non-spherical ice shapes and lower densities clearly slow the fall speeds of individual ice crystals.

Thick plates fall faster than thinner plates of equivalent diameter, due to their greater mass but equivalent drag forces. For a given size of

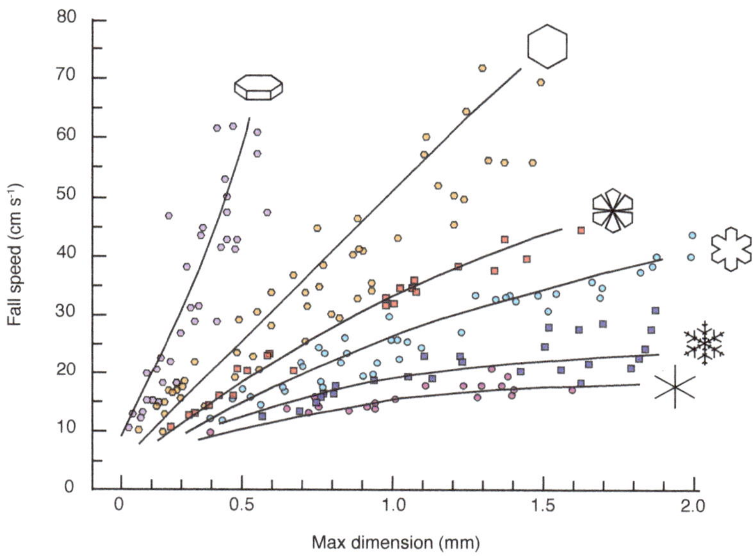

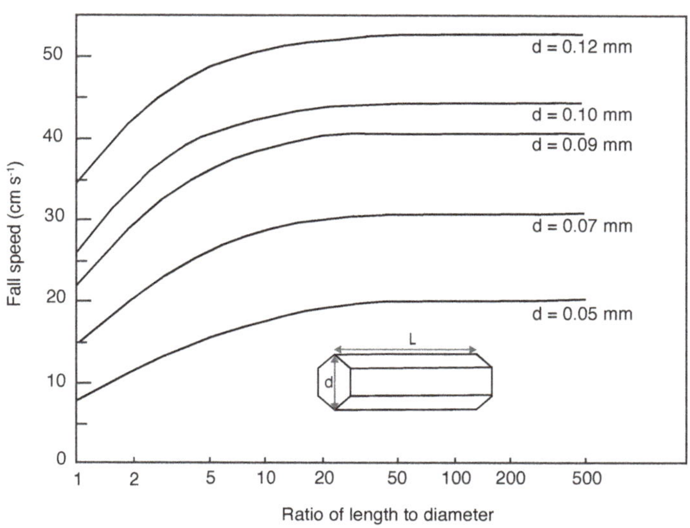

Fig. 4.17: *Top*: Observed fall speeds of plate-like and branched ice crystals (symbols), and estimates computed from drag data (lines). Picture near the end of each line shows the crystal habit for that set of data, from upper left to lower right: thick plates, thin plates, sector plates, broad branched plates, dendrites, and stellars. *After Kajikawa (1972). Bottom*: Observed fall speeds of columnar ice crystals versus their length L to diameter d ratio. *After Jayaweera and Cottis (1969).*

plate-like ice crystal, the fall speed decreases as the habit becomes more branched due to the increased surface area, which leads to an increase in the drag force.

Column-shaped ice crystals exhibit simplified trends in fall speeds (Fig. 4.17, bottom). The fall speed increases as the diameter of the column increases. Since columnar ice crystals generally fall with their maximum dimension nearly horizontal, a larger diameter increases the mass without a significant increase in drag force. For a given diameter, longer columns have greater fall speeds until the length is approximately 50 times the diameter, after which the increased drag force nearly compensates for the increase in mass, resulting in nearly constant fall speeds. The highest fall speeds for columns are approximately the same as those for sector plates, around 50 cm s^{-1}, but thick plates tend to fall slightly faster.

4.8.3 Aggregates

Aggregates, consisting of multiple individual ice crystals, can have intricate shapes with loosely joined segments, producing abundant spaces of air between some crystals. These characteristics greatly increase the drag force acting on the aggregate, even as the gravitational force increases due to the added mass. Observations of "small aggregates," consisting of only two or three crystals, exhibit lower fall speeds of around 40 cm s^{-1} over a wide range of aggregate sizes. Unrimed aggregates composed of more crystals tend to have fall speeds between 100 and 150 cm s^{-1} (Fig. 4.18). Rimed aggregates, which potentially have much more mass, can achieve higher fall speeds, up to several meters per second.

4.8.4 Graupel

Graupel particles have greater fall speeds than unrimed crystals of the same size, up to several meters per second, as shown in Fig. 4.19. The slopes of these curves are not as steep as those for raindrops, because the surfaces of the graupel tend to be rough from the rime, increasing the drag force acting on them, and because the rimed ice has a much lower bulk density than that of liquid water.

Conical graupel particles can generate more turbulent eddies around themselves as they fall, resulting in slightly more varied fall speeds for a given size compared to spherical graupel. Observations and numerical simulations indicate that conical graupel smaller than 1 mm in maximum dimension tend to fall with their flat bases oriented downward (Fig. 4.20, left panel). However, as they grow larger, they start to rotate from this orientation, and by the time they reach 3 mm in maximum dimension, they begin tumbling as they fall (right panel), similar to spherical graupel.

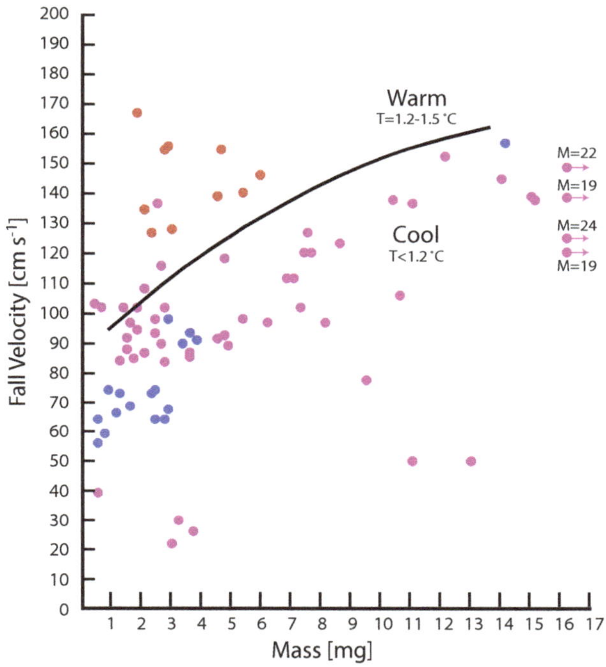

Fig. 4.18: Measured aggregate fall speeds as a function of aggregate mass, as observed in a natural snowfall over a single day in Japan, as temperatures fell from 2 to −0.5 °C. Fall speeds plotted above the solid curve indicate observations taken at temperatures above 1.2 °C, where the aggregates were beginning to melt during the measurements, and especially for red circles when the average temperature was 1.5 °C. Fall speeds below the solid curve indicate measurements taken at temperatures below 1.2 °C, with blue circles showing observations when the temperature was less than 0 °C. Fall speeds for four aggregates having masses (M) larger than the scale of the x-axis are plotted with arrows at far right. *After Sasyo and Matsuo (1980).*

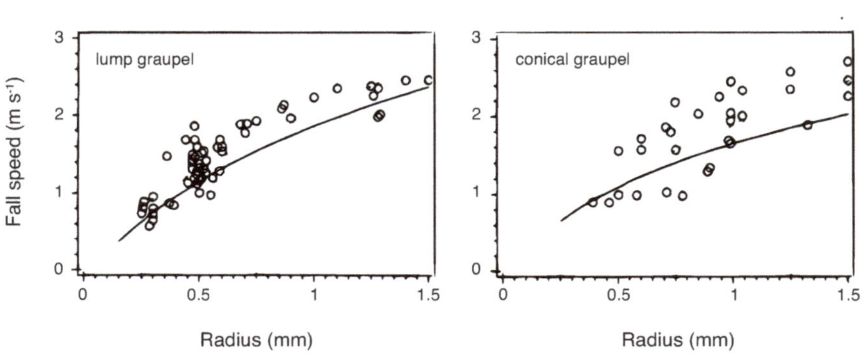

Fig. 4.19: Observed fall speeds of lump (quasi-spherical) graupel or conical graupel. *Adapted from Böhm (1989). © American Meteorological Society. Used with permission.*

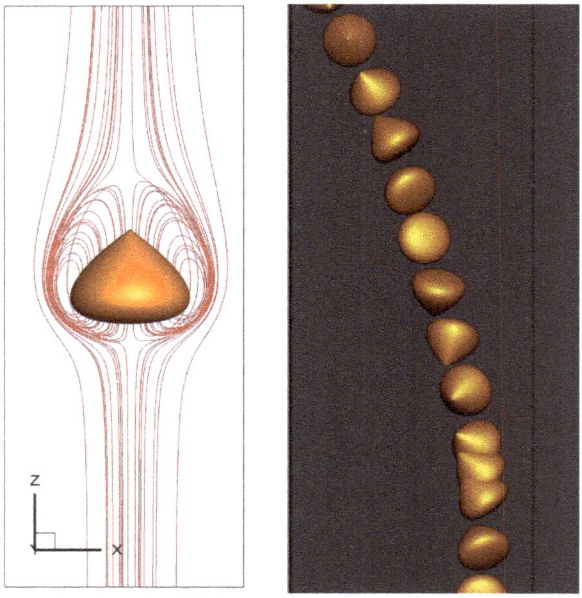

Fig. 4.20: Results of numerical simulations of falling conical graupel. *Left*: A 0.5-mm-diameter conical graupel particle falls with its base downward, with very little variation in its path. Streamlines (red) suggest counter-rotating vortices on the top of (in the wake of) the falling graupel particle. *Right*: As graupel particles grow larger than 3 mm in diameter, they start to tumble as they fall, possibly resulting in the loss of their conical shape with further riming. *From Chueh et al. (2018). ©2018, used with permission from Elsevier.*

4.8.5 Hailstones

The largest rimed hydrometeors, hailstones, achieve the highest fall speeds of all hydrometeors due to their significant sizes and densities. As shown in Fig. 4.21, their fall speeds—computed here using formulas for spherical hailstones—start around 10 m s^{-1} (22 mph) for the smallest hailstones and increase to 60 m s^{-1} (134 mph) or more for hailstones with diameters of 10 cm or greater. Hailstone density plays a crucial role: denser hailstones fall faster than less dense ones of the same size. Larger hailstones typically have higher densities.

However, larger hailstones may also have rougher surfaces and become markedly non-spherical, such as having "lobes" as shown in Fig. 4.6. These features result in a larger drag force compared to a smooth sphere of the same mass. Consequently, they may fall more slowly than more spherical hailstones, as depicted by the curves in Fig. 4.21, which consider different drag coefficients (proportional to the drag force). Like all other large hydrometeors, hailstones fall slightly faster higher in the at-

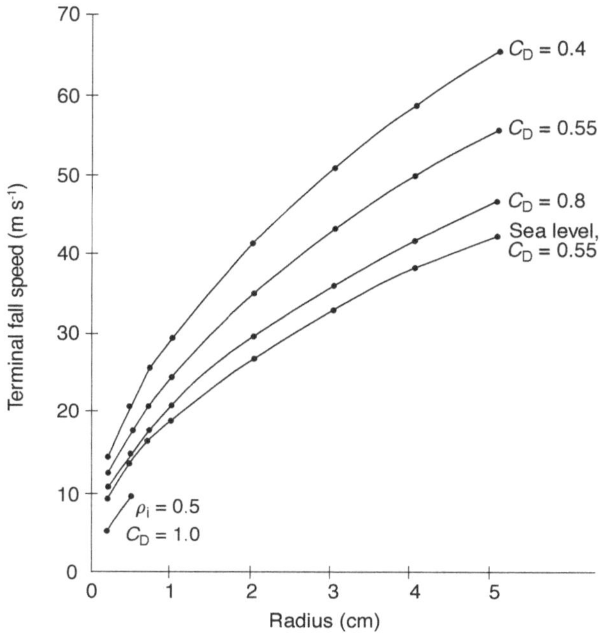

Fig. 4.21: Calculated fall speeds of hailstones. Different curves consider different unitless drag coefficients, with increasing values corresponding to rougher hailstone surfaces and thus a stronger drag force. Hailstone density is 0.9 g cm^{-3} for all curves except the bottom curve representing small hailstones with a density of 0.5 g cm^{-3}. All curves are assumed to be valid at a pressure of 500 hPa except the curve labeled "sea level."*From Knight and Knight (2001). © American Meteorological Society. Used with permission.*

mosphere, where the air density, and thus the drag force, is somewhat less than near sea level.

Because they combine large fall speeds with substantial masses, hailstones can easily flatten agricultural crops, damage cars and buildings, and injure livestock and humans. It is no wonder that a thunderstorm may be classified as "severe" based on the hail it produces!

4.9 Limits on Hydrometeor Growth

Time is the ultimate limiting factor on the maximum hydrometeor size that can be attained in a given cloud. It is restricted by (i) the **residence time** of a hydrometeor within the cloud according to its trajectory, and (ii) the overall **cloud lifetime**.

Within these time limits, however, the conditions within the cloud control the *rate* of hydrometeor growth. The trajectory of a hydrometeor de-

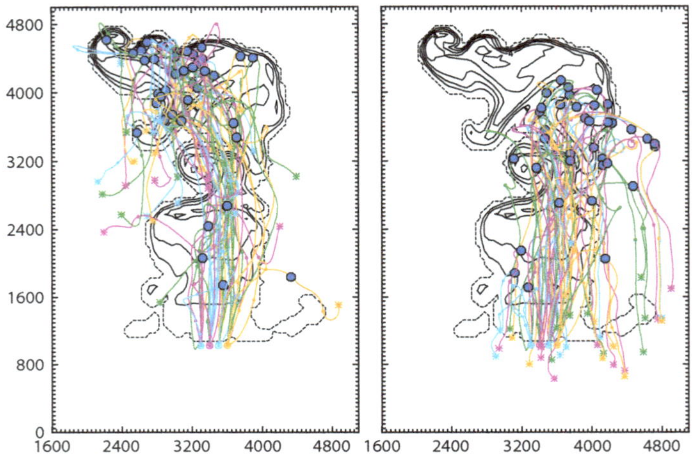

Fig. 4.22: Trajectories of smaller drops (left) and larger drops (right) through a simulated cumulus congestus cloud. Vertical cross-section of simulated cloud water contoured in black. Drop trajectories for the entire simulation are overlaid; current locations of drops at the time of the cross-section are blue dots, with asterisks marking locations at the end of the simulation. Distance in meters is labeled along horizontal and vertical axes. *Adapted from Lasher-Trapp et al. (2001). © American Meteorological Society. Used with permission.*

pends on its growth rate, as its increasing size can affect its fall speed and determine the point, if any, at which it begins descending out of the cloud.

4.9.1 Hydrometeor trajectories

The trajectory of a hydrometeor through a cloud ultimately depends on the cloud vertical velocity, which varies in space and time, and the hydrometeor's fall speed, which also changes over time as the hydrometeor grows or shrinks. *A hydrometeor can only remain suspended indefinitely within a cloud if its fall speed V_t is less than the local cloud updraft speed.* If the updraft speed within the region where the hydrometeor is located is less than its fall speed (or if the hydrometeor is within a downdraft), the hydrometeor will fall toward the ground.

Generally, for a given type of hydrometeor, smaller particles can be carried higher in the cloud updraft than larger ones due to their lower fall speeds. This tendency is illustrated in Fig. 4.22 for water droplets: within the same updraft, the trajectories of the smaller droplets extend much higher than those of the larger droplets. In the case shown here, the small droplets grow minimally over time, while the larger droplets grow much more rapidly, which further accentuates the difference in their trajectories.

We can make some rough estimates of hydrometeor trajectories using a simple mathematical model. First, the hydrometeor's upward or downward motion dz/dt (z being positive as it moves away from the ground) can be written as

$$\frac{dz}{dt} = w - V_{tr} , \tag{4.16}$$

where w is the updraft speed (negative for downdrafts) and V_{tr} is the hydrometeor fall speed relative to the air, which is a function of its size (here denoted by hydrometeor radius, r). We can then integrate (4.16) over some time to find the vertical position of the hydrometeor with respect to a chosen reference height z_0, such as the ground or cloud base, at any time t':

$$z_{t'} = z_0 + \int_0^{t'} w - V_{tr} \, dt, \tag{4.17}$$

and we account for the fact that the hydrometeor radius is likely increasing and/or decreasing with time, depending on whether it is inside or outside the cloud. The updraft speed is also a function of time along the hydrometeor trajectory; i.e., updrafts are never constant with height within a cloud. Depending on how complicated our assumptions are about how w and r, and thus V_{tr}, change over time, the integral can be evaluated either analytically or numerically.

Once the magnitude of the hydrometeor fall speed exceeds the local updraft velocity, the hydrometeor will fall through the cloud, generally continuing to grow by collecting other hydrometeors, so the maximum hydrometeor size has usually not yet been attained as it begins its descent. Frozen hydrometeors entering a warmer layer (> 0 °C) within the cloud also melt during their descent. Their size decreases as they either shed liquid drops during melting (hailstones, in particular) or collapse into a smaller spherical shape.

Once hydrometeors exit the saturated, water-laden (or icy) cloud, their residence time within the cloud has ended, and they can no longer grow. Instead, their size decreases due to sublimation, melting, or evaporation, or possibly a succession of these processes, as they fall through the unsaturated air, which typically increases in temperature as the hydrometeor approaches the ground.

4.9.2 Cloud lifetime

Of course, the total hydrometeor growth also depends on the **cloud lifetime**, which includes a developing stage, a mature stage (characterized by its maximum intensity), and a decaying stage. Even under very favorable hydrometeor growth conditions during the developing and mature stages, if these stages are brief, the maximum hydrometeor sizes will be constrained. During the decaying stage, when the cloud updraft ceases, the ascent of all hydrometeors will stop as well. As hydrometeors fall through the dissipating cloud, further growth is limited by the rate at which the cloud beneath is disappearing due to evaporation or sublimation.

Each cloud type discussed in Chapter 1 inherently possesses a range of possible cloud lifetimes, and sometimes this range is quite large and can thus be difficult to generalize. However, let us consider some examples.

The lifetimes of cumulus humilis or cumulus congestus clouds might be 30 minutes or less. Thus, if large hydrometeor growth rates are supported, they could not be maintained for too long a period. In contrast, a cumulonimbus cloud can have a lifetime of an hour, and a supercell thunderstorm may last for several hours, potentially producing high hydrometeor growth rates for much longer durations.

A nimbostratus cloud might last for several hours, whereas an orographic stratiform cloud could have a lifetime extending to a day or more. Both provide ample time for the highest growth rates these clouds can produce, which are generally lower than those found in many cumuliform clouds.

Cirrostratus can have a life cycle that extends several hours or longer, depending upon the duration of forcing by synoptic-scale weather patterns. Cirrocumulus, on the other hand, has developing and mature stages lasting approximately 30 minutes or less. However, these extremes in cloud lifetime are of less consequence than for other cloud types because the maximum possible hydrometeor growth rates are so small.

4.9.3 Implications for different cloud types

Based on our discussion thus far, we can assume that the maximum hydrometeor sizes occur in clouds with the strongest updrafts, the largest amounts of condensate (water and/or ice mass), and sufficiently long cloud lifetimes. Future chapters will explore in greater detail the cloud conditions necessary for the fastest hydrometeor growth rates. For now, we will consider large amounts of condensate as a proxy for this condition.

Let's revisit some of the general characteristics of different cloud types from Chapter 1 in light of these principles. Keeping these characteristics in mind, we can assess the likely maximum sizes and types of hydrometeors that each cloud type could produce. We consider cloud updraft speeds, hydrometeor fall speeds, and the impact of these factors on the trajectories (and thus residence times) of hydrometeors in the cloud, as well as the amount of condensate typical of each cloud type, which indicates possible maximum hydrometeor growth rates.

i) Stratiform clouds

Earlier, we learned that stratiform clouds exhibit the weakest updrafts, typically on the order of a few centimeters per second. Given the fall speeds shown in Figs. 4.14–4.18, we would expect the largest drops in these clouds to manifest as drizzle, while the largest ice particles are likely to be small aggregates or lightly rimed crystals. This is because they would fall out of the cloud before having the opportunity to grow larger.

In Chapter 3, we discussed how clouds that are deeper and/or have warmer bases usually have larger liquid water content that can potentially be converted to precipitation. Shallow stratiform clouds typically exhibit small liquid water contents and, therefore, support only minimal hydrometeor growth rates, producing limited or no precipitation. Very deep nimbostratus clouds not only are likely to have larger liquid water contents but also may extend well above the freezing level, supporting the production of a broad spectrum of liquid and frozen hydrometeors, precipitating small drizzle drops, ice crystals, aggregates, or lightly rimed ice.

Stratiform clouds, despite having some of the longest lifetimes of any cloud type, typically have weak updrafts and contain limited amounts of water or ice. As a result, they are unable to produce the largest hydrometeors—such as large raindrops, graupel particles, or hailstones—that are commonly seen falling from some cumuliform clouds.

ii) Cirriform clouds

Cirriform clouds can exhibit larger updraft speeds compared to stratiform clouds, typically ranging from 0.1 to 1 m s^{-1}. As these clouds are usually glaciated, only frozen hydrometeors are expected to form within them. Given their modest updraft speeds, one would primarily find single ice crystals and/or aggregates. While these updrafts can occasionally support rimed ice crystals or aggregates, the general scarcity of supercooled water within cirriform clouds renders such occurrences rare.

Cirriform clouds tend to have an extremely low ice mass content due to their formation in the cold upper troposphere, where little water vapor exists. Therefore, the largest hydrometeors formed within these clouds are typically single ice crystals or small aggregates. Cloud lifetimes for cirrus are generally moderate, with much of their duration spent in the decaying stage. While cirriform clouds can precipitate, none of the largest crystals or aggregates produced by these clouds reach the ground before sublimating.

iii) Cumuliform clouds

Finally, cumuliform clouds, having updraft speeds ranging from a few meters per second to 50 m s^{-1} or more, can contain the full spectrum of liquid and frozen hydrometeors discussed in this chapter. However, only the strongest cumulonimbus updrafts are capable of suspending large hailstones, as indicated by the hailstone fall speeds shown in Fig. 4.21.

Because deep cumuliform clouds with warm bases generally contain more condensate than shallower cumuli, cumulonimbus clouds in particular have the potential to produce the largest amounts of water and ice, which can then convert into precipitating hydrometeors such as the largest raindrops and hailstones. Their lifetimes tend to be the longest among cumulus clouds.

Conversely, shallow cumulus clouds contain much less water and/or ice mass and can have very short lifetimes. This limits their hydrometeors to cloud droplets and/or lightly rimed ice crystals that might never reach the ground as precipitation.

The constraints on hydrometeor growth imposed by the combination of all these factors in any of these different cloud types should be reflected in your own experience of the hydrometeors falling from these clouds (or lack thereof). In short, you would not worry about being caught in a heavy downpour of rain if the sky is overcast with thin stratus or cirrus clouds, nor would you worry about hailstones falling on your head! However, you would fret if you had forgotten to bring your umbrella when stepping outside with a nimbostratus cloud overhead or with a cumulonimbus cloud approaching.

Questions for Review

1. What are the different names given to liquid hydrometeors in Earth's atmosphere, and what units are typically used to quantify them by size and number concentration?
2. What are the different names given to frozen hydrometeors in Earth's atmosphere, and what units are typically used to quantify them by size and number concentration?
3. What is meant by the bulk density of a hydrometeor, how does it differ among various hydrometeor types, and why?
4. How does the shape of a raindrop differ between smaller and larger sizes?
5. In general terms, how do the fall speeds of various hydrometeor types differ, and how do these differences influence their residence time in a cloud?
6. How is a hydrometeor size distribution represented, and why is it important in cloud physics?

7. How do the distributions of hydrometeor surface area and volume compare to hydrometeor size distributions, and for what purposes might these other representations be useful?
8. What analytical size distributions are typically used for smaller versus larger hydrometeors?
9. How do the typical characteristics of cumulus, stratus, and cirrus clouds determine which hydrometeor types they produce?

Homework problems and other resources for students and instructors can be found at bit.ly/4dYBwB5, or scan this code:

Further Reading

For tabulations of hydrometeor characteristics from past studies [advanced]:

Pruppacher, H. R., and J. D. Klett, 2010: *Microphysics of Clouds and Precipitation*. 2nd ed., Academic Press

For more on meteorologists' usage of hydrometeor terms:

Bluestein, H. B., 1992: *Synoptic-dynamic Meteorology in Midlatitudes: Observations and Theory of Weather Systems*, Vol. 2. Taylor & Francis

CHAPTER 5

Cloud Microphysical Measurements

Because many hydrometeors formed within clouds never reach the ground as precipitation,[1] cloud physicists must make detailed measurements of different stages of hydrometeor growth. This can be achieved in several ways: in the laboratory using cloud chambers, by strategically placing instruments at various altitudes on high terrain to immerse them in clouds, or by mounting instruments on research aircraft to collect measurements within clouds. These diverse approaches to collecting data pertinent to clouds, or actually within them, are discussed in this chapter, with an emphasis on airborne measurements because they are often not presented to students. We then examine general types of microphysical instrumentation used to take measurements with any of these approaches.

If detailed information on the small-scale spatial variability of the hydrometeor population is less important than obtaining larger samples with less detail, then remote sensing (e.g., radar and satellite measurements) is more useful. This chapter concludes with a basic review of the theory and use of weather radars in characterizing hydrometeors within a cloud or the precipitation that falls from it.

Often, we want to know both the hydrometeor details within the cloud and the larger-scale aspects of its coverage, overall morphology, and evolution. Many field studies use both kinds of instrumentation in tandem.

[1]Measurement of precipitation at ground level is discussed in Chapter 13.

5.1 Hydrometeor Measurement Approaches

5.1.1 Cloud chambers

The field of cloud physics has long been rooted in laboratory experiments, as measurements made in real clouds cannot track specific hydrometeors over time to observe the microphysical processes acting on them. Within a laboratory device known as a **cloud chamber**, a cloud can be artificially created under controlled conditions, allowing the hydrometeors within it to be monitored and measured in much more detail. Cloud chamber studies have been, and continue to be, invaluable for understanding processes occurring at the smallest scales. The details of many processes discussed throughout this textbook have been acquired or verified using experiments performed within cloud chambers.

There are many sizes and types of cloud chambers worldwide, each designed for specific measurement goals. Cloud chambers cannot replicate *all* the features of real clouds but instead help to isolate specific conditions found in real clouds to study their individual effects. In actual clouds, these conditions are always combined, making it challenging to examine them separately.

For example, some small cloud chambers are designed to measure the nucleation and growth of cloud droplets or ice crystals from water vapor. The small size of the chamber makes it easier to create homogeneous conditions across its volume. They can also be used to study hydrometeor evaporation and sublimation.

Others of a little larger size may be used to study the influence of specific types of wind fields (laminar or turbulent) on hydrometeor growth or decay or even the interactions among hydrometeors under such conditions. However, the range of turbulent motions that can be generated in a chamber is always less than what can occur in real clouds.

Some cloud chambers can be quite large—even up to several stories tall within a building—and are designed to study microphysical processes over time as hydrometeors initiated at the top of the chamber fall to the bottom of the chamber (Fig. 5.1). These chambers aim to replicate some of the conditions that hydrometeors experience as they fall through a real cloud, or even outside a cloud.

Within cloud chambers, instrumentation can be used to count and size hydrometeors and observe their phase changes. All our knowledge of what occurs *on the scale of individual hydrometeors* has been acquired through the use of cloud chambers and related laboratory experiments due to the impossibility of observing these scales in real clouds.

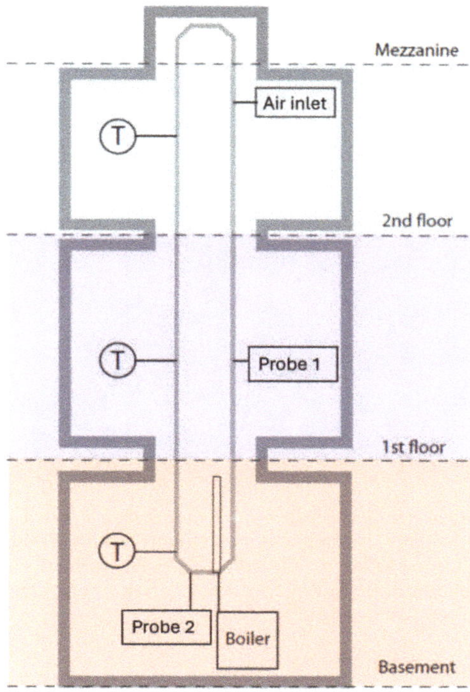

Fig. 5.1: Schematic of the Manchester Ice Cloud Chamber at the University of Manchester, UK. It consists of a one-meter-wide cylinder that spans 10 meters vertically, extending across three separate floors of the building it occupies. Each segment (shaded regions) can be maintained at a separate temperature between +20 and −50 °C (monitored at each location 'T'). In the particular setup shown, an air inlet introduces particles that can nucleate ice at the top of the chamber, and a boiler at the bottom of the chamber introduces cloud droplets there. Different probes (two are shown in this example) can be mounted along the chamber to photograph, count, and/or size hydrometeors as they fall through the chamber and grow. *Adapted from Connolly et al. (2012). (CC-BY)*

5.1.2 Mountaintop observatories

Another method by which scientists can gather data on hydrometeors within clouds, as well as atmospheric aerosol particles, is through mountaintop observatories. These high-altitude research stations (a dramatic example is shown in Fig. 5.2) provide information on aerosol particles that are less influenced by day-to-day human activities and reflect more seasonal and multi-year trends. Occasionally, these observatories can be immersed within a cloud at the mountaintop, allowing "ground-based" instruments at the observatory to sample the clouds without the need for aircraft.

Fig. 5.2: The Sphinx Observatory, a part of the Jungfraujoch Research Station run by the Swiss government in the Alps. Numerous kinds of high-altitude research can be performed at the laboratory, including aerosol and cloud studies. Researchers from all over the world can install their instruments, take measurements, and even sleep in the researchers' quarters on site. *Photo by Peter Stein/stock.adobe.com*

5.1.3 Research aircraft

Except perhaps when one is observing clouds from a mountaintop, direct—or *in situ*—measurements of the various hydrometeors within clouds require the use of research aircraft or, in some rare cases, instrumented balloons. Sampling clouds from mountaintop observatories necessitates the presence of a cloud at the mountaintop for sampling. A more straightforward approach is to bring the instruments to the clouds, which entails using a research aircraft. Although this may seem simpler in principle, numerous challenges arise.

Research aircraft (Fig. 5.3) are outfitted with microphysical *probes* (described in detail in Sec. 5.2), usually mounted on the underside of the aircraft wings. An aircraft flies through a cloud and essentially samples a one-dimensional horizontal line across the cloud. The aircraft may repeat this process at different altitudes within the cloud to understand how the hydrometeor population changes as the air rises or descends inside the cloud. It may also collect data on how the hydrometeors evolve over time as the cloud progresses through its life cycle. Alternatively, the aircraft can fly at the same altitude through multiple small cumuliform clouds or for an extended period across the same stratiform or cirriform cloud, acquiring more measurements to better understand the spatial variability in

Fig. 5.3: The 2024 fleet of U.S. National Science Foundation-sponsored research aircraft that can carry a range of sophisticated meteorological and microphysical instrumentation for taking measurements in clouds.
Top: The NSF-National Center for Atmospheric Research (NCAR) C-130 propeller aircraft. *Photo from NSF Multimedia Gallery.*
Center: A close-up of some of the microphysical probes (gold and silver instrumentation on the left and right sides of the image) attached beneath a wing of the NSF-NCAR High-performance Instrumented Airborne Platform for Environmental Research (HIAPER) jet aircraft. The large silver-tipped enclosure in the center hosts the HIAPER cloud radar, which allows for 240° cross-track scanning as well as fixed profiling at zenith or nadir. *Photo by S. Lasher-Trapp.*
Bottom: Scheduled to begin research flights in 2024, this new University of Wyoming King Air (UWKA) atmospheric research aircraft is a replacement for an older version that operated from 1977 until 2022. The new UWKA will carry instruments at several locations, including under the wings, along the top of the main fuselage, and on the nose. *Photo courtesy of J. French.*

hydrometeor characteristics—among a group of clouds or within a single, extensive cloud.

These aircraft also typically measure temperature, pressure, humidity, and winds as they pass through the cloud, providing critical thermodynamic and dynamical information about the local environments in which the hydrometeors are located at that instant. Only a few scientific agencies around the world possess the research aircraft and specialized microphysical probes required for conducting these airborne measurements.

i) Aircraft sampling

The engineering behind airborne cloud measurements is impressive: flying at speeds of 100 m s^{-1} (224 mph) or greater, the aircraft probes measure particles that may be as small as a few micrometers ($\sim10^{-6}$ m), across a cloud that is tens or even hundreds of kilometers wide ($\sim10^5$ to 10^6 m). Data are thus being collected over length scales spanning up to twelve orders of magnitude! This presents a significant data collection challenge: some probes measure millions of particles per second, creating a large volume of data that must be stored during a flight lasting hours. After landing, the information is transferred to larger computers for processing and careful quality control before it can be released to scientists for study.

Certain aspects of research aircraft pose challenges in data collection. Larger or faster aircraft have a significant "turning radius," making it difficult to sample narrow clouds like small cumuli efficiently. Much of the flight time and fuel might be expended on turning the aircraft to make additional cloud passes. For cumulus clouds, vertical variability is typically much greater than horizontal variability, necessitating multiple passes at different altitudes to fully characterize the cloud. Given these considerations, the overall flight pattern plotted at the end of a session sampling small cumuli can appear convoluted and haphazard (Fig. 5.4).

In addition, when aircraft sample a cloud multiple times, their passage through the cloud can potentially alter some hydrometeors. For example, the aircraft fuselage may collect and shed water drops, or the airflow around the aircraft propellers may nucleate new ice particles (referred to as aircraft-produced ice particles, or APIPs). Usually, it is assumed that if the aircraft passes through the cloud only once, any effects caused by the aircraft will be observable only in its wake, thus avoiding contamination of the measurements made by that aircraft. When the same cloud is sampled multiple times, the possible influence of the passage of the aircraft through the cloud upon the measurements must be considered in the data analysis.

In contrast to cumuliform clouds, stratiform clouds necessitate different flight patterns and sampling strategies due to their greater horizontal extent but smaller depth. The horizontal variability within these clouds

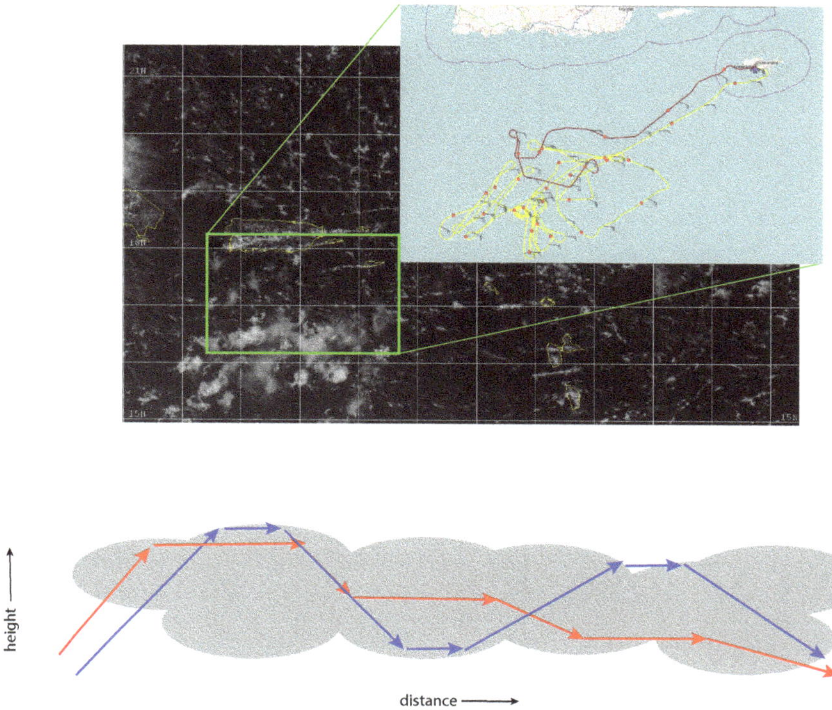

Fig. 5.4: *Top*: Visible satellite image of a field of cumulus over the Caribbean Ocean, with inset showing plan view of the aircraft flight track (yellow, with wind barbs of local winds overlaid) used to sample them. Sampling would also occur at different altitudes (not shown). *Bottom*: Hypothetical vertical cross-section of a deep stratiform cloud with two possible flight tracks overlaid to sample both horizontal and vertical variability (red is a descending pattern; blue is a sawtooth pattern).

is significantly larger than their vertical variability. Back-and-forth flight patterns traversing long distances at a near-constant altitude can be employed to sample this horizontal variability. Conversely, deeper stratiform clouds may require a "sawtooth" or other flight pattern to accurately characterize both the vertical and horizontal variability in the cloud (Fig. 5.4, bottom).

Cumulonimbus clouds were once sampled using specially armored storm-penetrating aircraft capable of withstanding high wind shear, lightning strikes, and particularly hailstones up to 7.5 cm (3 inches) in diameter. Unfortunately, the last of these aircraft (Fig. 5.5) was retired in 2004, and no similar aircraft currently exists anywhere in the world. Thus, for nearly two decades, atmospheric scientists have been unable to make in situ observations of the hydrometeors within the most intense portions of

Fig. 5.5: The U.S. NSF-sponsored armored T-28 storm-penetrating aircraft formerly operated by the South Dakota School of Mines and Technology. *Photo courtesy of T. Warner.*

thunderstorms, aside from some measurements within hurricanes where hailstones are rare.[2]

For any type of cloud, the geographical location can make aircraft sampling difficult. If the clouds of interest are over large urban areas or within the flight corridors of major airports, research aircraft may not be permitted to operate in that airspace. Conversely, when clouds are located over oceans, far from land and urban areas, aircraft must have ample fuel capacity to ensure they can fly to the clouds, execute the necessary flight patterns to take measurements, and then return safely to the airport with some fuel to spare. It is not unusual for the ferry time (transit time between the home base and the cloud area) to greatly exceed the time actually spent making measurements in clouds over the ocean.

ii) Hydrometeor splashing and shattering

Measurements of large hydrometeors, whether liquid or solid, present additional challenges. As the aircraft flies through a cloud at considerable speed, large hydrometeors may splash or shatter upon impact with the probe, creating numerous smaller particles that are recorded by the probe,

[2]Some unarmored research aircraft continue to make measurements inside mesoscale convective systems but remain a safe distance outside the more intense convective cores. Such measurements cannot represent the full range of updraft speeds and hydrometeors associated with these cloud systems. However, aircraft-mounted weather radar, and especially Doppler radar, can help fill in some of the gaps.

introducing artifacts into the measured hydrometeor population. To correct for these biases, statistical algorithms have been developed that can be retroactively applied to the collected data. These algorithms are based on our best knowledge to date but are still considered imperfect.

iii) Spatially and temporally evolving clouds

Aircraft microphysical probes are carefully calibrated in the laboratory before data collection, and tests are conducted to rigorously assess the uncertainty in their measurements while deployed during field campaigns. However, *the greatest source of uncertainty in aircraft microphysical measurements arises from the need to extrapolate general cloud properties and behavior from measurements collected over an extremely small area and time relative to the total area and time over which the clouds exist.*

The challenge thus lies in the fact that the interior characteristics of clouds can vary greatly in space and rapidly evolve over time. During a single pass through a cloud, a research aircraft collects data along a relatively narrow path. It is important to understand that these data are only representative of the particular location and time at which they were collected. An aircraft flying the same pattern but offset in a slightly different horizontal location, or at the same location but later in the cloud lifetime, can sample very different hydrometeors because clouds transition into different stages of development or even decay. Multiple data sets from similar clouds are used to understand the generality of the cloud measurements collected on a given day and at a given location.

When aircraft sampling is combined with simultaneous radar or satellite measurements, additional information concerning the cloud stage of development or the location of maximum hydrometeor amounts or sizes may be accessible. This information can also help verify the stage of cloud evolution at the time of the aircraft measurements. Sometimes, despite best efforts, the aircraft may be unable to be in the right place at the right time to capture a particular stage in cloud evolution. Therefore, these and other practical limitations must be considered when interpreting these data to address specific research questions.

5.2 In-situ Microphysical Probes

As discussed in Chapter 4, hydrometeors within clouds exhibit a vast range of sizes and number concentrations. Currently, no single instrument can measure this entire range simultaneously, nor is one expected to in the near future. Therefore, a suite of different microphysical probes is employed to capture the complete spectrum of hydrometeors present in a cloud. Regardless of the observational approach—be it cloud chambers,

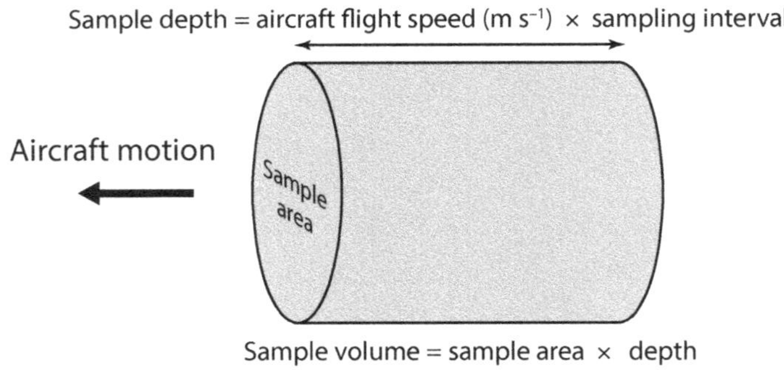

Fig. 5.6: Illustration of a hypothetical probe sampling area (circle) and the corresponding probe sampling volume (cylinder). The probe opening is oriented into the airflow as the aircraft flies toward the left. The depth of the sampling volume is proportional to the airspeed of the airplane and to the sampling interval.

mountaintop observatories, or research aircraft—the fundamental operational principles of in-situ microphysical probes remain similar, with specific adjustments made for the platform in use.

5.2.1 Basic design considerations

Of critical importance among the many different microphysical probes in use is a quantity called the **sample volume** (Fig. 5.6). Essentially, this is the volume of air within which a given population of hydrometeors has been measured. The *sample volume rate* is simply the *sample area* of the probe, which is generally a function of the probe opening or *inlet* that ingests the air containing the hydrometeors, multiplied by the aircraft speed. The sample volume is then the sample volume rate multiplied by the sampling interval Δt, i.e., the reciprocal of the sampling frequency of the probe. Typically, the sampling interval/frequency can be easily adjusted so that the sample volume can be made larger or smaller, but only within the range possible for the design and operational principles of a given probe.

An unavoidable trade-off exists in probe design that necessitates a variety of instrumentation for measuring the full spectrum of hydrometeors possible in a cloud. Measuring larger hydrometeors, which are always present in smaller number concentrations, requires a large sample volume combined with a relatively low-resolution sensor (for example, it is not necessary to measure graupel particles with micrometer precision). For the smaller and far more numerous aerosol particles and cloud droplets, they are best measured using a small sample volume and a much higher resolution sensor. Thus, in Table 5.1, as the maximum sizes of hydrometeors measured by an instrument increase, so does the sample volume, while the resolution at which their sizes are measured decreases.

Table 5.1: Examples of typical probes, their sizing range and resolution, and their sample volumes per one-second sample interval. Sample volume is a function of flight speed and, in some cases, the size of the particles, so approximate values are listed.

Probe	Sizing range (μm)	Resolution (μm)	Sample Volume
CDP	2–50	2	50 cm^3
PHIPS	30–700	6	30 cm^3
2D-S	10–1,280	10	15 L
2D-C	25–1,500	25	8 L
PIP	100–6,400	100	160 L
HVPS	1,000–50,000	200	400 L

In the following subsections, four major categories of microphysical probes will be discussed: bulk probes, optical particle counters, optical array probes, and camera-based probes.[3] These subsections will cover their individual operating principles, the types of data they collect, the scientific needs that each probe type addresses, and their limitations.

5.2.2 Bulk probes

Some probes only provide bulk (usually mass) measurements of the smaller liquid and/or frozen hydrometeors in a cloud. Several varieties of these probes exist, a few of which are discussed here. So-called *hot-wire probes* contain a wire stretched between the two forward-facing arms of the probe (Fig. 5.7). This wire is maintained at a constant temperature, high enough to evaporate any cloud droplets that collide with it. As the droplets are intercepted by the wire and evaporate, they extract heat energy from the wire (i.e., start to decrease its temperature by latent cooling). The probe electronics detect any small decrease in the wire temperature and, in response, increase the power sent to the wire to maintain its temperature. The amount of power required to maintain the wire temperature can be directly related to the mass of liquid water evaporated. Knowing the sample volume of the wire, that mass can then be converted to the liquid water content L in units of grams of liquid water per cubic meter of air ($g\ m^{-3}$) as discussed in Chapter 3. Current versions of these probes wrap the wire around a thin bar a few millimeters in diameter.

---

[3]Information about other measurement technologies, current manufacturers of the probes discussed in this section, and a longer discussion of measurement challenges can be found in Baumgardner et al. (2017).

Fig. 5.7: A hot-wire probe mounted on the underside of an airplane wing. The pointed arms face toward the front of the aircraft. As the aircraft flies through a cloud, droplets collide with the heated wire that is stretched between the two arms. The exposed wire between the two arms is just a little over 2 cm in length and has a diameter of approximately 0.2 cm.

Hot-wire probes present certain complications during their operation. They may miss the smallest droplets, as they can be deflected by the airstream around the wire instead of directly colliding with it. These probes are also known to significantly underestimate the contributions from drops larger than approximately 50 μm in diameter, because part of the drop can be shed upon impact with the wire rather than evaporating as intended.

In clouds that may contain supercooled liquid droplets or ice, other bulk measurements become more useful. The *Rosemount Icing Detector* is a simple metal rod mounted on the aircraft fuselage designed to capture supercooled cloud droplets ranging from about 50 to nearly 150 μm in diameter. As supercooled droplets collide with the rod, they freeze upon contact. The rod has a natural vibration frequency during flight, which changes in proportion to the amount of ice mass accumulated on it. This frequency change is then converted to a mass equivalent of supercooled liquid water per sample volume of the rod. When the accumulated ice reaches a threshold where the vibration frequency can no longer be used reliably to determine its mass, a heater activates to melt the ice.

The *Nevzorov probe* measures both bulk liquid and ice mass in a cloud by determining the total mass content and then subtracting the liquid mass content, computing the ice content as the residual. It features two sensors: one with a rounded end to collect cloud droplets and another with a concave end to capture both liquid and ice particles. This setup provides separate estimates of cloud droplet mass and small ice mass.

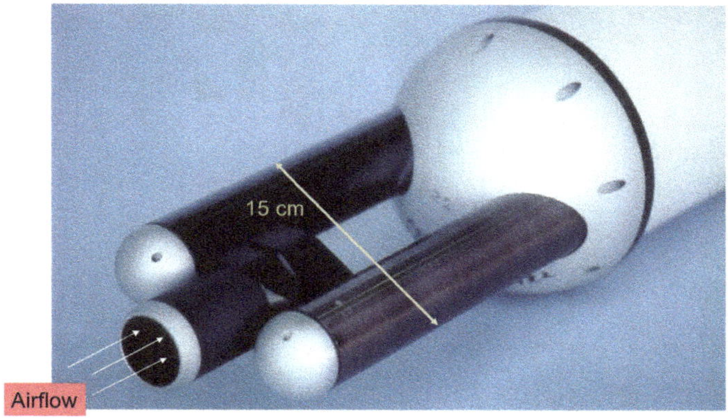

Fig. 5.8: An optical particle counter known as the FSSP. The probe is typically mounted on the underside of the airplane wing, with the two arms facing forward. Droplets enter the tube and pass through a laser that points across the arms. *Adapted from Baumgardner (1989).*

5.2.3 Optical particle counters

As discussed in Chapter 4, the size distribution of hydrometeors is more useful than only knowing the bulk hydrometeor mass for some scientific problems. This class of hydrometeor probes detects single cloud droplets and sizes them by measuring the intensity of light that *each water droplet* scatters as it passes through a laser beam directed between the two leading arms of the probe. Part of the scattered light is used to count how often a droplet has been detected, thus tracking the number concentration of droplets. The other part is used to relate the strength of the scattered light to the droplet size. Examples of these probes are the *Forward Scattering Spectrometer Probe* (FSSP; Fig. 5.8) and the *Cloud Droplet Probe* (CDP).

Optical particle counters provide far greater quantities of data than the bulk probes discussed in the previous section. Here, individual droplets are counted, sized, and then assigned to size "bins," so that an entire droplet size distribution like that shown in Fig. 4.7 can be produced. The sampling volume of optical particle counters is very small, approximately 50 cubic centimeters, to minimize biases caused by multiple droplets being within it simultaneously. Any such biases must be corrected during data processing. Sampling typically occurs at a rate of 1 to 10 Hz, although some probes are designed to operate at rates up to 1000 Hz.

These probes are specifically used to measure cloud droplets with diameters up to nearly 50 μm, and their design is optimized for measuring liquid spheres. The functioning of these probes relies on the principles of light scattering from spheres, as described by *Mie theory*. This theory is applicable when the wavelength of the light is comparable to, or slightly smaller than, the size of the scattering object—in this case, the droplets.

Because the Mie scattering curve can produce multiple droplet diameter values for a given scattering intensity, the probe is designed to consolidate these multiple values into a single, broader bin to maintain high accuracy. Consequently, this design decision results in some droplet bins being inherently wider than others, which means the resolution of the measurement can vary across different size ranges. Therefore, the number concentrations must be normalized by the bin width, as discussed in Sec. 4.6 and shown in Fig. 4.7.

5.2.4 *Optical array probes*

The optical particle counters described in the previous section are effective for measuring liquid droplets smaller than approximately 50 μm. Their method of correlating scattered light to hydrometeor size is accurate only for perfectly spherical hydrometeors, such as cloud droplets, and they are therefore unsuitable for ice crystals or larger drops. Additionally, their sample volume is insufficient for measuring larger hydrometeors, which occur in much lower number concentrations than cloud droplets.

A different class of instruments, called *optical array probes*, is designed to measure larger hydrometeors than optical particle counters and to distinguish between liquid and frozen hydrometeors. Optical array probes detect single hydrometeors and size them by recording the "shadow" (Fig. 5.9) that a hydrometeor casts upon an array of photodiodes as it passes through a laser beam directed across the two leading arms of the probe. The more photodiodes that are shadowed as a hydrometeor passes through the instrument, the larger the size of the hydrometeor.

Examples include the *2D Cloud Probe* (*2D-C*) and the *2D Precipitation Probe* (*2D-P*). Newer versions are known as the *Cloud Imaging Probe* (*CIP*), the *Precipitation Imaging Probe* (*PIP*), the *2D-Stereo probe* (*2D-S*), and the *High-Volume Precipitation Spectrometer* (*HVPS*). The sample volumes of these probes range from approximately 5 L s^{-1} for the 2D-C/CIP to 400 L s^{-1} for the HVPS.

Because they depend solely on the shadows cast by hydrometeors, these probes can be used for both liquid and frozen hydrometeors. They are capable of measuring the sizes of larger hydrometeors (sometimes much larger) due to their larger sample volumes. The probe sample volume must increase to measure larger hydrometeors since larger hydrometeors exist in lower number concentrations.

Optical array probes generate even more data than the optical particle counters discussed previously: here, the instrument additionally saves *images* of the hydrometeors as they are measured. The images can subsequently be processed to identify hydrometeor phase based on shape,

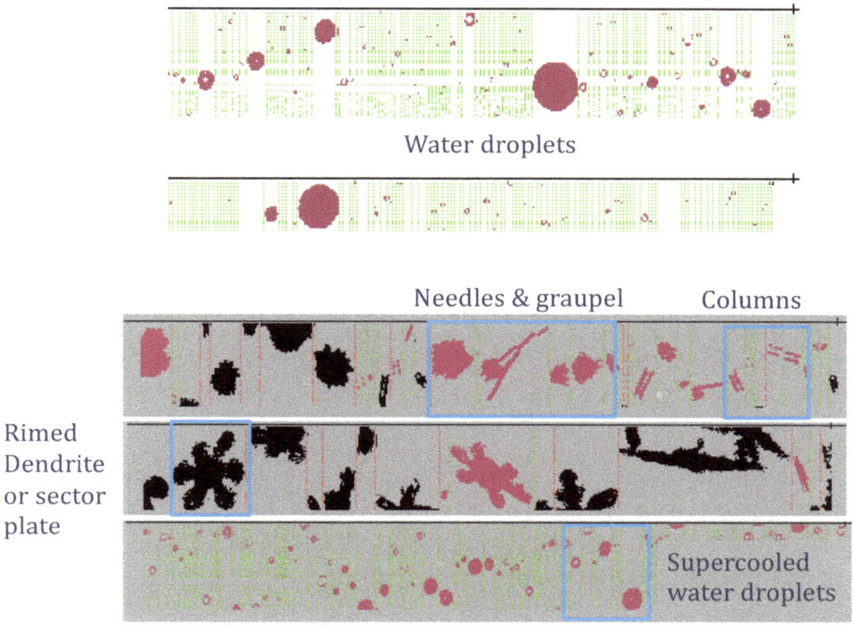

Fig. 5.9: Images from a 2D-C optical array probe. These instruments are particularly useful for distinguishing between liquid hydrometeors (top and extreme bottom) and frozen hydrometeors (middle). *Images courtesy of A. Johnson.*

including specific ice crystal habits, as shown in Fig. 5.9. Originally, the particle images had to be classified manually, but over the last several decades, automated algorithms have proven useful. More recently, artificial intelligence algorithms are being developed for this purpose.

5.2.5 Camera-based probes

While optical array probes can be useful for distinguishing between liquid and frozen hydrometeors, the shadow images they collect can sometimes leave the hydrometeor phase ambiguous, especially for smaller hydrometeors. Additionally, discerning the ice crystal habit from shadow images can be particularly challenging.

A newer class of probes uses digital cameras to photograph hydrometeors as they pass through the instrument, providing even greater detail in determining the phase. For ice particles, these probes can also identify the habit and the degree of riming. The images can be stunning in their detail and are starting to rival those produced in laboratories.

The earliest instrument of this kind was the *Cloud Particle Imager* (CPI) (see Fig. 4.3). Several lasers in the instrument are directed across opposite corners, crossing at the center. A camera is focused on the intersection where the lasers meet, and the multi-angle illumination of the hydrometeor helps capture detailed structural information. This instrument is useful for distinguishing between liquid and frozen hydrometeors in a cloud, as well as determining their size. It can potentially measure hydrometeors ranging from 15 to 2,500 μm with a nominal resolution of 2.3 μm. However, due to the design of the instrument, the sample volume is difficult to define. Consequently, hydrometeor size distributions are not commonly evaluated using the CPI. This probe is most effective for measuring hydrometeors with number concentrations below approximately 1 cm^{-3}. At this concentration, the likelihood of multiple hydrometeors appearing simultaneously in the camera's field of view is minimized, preventing overlap and ensuring clear images of individual hydrometeors.

A newer camera-based probe with a well-defined sample volume can be used to create separate size distributions for smaller liquid and frozen hydrometeors, ranging in size from about 30 to 700 μm. The *Particle Habit Imaging and Polar Scattering (PHIPS)* probe combines light-scattering measurements with digital photography (Fig. 5.10) of hydrometeors illuminated within its sample volume. It provides new information on hydrometeor phase, even for hydrometeors as small as a few micrometers. When a hydrometeor passes into the PHIPS probe and intersects a polarized laser, the light scattered at different angles is measured. The scattering patterns differ significantly between liquid and ice, and among different ice habits, allowing algorithms to automatically categorize the hydrometeor as liquid or ice, identify its habit, and measure its size. Two digital cameras are also located inside the instrument to document the habits and sizes.

A notable limitation of the PHIPS probe is its small sample volume rate, a few tens of centimeters per second, the smallest of all the probes discussed in this chapter (Table 5.1). While a small sample volume allows for highly detailed measurements of individual liquid and frozen hydrometeors, it also means that the limited statistical representativeness of these measurements within a broader cloud context is a key consideration when analyzing the data.

5.3 Weather Radar Measurement of Hydrometeors

Weather radars are indispensable tools for assessing and quantifying clouds and precipitation. Weather radar technology was developed after World War II, when the military noticed that the radars used to detect enemy airplanes also sometimes detected larger clouds and storms. The targets (hydrometeors) detected by weather radars are, of course, much smaller than aircraft, so radar designs were optimized to increase their sensitivity to hydrometeors.

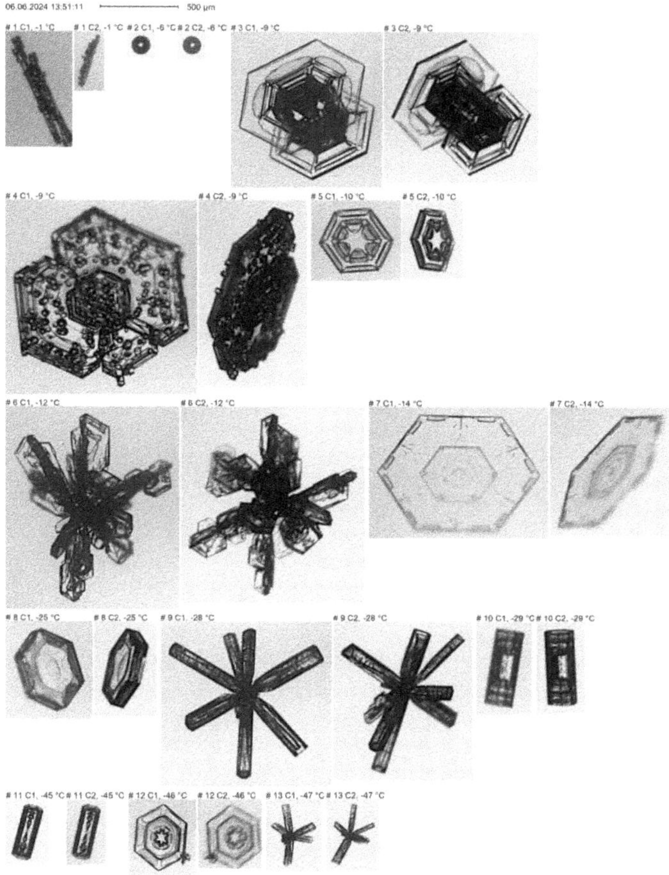

Fig. 5.10: Hydrometeor images from the PHIPS probe, including water drops, ice crystals, and rimed ice crystals. A scale for reference is shown at the top of the figure. Each hydrometeor is shown from the different viewing angles of the two cameras. *Images courtesy of E. Järvinen and M. Schnaiter.*

Radars can be versatile in their deployment. They can be mounted on aircraft, functioning similarly to the in situ measurements discussed earlier in this chapter by collecting data as the aircraft passes through the cloud. Additionally, radars can be flown on satellites, providing a space-based perspective of clouds and storms. For ground-level observations with flexibility of movement, radars can be mounted on trucks, essentially bringing the radar to the clouds to be measured. Their deployment on the ground, either as mobile or fixed sites, is generally much less costly than aircraft measurements and certainly less costly than space-based platforms, so the majority of radar measurements are taken from the ground.

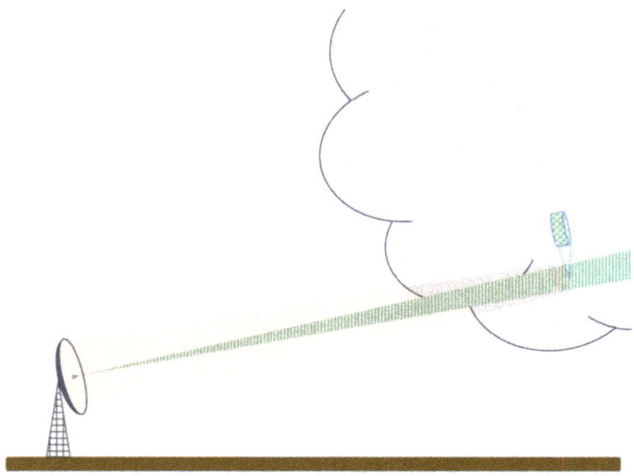

Fig. 5.11: Illustration of the basic geometry involved with a radar scanning a cloud. The radar sends a pulse of energy outward toward the cloud (green beam). The parabolic dish then receives any backscattered energy (red shading) from hydrometeors within successive volumes as the radar pulse intercepts them. A single volume along the beam is enlarged as an example.

Radars have both advantages and disadvantages compared to the microphysical measurements we have discussed. In contrast to detailed microphysical probes mounted on aircraft that can count and size individual hydrometeors, radar data are truly "bulk" measurements, even compared to bulk probes. Radars cannot provide detailed information on the size or phase of individual hydrometeors but can characterize a population of hydrometeors over a much larger volume, either within or falling below the cloud. Thus, a notable advantage of radar is that it overcomes a key limitation associated with aircraft measurements: radars do not suffer from the problem of only being able to sample a tiny fraction of a cloud, as discussed on page 98.

As a result, combining aircraft-based microphysical measurements with simultaneous radar scans of a cloud proves to be a powerful method of scientific study. The radar data provide the broader context for the cloud, including its width, depth, morphology, and stage of precipitation development. Meanwhile, the in-situ microphysical probe data from the aircraft offer detailed insights, aiding the interpretation of microphysical developments at finer scales.

Several excellent textbooks cover the details of weather radar operations and the various types of analyses that can be performed using radar data. Here, we provide only a brief overview, enabling us to refer to radar data as needed throughout this book. The reader is encouraged to explore the recommended texts listed at the end of this chapter for more in-depth information.

5.3.1 General operating principles

The general working principles of weather radars are shown in Fig. 5.11. The radar first emits a very short pulse (typically $\sim$1 μs or 100 m) of microwave energy into a narrow beam along the axis of the parabolic dish antenna. This energy travels at the speed of light. The radar then pauses briefly between pulses to "listen" using the same antenna as a receiver. If the transmitted pulse encounters hydrometeors within or falling beneath a cloud, a portion of its energy is scattered backward toward the radar. The received power is recorded as a function of the time t elapsed since the transmitted pulse, which is interpreted as a distance, or *range*, from the radar according to the relationship

$$d = \frac{ct}{2}, \tag{5.1}$$

where c is the speed of light. The factor of two in this expression accounts for the fact that the radar energy must make a two-way trip before it is detected by the receiver. The radar display system further processes the amplitude of the returned power from the hydrometeors to calculate a quantity known as the *radar reflectivity factor*, which depends only on the properties of the hydrometeors intercepted by the radar beam.

Following each pulse, the radar antenna makes a slight rotation. This is typically done in azimuth, meaning around a vertical axis, but it can also involve a change in elevation, either ascending or descending. Following this adjustment, the radar sends out a new pulse of energy. This process allows the radar to sample a new volume immediately adjacent to the previously sampled one. After the radar antenna has completed its scan across the entire desired range of azimuth or elevation, the results are composited into an image. When a cloud is present, this image, which visually represents the distribution and intensity of hydrometeors along a two-dimensional vertical or horizontal slice through the cloud, is referred to as a **radar echo**.

The scanning strategy of a weather radar is designed to meet the scientific objectives of the data collection. Typical weather radars used operationally by the U.S. National Weather Service use a series of PPI (Plan Position Indicator) scans that form 360° cones at increasing elevations around the radar (Fig. 5.12a). Research radars can also perform RHI (Range-Height Indicator, Fig. 5.12b) scans to collect data in vertical slices, which are useful for observing the microphysical development within an updraft in the cloud core. Aircraft-mounted research radars can scan below, above, or alongside the aircraft, or be fixed to point straight up or down (or both) as the aircraft flies through a cloud (Fig. 5.12c). The advantage of a fixed vertical radar beam orientation is that it allows detailed tracking of precipitation development, particularly within updrafts or downdrafts.

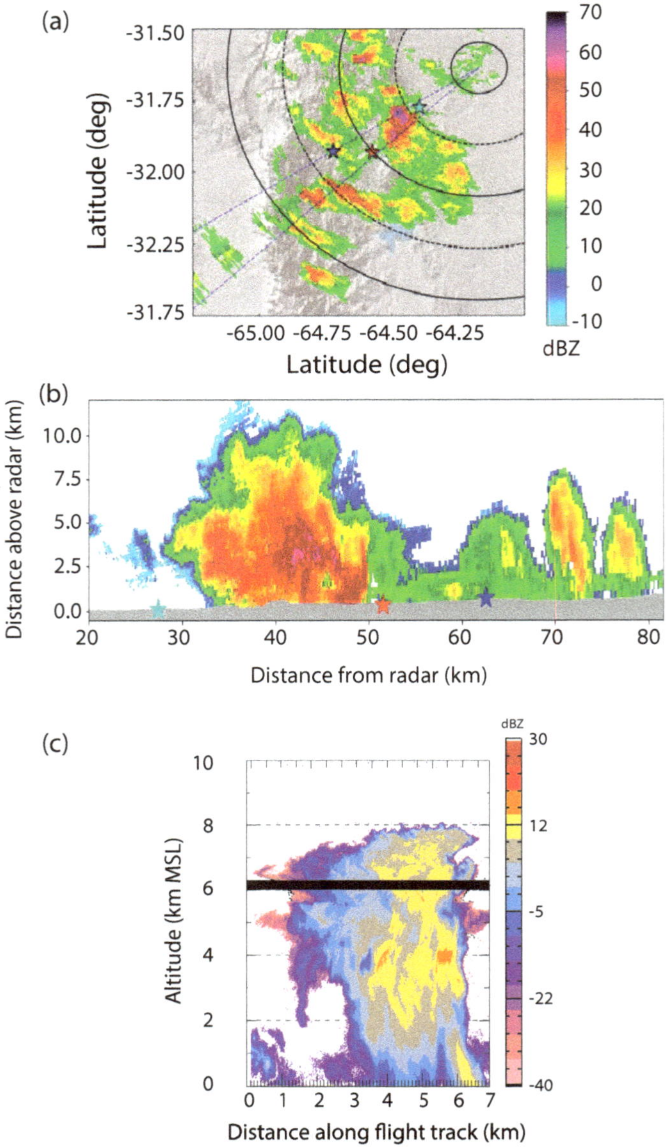

Fig. 5.12: Examples of radar images derived from different radar scanning strategies: (a) low-level PPI from a ground-based radar (only southwest quadrant is shown); (b) RHI from a ground-based radar along the southernmost dashed line in (a); (c) image from an upward-downward-looking radar mounted on an aircraft (black line in center denotes flight path of aircraft). *Images in (a, b) courtesy of T. Ross; image in (c) courtesy of D. Leon.*

Table 5.2: Common radar wavelengths.

Wavelength	Radar Name	Example	Hydrometeors Measured
10 cm	S-band	NOAA WSR-88D	Drizzle, raindrops, snow, graupel, hailstones
5 cm	C-band	DOE-ARM C-SAPR	Drizzle, raindrops, snow, graupel
3 cm	X-band	Univ. of Illinois Doppler on Wheels	Drizzle, raindrops, snow, graupel
~1 cm	Ka-band	DOE-ARM KAZR	Cloud drops, ice crystals, drizzle, small aggregates
1 mm	W-band	Univ. of Wyoming Cloud Radar	Cloud drops, ice crystals, drizzle

5.3.2 Radar wavelength

Weather radars are designed to operate in the regime where *Rayleigh scattering* occurs, meaning the hydrometeors to be quantified have diameters much smaller than the wavelength of the emitted radar energy. When this condition is met, the amount of energy scattered back to the radar by the hydrometeors increases with hydrometeor size. If the hydrometeor diameters were comparable to or larger than the wavelength of the transmitted radar energy, the scattered energy would no longer increase monotonically with increasing hydrometeor size, resulting in ambiguous information.

Table 5.2 lists the typical wavelengths used by radars in atmospheric science and the main classes of hydrometeors they are designed to quantify. Radars using longer wavelengths are usually referred to as weather radars and are used for either operational meteorology or research. The shortest (1-mm) wavelength radars are often referred to as "cloud radars" and perform best at detecting and measuring smaller hydrometeors down to cloud droplet size; these are used primarily for research purposes. This does not imply that larger hydrometeors not listed in the table cannot be measured at all by the shorter-wavelength radars, but their quantification will be less accurate.

Fig. 5.13: *Left*: A WSR-88D 10-cm operational weather radar. The dish inside the radome (white globe) has a diameter of 10 m. This is a fixed-site radar. *Photo courtesy of the U.S. National Weather Service*. *Right*: A mobile radar, the 3-cm Doppler on Wheels (DOW), a research radar at the University of Illinois. Its dish is less than 2.5 m, allowing it to be easily mounted on a truck and transported to regions of interest. *Photo courtesy of E. Wolff.*

5.3.3 Antenna size

The radar wavelength is also an important factor in determining the size of the parabolic dish used to transmit and receive microwave energy. This is because the sharpness of the radar beam, measured as an angular *beamwidth*, depends on the ratio of the wavelength to the dish diameter. If a longer wavelength is used, the dish must be proportionally larger in diameter to achieve an acceptably narrow beamwidth; otherwise, the radar will lack the ability to resolve fine details in directions perpendicular to the beam, especially at greater distances.

For example, the operational NOAA WSR-88D radar used in the U.S. (10 cm wavelength, Fig. 5.13, left panel) requires a substantial 10 m diameter dish to achieve its target beamwidth of approximately 1°. In contrast, a Doppler on Wheels (DOW, right panel) operates at a much shorter wavelength of 3 cm and therefore can utilize a more easily transported 2.5 m dish with only slightly poorer beam resolution.

5.3.4 Attenuation by hydrometeors

As the wavelength of the energy transmitted by the radar decreases, another issue becomes increasingly problematic: *attenuation* (losses of energy) of the radar beam. Precipitation-sized hydrometeors can sometimes absorb and scatter significant amounts of energy within the beam, depleting it before it can reach deeper into the cloud or storm. This results in a partial or even complete loss of information about the hydrometeors at those locations.

Generally speaking, the stronger the radar echo, the greater the attenuation within that echo region, increasing the likelihood that more distant radar echoes will appear weaker than they actually are, or potentially disappear altogether. This is another reason why the measured hydrometeors listed in Table 5.2 decrease in size as the radar wavelength decreases. A large number of hydrometeors relative to the radar wavelength will deplete the radar signal more quickly and result in inaccurate measurements. Attenuation by precipitation is rarely a problem for 10 cm wavelength radars, but it is almost always an issue with shorter wavelengths and higher rain rates.

5.3.5 *Radar reflectivity factor*

The **radar reflectivity factor**, denoted as Z, expresses the dependence of backscattered radar energy on the size distribution of spherical, liquid hydrometeors, assuming that the hydrometeors are much smaller than the radar wavelength. In this case, the criterion for Rayleigh scattering is met, and the backscattered radiation is proportional to the sum of the sixth powers of the hydrometeor diameters in a unit volume of air:

$$Z = \sum_i N_i D_i^6, \tag{5.2}$$

where we sum over the different sizes D_i present in the volume, and N_i is the number concentration of hydrometeors corresponding to each size D_i. By convention, the units of Z are $\mathrm{mm^6\ m^{-3}}$.

Note the weighting given to the size of the hydrometeors: for each size bin, their diameter is raised to the sixth power, while their number is only raised to the first power. Thus, *the radar reflectivity factor is dominated by the largest hydrometeors in the cloud*. For this reason, Z is always orders of magnitude larger for volumes that contain precipitation than for volumes containing only cloud droplets or ice crystals, a fact that makes radar especially useful for detecting precipitation and assessing its relative intensity and evolution.

Because the numerical value of the radar reflectivity factor Z can range from as little as $10^{-5}\ \mathrm{mm^6\ m^{-3}}$ for non-precipitating clouds to as much as $10^7\ \mathrm{mm^6\ m^{-3}}$ in the heaviest precipitation, it is conventional to instead express the radar reflectivity factor on a logarithmic scale:

$$\mathrm{dBZ} = 10 \log_{10}\left(\frac{Z}{1\ \mathrm{mm^6\ m^{-3}}}\right) \tag{5.3}$$

where Z is given by (5.2). The units are now in decibels of reflectivity, denoted as dBZ. This transformation allows the radar reflectivity factor to be displayed on a scale ranging from less than −20 to +70 dBZ. The higher the value, the larger the hydrometeors in that part of the cloud. All weather radar displays depicted by the media and on weather websites, as well as those used in formal scientific articles, are shown in units of dBZ. Therefore, we often refer to the quantity dBZ simply as the *radar reflectivity*.

In reality, the backscattered radar power attributable to a real population of hydrometeors depends on additional factors. For example, these equations do not include the influence of hydrometeor phase on the reflectivity. In fact, liquid drops scatter about five times more energy—equivalent to about 7 dBZ—back to the radar than a sphere of pure ice of the same size. More generally, the displayed radar reflectivity values from falling snow will always be lower than those from raindrops of similar size and number concentration.

In addition, these equations assume that all the hydrometeors in the radar beam are perfect spheres that are small compared to the radar wavelength. As discussed in Chapter 4, the assumption of a spherical shape is usually not valid for frozen hydrometeors and is not entirely accurate for the largest raindrops. For radars utilizing shorter wavelengths—for example, X- and W-band radars—the assumption of Rayleigh scattering can also break down for these large hydrometeors, in which case the assumed D^6 dependence of radar reflectivity on particle size in (5.2) becomes invalid.

Other aspects of the specific radar hardware must also be considered, including the beam width, the gain of the antenna, and attenuation losses within the radar beam. The actual energy that the hydrometeors scatter back to the radar is calculated using the *radar equation*. This equation is beyond the scope of this text but is derived and explained in the references at the end of this chapter.

5.3.6 Doppler radar

Certain radars can also distinguish whether the hydrometeors are themselves moving toward or away from the radar. They achieve this by measuring the relative phase of the backscattered wave, which is related to its wavelength or frequency. A wave crest backscattered by an object moving toward the radar takes less time to make the round trip (see Eq. 5.1) than the previous wave crest; thus, the period of the backscattered wave is reduced and the frequency increased relative to the originally transmitted wave. For objects moving away from the radar, the backscattered frequency is reduced. This phenomenon is known as the *Doppler effect*, which also describes the shift in the pitch of the sound of cars speeding around a racetrack.

It is important to note that the so-called Doppler shift depends solely on the component of motion of the scatterer directly toward or away from the radar, known as the *radial velocity*. The radar cannot detect hydrometeor motions in any direction perpendicular to its beam.

In the horizontal direction, the motion of hydrometeors toward or away from the radar results from air motions carrying them. Thus, we can infer one component of the horizontal wind field within the cloud or storm from Doppler radar measurements of the radial velocity. Because a single radar cannot measure the velocity components perpendicular to its viewing direction, two Doppler radars observing the same air volume from different directions are necessary to infer the complete horizontal wind vector. This technique is called *dual-Doppler radar analysis* and has been successfully used to assess the horizontal motions transporting hydrometeors within a cloud or storm.

Vertically pointing Doppler radars, whether mounted on the ground or on aircraft, can measure the vertical velocity of the air, including updrafts and downdrafts, since in this case these air motions are towards or away from the radar. However, it is important to note that the observed hydrometeor motion is a combination of the air motion *and* the fall speeds of the hydrometeors. Therefore, estimates of the hydrometeor fall speeds must be subtracted from the measured Doppler velocities to isolate the true vertical component of the air motion.

5.3.7 Radar estimation of precipitation rate

The radar reflectivity factor Z depends on the raindrop size distribution within a volume of the radar beam. For any given distribution—i.e., a specific set of parameters in a gamma or exponential distribution—it is possible to derive the value of Z as a function of the rain rate R. Such a function, often represented as a power law of the form

$$Z = aR^b \tag{5.4}$$

with constants a and b, is called a **Z–R relationship**. In theory, one can specify an observed radar value of Z (in mm^6 m^{-3}) and solve for the rain rate R (normalized by 1 mm hr^{-1}), allowing surface rainfall to be mapped by radar for climatological, hydrological, and agricultural purposes.

One such relationship, given by

$$Z = 200R^{1.6}, \tag{5.5}$$

has long been used based on the assumption of a Marshall-Palmer distribution (Sec. 4.6.4). Numerous other relationships, using different values for a and b, have been developed over the years to account for regional, seasonal, and even storm by storm variations in raindrop size distributions. Careful field observations have shown that each type

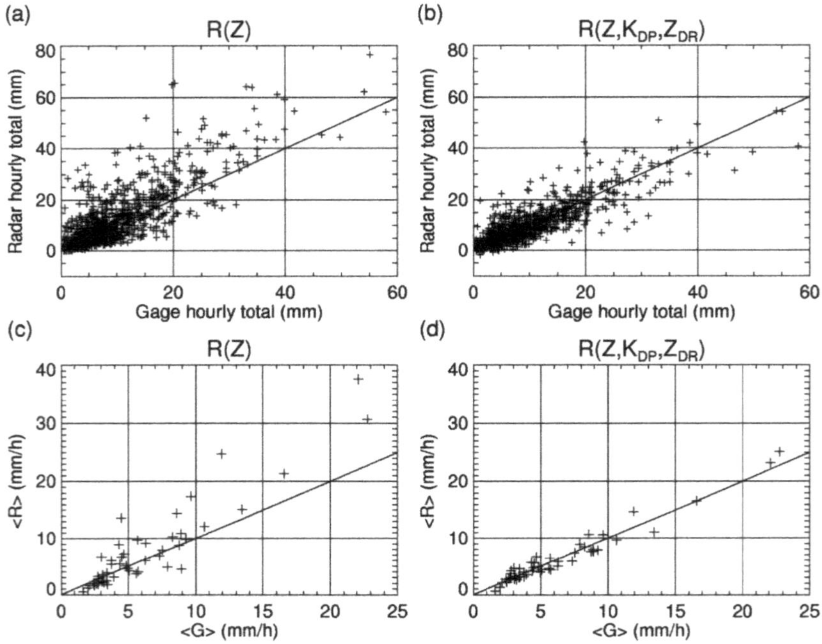

Fig. 5.14: Comparison of rain gauge measurements (x-axis) versus radar-derived rainfall estimates (y-axis) from an older Z–R relation (left column) and a newer polarimetric radar algorithm (right column). Hourly accumulations are shown in the top row and area-averaged rainfall rates are shown in the bottom row. One-to-one line (indicating a perfect match of the two different measurements) is shown as a solid black line. *From Ryzhkov et al. (2005). © American Meteorological Society. Used with permission.*

of precipitation—e.g., showery from smaller clouds, heavy precipitation from convective storms, light precipitation from stratiform clouds—typically requires different constants in (5.4).

Unfortunately, because of the high sensitivity of Z–R relationships to unknown variations in the raindrop size distribution, rain rate estimation based on radar reflectivity alone has been prone to rather large errors, as seen in Fig. 5.14a,c.

5.3.8 Dual-polarization radar

In 2011, the National Oceanic and Atmospheric Administration (NOAA) upgraded its U.S. operational weather radar network to **dual-polarization**. Now, each radar emits two pulses of electromagnetic energy: one that is horizontally polarized and one that is vertically polarized, where the polarization refers to the direction of vibration of the electric field component of the traveling wave.

Hydrometeors with a horizontal dimension greater than their vertical dimension (such as raindrops larger than about 2 mm in diameter, Fig. 4.15) scatter more horizontally polarized energy back to the radar compared to vertically polarized energy. By measuring this difference, the radar can estimate the size of the largest raindrops from the **reflectivity depolarization ratio,** Z_{DR}:

$$Z_{DR} = 10\log_{10}\left(\frac{Z_h}{Z_v}\right) \tag{5.6}$$

where Z_h is the radar reflectivity given by (5.2) for the horizontally polarized signal, and Z_v is the reflectivity for the vertically polarized signal. The units of Z_{DR} are decibels (dB). Large raindrops are flattened compared to perfect spheres, and they will have a Z_{DR} greater than zero. For example, a 2.5-mm-diameter raindrop will have a Z_{DR} of 1 dB, a 3.5-mm raindrop will have a Z_{DR} of around 2 dB, and a 6-mm raindrop will have a Z_{DR} of approximately 4 dB. A perfectly spherical raindrop (typically less than 1 mm in diameter) will have a Z_{DR} of exactly zero.

Using this additional information, along with the radar reflectivity factor and other polarimetric radar-derived variables, much more robust estimates of the precipitation rate, R, can be produced (Fig. 5.14b,d). However, the agreement between rain gauges and radar estimates is not always as good as shown here, so research on improving dual-polarization rainfall estimation continues.

5.3.9 Hydrometeor identification

Hydrometeor identification algorithms use a combination of dual-polarimetric radar-derived variables and local atmospheric soundings (which provide the approximate temperature versus height in the cloud) to "map" the different kinds of hydrometeors present in clouds and storms as the radar scans them. Two examples are shown in Fig. 5.15.

In panel (a), a vertical scan of a stratiform cloud shows a horizontal layer of higher reflectivity near 2 km altitude, known as the **bright band** (discussed further in Chapter 12). This phenomenon occurs because the snow starts to melt as it falls below the height of the 0 °C isotherm. As melting begins, the liquid coating on the aggregates initially makes the hydrometeors resemble larger raindrops at that altitude, increasing the radar reflectivity. Once fully melted at lower altitudes, they become more compact and the radar reflectivity reduces accordingly. The hydrometeor identification algorithm uses different polarimetric variables to assess the phases of the hydrometeors and thus can show more clearly where this

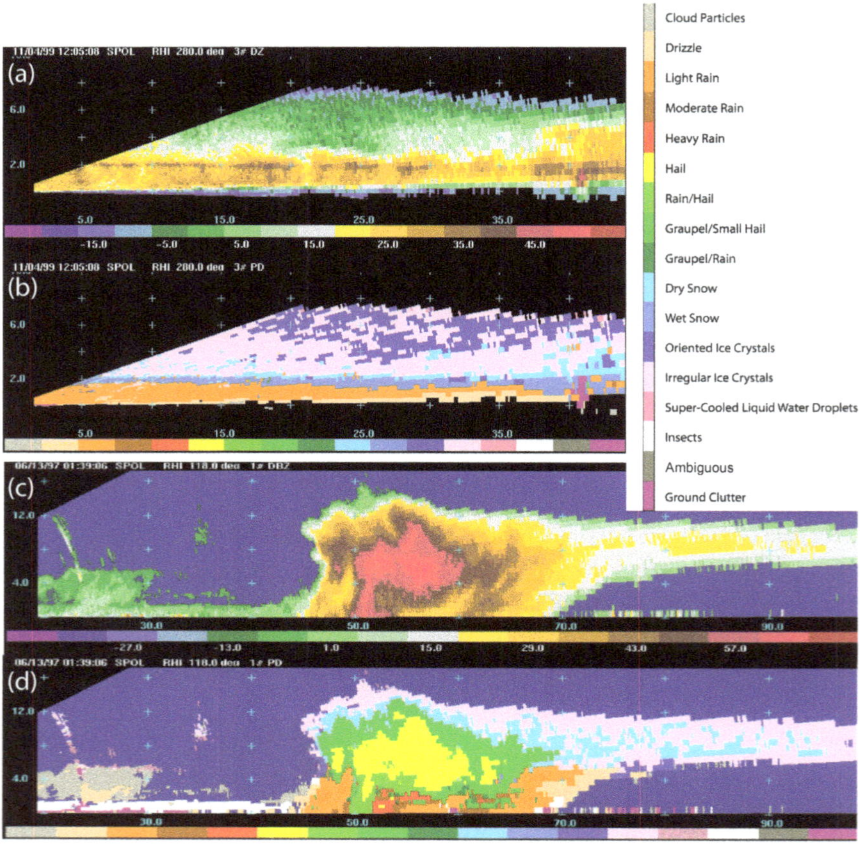

Fig. 5.15: RHIs from a cold-season stratiform cloud showing (a) radar reflectivity factor (scale in dBZ at bottom of plot) and (b) hydrometeor types colored according to legend, upper right; (c) and (d) are the same but for a warm-season deep convective cloud. *Images courtesy of J. Vivekanandan.*

phase transition is occurring. Combined with an atmospheric sounding, the algorithm appropriately labels the area immediately above the bright band as "dry snow," then "wet snow" within the bright band, and finally "light rain" near the base of the cloud in panel (b).

In Fig. 5.15d, the hydrometeor identification algorithm used with a vertical scan through a strong thunderstorm (panel c) reveals extensive areas of potential graupel and hail within much of its interior, with mainly snow and ice in the anvil region at the top, and some hail or graupel possibly even reaching the ground along with heavy rain. However, these results should be interpreted with caution, especially in deep convective clouds where aircraft cannot fly inside to verify the accuracy of the algorithm.

Research is ongoing to enhance hydrometeor identification algorithms as well.

Questions for Review

1. What are some different direct approaches to hydrometeor measurement, what are the advantages and limitations of each approach, and what additional information can remote sensing provide?
2. What are the main types of microphysical probes, what quantities do they measure, and what are their limitations?
3. What strategies can be employed to mitigate the challenges of in-situ microphysical measurements?
4. What quantities are bulk probes used to measure in clouds?
5. What additional details can optical particle counters provide that bulk probes cannot?
6. Which instruments are useful in distinguishing between liquid and frozen hydrometeors?
7. How do weather radars quantify hydrometeors in clouds and the precipitation they produce?
8. What factors must be considered when choosing radar wavelengths for operational meteorology versus research applications?
9. What is the relationship between radar reflectivity factor Z and the raindrop size distribution?
10. What are some newer developments in the use of weather radar to estimate rainfall and to detect different kinds of hydrometeors?

Homework problems and other resources for students and instructors can be found at bit.ly/4dYBwB5, or scan this code:

Further Reading

For a review of the evolution of atmospheric observations, including microphysical instrumentation [advanced]:

Stith, J. L., and Coauthors, 2018: 100 years of progress in atmospheric observing systems. *Meteorological Monographs*, **59**, 2.1–2.55

For more on the principles and applications of weather radar:

Rinehart, R. E., 2010: *Radar for Meteorologists*. 5th ed., Rinehart Pub

For a deeper dive into weather radar [advanced]:

Rauber, R. M., and S. W. Nesbitt, 2018: *Radar Meteorology. A First Course*. John Wiley & Sons

CHAPTER 6

Homogeneous Nucleation of Droplets and Ice

In this chapter, we embark on the journey of understanding how precipitation forms in clouds. We start with the basic assumptions about the possible phase changes of water and discover that some of these changes are impossible in Earth's atmosphere without external agents. However, the concepts and equations discussed will be useful in scenarios often relevant to Earth's clouds.

6.1 Thermodynamics Related to Phase Changes

6.1.1 Thermodynamic equilibrium for a water–vapor system

We begin with a brief review of the concept of thermodynamic equilibrium as it applies to phase changes. Consider a sealed glass jar half-filled with pure water (Fig. 6.1). Molecules in the liquid are constantly agitated, with some gaining enough energy to break free and escape into the gas phase above. Simultaneously, some water vapor molecules in the air, similarly in constant motion, return to the liquid and are captured by the attractive forces exerted by the other water molecules.

At a constant temperature, equilibrium is reached when the rate of molecules leaving the liquid equals the rate of molecules returning (Fig. 6.1, center). The vapor pressure, e, at this equilibrium is defined as the saturation vapor pressure, e_s, discussed in Sec. 2.3.1. This equilibrium does not mean that phase changes cease; it simply indicates that the two-way exchange of molecules between phases is balanced.

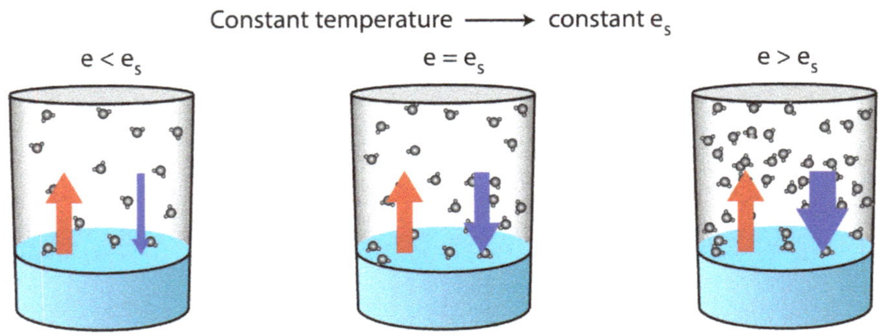

Fig. 6.1: Water molecules in the vapor state and in the liquid state under identical temperature conditions, so that the saturation vapor pressure is identical in all three cases. Arrows indicate the overall direction of the phase change, with bigger arrows denoting a larger imbalance in the number of molecules changing phase in each direction.

Recall that e_S depends only on temperature. If the temperature remains constant, this balanced exchange of molecules persists indefinitely. The air is then *saturated*, having a relative humidity of exactly 100% as calculated from (2.7).

Now consider the scenario where the initial vapor pressure is less than e_S (Fig. 6.1, left), but the saturation vapor pressure remains the same due to identical temperatures in the jars. The relative humidity of the air in the jar will be less than 100%, and we could call the air *subsaturated*. The system will adjust toward saturation by favoring **evaporation**: more water molecules leave the liquid phase to join the vapor phase. In this case, molecules leave the liquid phase at the same rate as in the first case, but there are fewer molecules in the vapor phase and therefore fewer to join the liquid phase. The two-way exchange is no longer in balance, and e increases until it reaches e_S, at which point it is back in equilibrium, and the air is saturated.

Lastly, consider the case on the far right of Fig. 6.1, where the air initially contains more water vapor than that in the other jars. Although the temperature—and thus e_S—is the same, the air is initially *supersaturated* because $e > e_S$. Here, **condensation** is favored: more water molecules leave the vapor phase to join the liquid phase, reducing e to e_S. Once again, equilibrium is achieved when $e = e_S$, and the rates of transfer between phases are balanced.

6.1.2 Dependence on temperature

Consider a different set of three jars, shown in Fig. 6.2, where now the air within them initially contains identical amounts of water vapor, but each

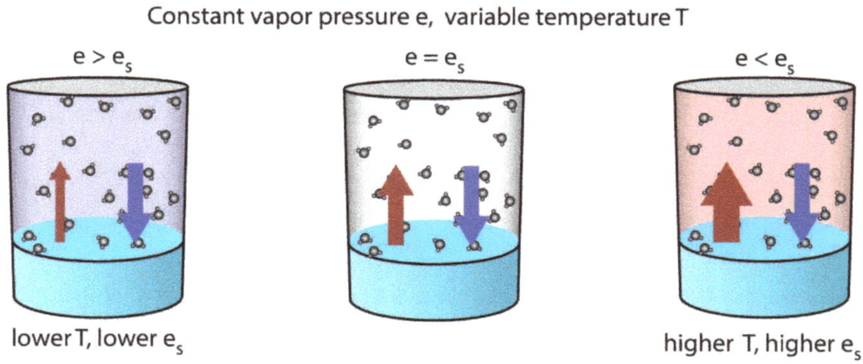

Fig. 6.2: As in Fig. 6.1, except the jar on the left has a lower temperature (air shaded blue) and thus lower saturation vapor pressure than the center jar, and the jar on the right has a higher temperature (air shaded red) and thus greater saturation vapor pressure than the center jar. The jar in the center is in equilibrium. The amount of water vapor in the air within the three jars is initially identical.

jar of water and air is at a different temperature. We will assume the center jar is again in equilibrium, i.e., saturated, where $e = e_S$. The jar to the left in the figure is at a lower temperature than the other jars, and the jar to the right is at a higher temperature than the others.

The cooler jar will have a lower saturation vapor pressure compared to the other jars, because molecules in the liquid phase have lower average energy and therefore escape to the vapor phase at a reduced rate. However, the actual vapor pressure was specified to be at equilibrium (saturation) for the middle warmer jar. Thus, that same initial vapor pressure in the cooler jar produces a condition of supersaturation because $e > e_S$.

For similar reasons, the warmer jar on the right has a higher saturation vapor pressure compared to the jars at lower temperatures. The higher temperature causes more liquid to evaporate, thus increasing the vapor pressure and bringing the system to equilibrium.

The dependence of the Clausius-Clapeyron equation (2.4) and thus of e_S on temperature had previously been noted; we now have a qualitative understanding of *why* this is the case.

In short, when the vapor in the system is initially saturated but then cooled, condensation will occur to bring it back into equilibrium. This situation is exemplified by an *ascending cloudy air parcel* in the troposphere: as it ascends, it cools adiabatically, its saturation vapor pressure decreases, and water vapor condenses to move the system back toward equilibrium.

Conversely, when an initially saturated system is warmed, evaporation will occur to increase the vapor pressure until it equals the greater saturation vapor pressure, thus restoring equilibrium. This scenario is exemplified by a *descending cloudy air parcel* in the troposphere: as it descends, it warms adiabatically, which increases its saturation vapor pressure, causing the liquid droplets to evaporate in an attempt to attain equilibrium.

We will see in this and the next chapters, however, that the change of phase associated with *cloud droplets*, as opposed to a flat surface of pure water, is more complex than shown in the above thought experiments.

6.1.3 Thermodynamic equilibrium for an ice–vapor system

Now suppose that the jars in Figs. 6.1 and 6.2 contain a block of pure ice in the lower part of each jar, instead of pure water. The same arguments hold for the ice–vapor system as for the liquid–vapor system, except now the saturation vapor pressure with respect to ice, e_{si}, must be considered and compared to the vapor pressure e.

In Fig. 6.1, all the jars are at the same temperature, which we now assume to be below 0 °C. Over time, an equilibrium state is reached, as shown in the center jar, where the vapor pressure exactly equals the saturation vapor pressure over ice at that temperature. There is a continuous but equal exchange of water molecules between the surface of the ice and the air above it, i.e., the air is *saturated with respect to ice*.

If, however, the jar on the left initially contains less water vapor than the center jar, its vapor pressure will be less than the saturation vapor pressure over ice ($e < e_{si}$, i.e., *subsaturated with respect to ice*). In response, more molecules leave the ice phase and enter the vapor phase—a process called **sublimation**—to bring the system into equilibrium. Consequently, the mass of the block of ice would be reduced.

On the other hand, if there is initially more water vapor in the jar on the right than in the other jars, its vapor pressure would be greater than the saturated vapor pressure over ice ($e > e_{si}$, i.e., *supersaturated with respect to ice*). As a result, more molecules will leave the vapor phase and join the surface of the ice—a process called **deposition**—until the system reaches equilibrium. Consequently, the ice block will grow larger.

We can also reinterpret Fig. 6.2 if the jars, held at different temperatures (but now all less than 0 °C), contain a block of ice, and all have the same amount of water vapor in the overlying air. Because e_{si} is also an increasing function of temperature, the air in the leftmost jar will have a lower saturation vapor pressure than in the other jars, and thus the vapor pressure will exceed it. To reach equilibrium, deposition of water vapor onto the ice surface will increase, bringing the system into equilibrium. This process occurs in the atmosphere within an ascending (and thus cooling) cloudy air parcel that contains ice. On the other hand, the rightmost jar will have a higher saturation vapor pressure than the other jars due to its higher temperature. The vapor pressure will initially be less than the saturation vapor pressure, and sublimation of the ice will increase, bringing the system into equilibrium, much like a descending (and thus warming) cloudy air parcel that contains ice.

6.1.4 Thermodynamic equilibrium for a water–ice system

The final system to be examined can be visualized considering two jars that contain no air or water vapor, only pure liquid water or pure ice. Therefore, vapor pressure and saturation vapor pressure are irrelevant in this scenario.

As the temperature of the jar filled with pure water is initially decreased to 0 °C and below, some water molecules in the liquid can start to organize into the molecular lattice characteristic of ice. Once initiated, **freezing**, the conversion from the liquid to the solid phase of water, will continue until all the liquid water is converted to ice, provided that the latent heat released does not increase the temperature of the system above 0 °C. As we will discuss later in this chapter, if the liquid water is pure, it does not freeze when the temperature first reaches 0 °C. We call water in the liquid state at temperatures below 0 °C **supercooled**.

On the other hand, if a jar initially at a temperature below 0 °C and containing only pure ice is warmed to 0 °C, it will begin **melting**: water molecules in the ice phase will break free from the ice lattice and transform into the liquid phase. Although melting occurs immediately at 0 °C in our simple system within the jar, we will see in Chapter 12 that the onset of melting can sometimes be delayed for ice falling into warm, unsaturated air.

6.1.5 Latent heating/cooling

In atmospheric processes, *latent heating* or *latent cooling* refers to the exchange of heat energy that occurs when water undergoes a phase change. This process involves either the release or absorption of energy. When water molecules transition from a higher energy phase to a lower energy phase (for example, from vapor to liquid or liquid to ice), they release energy. In this sequence, the energy of water molecules is highest in the vapor phase, intermediate in the liquid phase, and lowest in the ice phase. Conversely, when water molecules change from a lower energy phase to a higher energy phase (such as from ice to liquid or liquid to vapor), they absorb energy from their surroundings.

The **latent heat of vaporization**, L_v, is the amount of energy required for the phase change between vapor and liquid, which must be supplied by the water molecules or their environment. The **latent heat of sublimation**, L_s, is the amount of energy required for the phase change between water vapor and ice. Finally, the **latent heat of fusion**, L_f, is the amount of

Table 6.1: Approximate values of latent heat of vaporization, fusion, and sublimation versus temperature.

Temp (°C)	L_v (J kg^{-1})	L_f (J kg^{-1})	L_s (J kg^{-1})
40	2.406×10^6		
30	2.430×10^6		
20	2.453×10^6		
10	2.477×10^6		
0	2.500×10^6	3.34×10^5	2.834×10^6
−10	2.524×10^6	3.13×10^5	2.837×10^6
−20	2.547×10^6	2.91×10^5	2.838×10^6
−30	2.575×10^6	2.64×10^5	2.839×10^6
−40	2.603×10^6	2.36×10^5	2.839×10^6

energy required for the phase change between liquid and ice. These latent heat values for water are weak functions of temperature (Table 6.1).

An inspection of the values in this table reveals that L_f is about an order of magnitude smaller than both L_v and L_s, with L_s being the greatest. In fact, the conservation of energy requires that $L_s = L_v + L_f$.

The difference in these latent heat values has important implications for clouds. Consider a rising air parcel containing cloud droplets, where water vapor is condensing onto these droplets. In this scenario, significant latent heat is released into the air within this parcel, thereby increasing its buoyancy. This release of latent heat is greater than what would occur if the same mass of water, in the form of cloud droplets, were freezing to become ice crystals within the air parcel. Notably, the maximum release of latent heat would occur in an air parcel containing only ice crystals that are growing through deposition directly from the vapor phase or, equivalently, condensing into supercooled water and then immediately freezing.

The saturation vapor pressures over liquid or ice, given by (2.4) and (2.5) respectively, result when one assumes the latent heats of vaporization and sublimation remain constant with temperature. However, as shown in Table 6.1, this assumption is not accurate. For higher accuracy in saturation vapor pressure values, the following curve fits[1] are used, taking into account the temperature dependence of latent heat. These equations are accurate to within 0.001–0.006% of observed values over the temperature

[1]See Huang (2018).

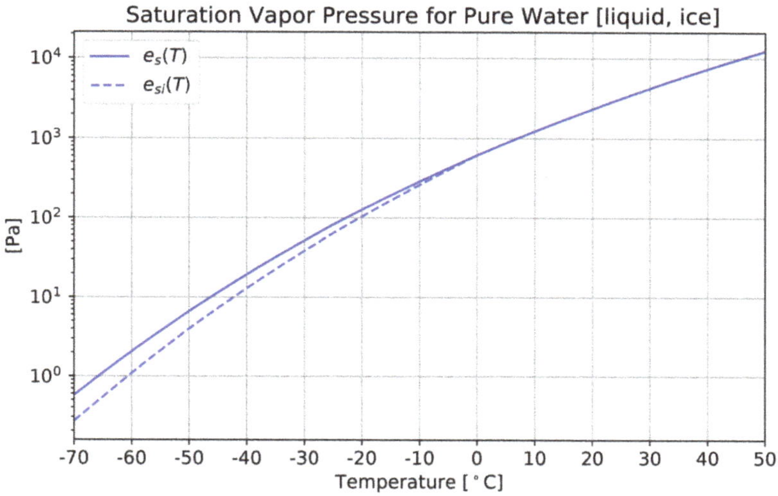

Fig. 6.3: The saturation vapor pressure with respect to pure liquid water (solid curve) and ice (dashed curve), from Eqs. (6.1) and (6.2), respectively.

range from −100 to 100 °C. For saturation with respect to liquid water,

$$e_s = \frac{\exp\left(34.494 - \frac{4924.99}{T+237.1}\right)}{(T+105)^{1.57}}, \tag{6.1}$$

and for saturation with respect to ice ($T < 0$ °C),

$$e_{si} = \frac{\exp\left(43.494 - \frac{6545.8}{T+278}\right)}{(T+868)^2}, \tag{6.2}$$

where in each case the temperature T is in degrees Celsius, and the saturation vapor pressure e_s or e_{si} is in pascals. These relationships are plotted in Fig. 6.3 for the meteorologically important range of temperatures. As previously inferred from the approximate expressions (2.4) and (2.5), we see that e_{si} decreases more rapidly than e_s with decreasing temperature below 0 °C. We will later discuss that the difference between e_s and e_{si} is crucial for the growth of ice particles in cold clouds.

6.2 Homogeneous Nucleation

Homogeneous nucleation refers to a phase change occurring in a pure substance without the aid of foreign particles or surfaces. We will consider three possible cases: the nucleation of cloud droplets from water vapor, the nucleation of ice crystals from water vapor, and the nucleation of ice crystals within pure water drops. These phase transitions are indeed more complex than the simple systems depicted in Figs. 6.1 and 6.2.

The ability of these nucleation events to occur spontaneously depends on both the energy required for the phase transition and the change in entropy associated with the transition. Specifically, a spontaneous phase transition is possible if there is a decrease in the **Gibbs free energy** of the system.

The Gibbs free energy is a thermodynamic quantity used to predict the direction of chemical reactions and phase changes under constant temperature and pressure conditions. It measures the maximum reversible work that a thermodynamic system can perform. For nucleation to occur spontaneously in nature, it must result in a decrease in the Gibbs free energy of the overall system, which includes both phases of water relevant to our application: liquid–vapor, ice–vapor, or liquid–ice.

6.2.1 Homogeneous nucleation of water droplets from water vapor

Let us start with the nucleation of cloud droplets from water vapor. Consider a group of water molecules in the vapor state in the air, coming together by chance (as the molecules are always in motion) to temporarily form a very tiny spherical liquid droplet. We will assume that the air is at some constant temperature T and pressure p.

The Gibbs free energy *required* to "build" the surface of a droplet of radius r depends upon its surface area $4\pi r^2$ and the surface energy (also called the *surface tension*), which we will denote by σ_{lv}, with the "lv" subscript referring to the liquid–vapor system. The surface energy per unit area at 0 °C, for example, is 0.075 J m^{-2}, and at 10 °C, it is 0.074 J m^{-2}.

The Gibbs free energy *released* due to the formation of the droplet, resulting from the phase change of water molecules from a state of higher energy (vapor) to a state of lower energy (liquid), depends on the difference in Gibbs free energies of the molecules in these phases. It is highest for the gas phase and lowest for the solid phase, with the liquid phase being intermediate. The Gibbs free energy released can be expressed as a function of the vapor pressure e and the saturation vapor pressure e_s of the air surrounding the droplet, as well as the total droplet volume $\frac{4}{3}\pi r^3$, because every molecule within the droplet must change phase, not just those at the surface.

Putting all of this together, we can express the total change in Gibbs free energy in the system (ΔG) due to the formation of a temporary spherical

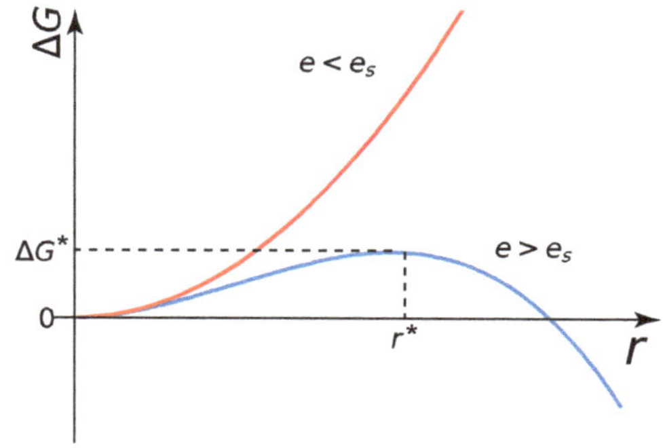

Fig. 6.4: The change in Gibbs free energy ΔG resulting from a drop nucleation event versus the droplet radius r, based on Eq. (6.3), within air that is subsaturated (red curve) or supersaturated (blue curve). Critical values discussed in the main text are denoted with an asterisk.

droplet with radius r as

$$\Delta G = 4\pi r^2 \sigma_{lv} - \frac{4}{3}\pi r^3 nkT \ln\left(\frac{e}{e_s}\right), \tag{6.3}$$

where ΔG is measured in joules, n is the number of water molecules per unit volume of liquid (approximately 3.35×10^{28} m^{-3}), k is the Boltzmann constant (1.38×10^{-23} J molecule^{-1} K^{-1}), and T is the air temperature in kelvin.

So, for this potential droplet nucleation event, there is free energy required to form the droplet surface (first term on the right-hand side) and free energy released as water molecules transition from the vapor phase to the liquid phase (second term on the right-hand side). Both terms depend on the radius of the nascent droplet, but the energy released has a stronger dependence on the radius (r^3 versus r^2).

The free energy released according to (6.3) also depends on the ratio of the vapor pressure e to the saturation vapor pressure e_s at that temperature. If the air is subsaturated ($e < e_s$), then $\ln(e/e_s) < 0$, and the free energy change is always positive, meaning that *additional* energy is needed to achieve the phase transition. Conversely, if the air is supersaturated ($e > e_s$), then $\ln(e/e_s) > 0$, and ΔG is reduced. Finally, if the air is exactly saturated, then ΔG results solely from the increased surface area of the droplet and is always positive.

The relationship between ΔG, the droplet radius r, and the saturation of the air is shown graphically in Fig. 6.4. Earlier, it was stated that nucleation events can occur spontaneously in nature if they result in a *decrease in the free energy of the system*, i.e., the energy released exceeds the energy required for nucleation to occur. Adding a single water molecule to the temporary droplet increases its radius by a tiny amount, thus moving toward the right along either curve. If ΔG *increases*, then the droplet is *not* energetically favored to grow by condensation and will instead evaporate. This is *always* the case when the air is below saturation (red curve).[2]

If, however, by adding a single water molecule to the droplet, ΔG decreases, the droplet will persist, and we say that it is *nucleated*. This is *sometimes* the case when the air is above saturation (blue curve), at least for droplet sizes larger than r^*. What then remains is to find the value of r^*, the minimum size of a temporary droplet required for it to nucleate, as determined by the local environment (temperature and degree of supersaturation). Then we will know that any group of water molecules forming a droplet of radius exceeding r^* will be nucleated.

6.2.2 Kelvin equation

We can use basic calculus to determine the location of the maximum (peak) in (6.3) when the air is supersaturated (blue curve in Fig. 6.4), i.e., when $e > e_s$. Starting from (6.3), we take the first derivative and set it equal to zero to solve for $r = r^*$, obtaining

$$r^* = \frac{2\sigma_{lv}}{nkT \ln\left(\frac{e}{e_s}\right)}. \tag{6.4}$$

This equation, called the **Kelvin equation**, is fundamental in cloud microphysics. It describes the equilibrium conditions for a pure water droplet to exist in a particular supersaturated environment. We can plot curves of r^* versus $S = e/e_s$ at different temperatures, as shown in Fig. 6.5.

From this figure, we observe that the dependence of the critical droplet radius r^* on temperature is weak, but the dependence on the ratio e/e_s, i.e., the saturation ratio S as defined by (2.13), is significantly greater. Regardless of temperature, for droplet nucleation to occur, the smallest droplets must be in a *very* supersaturated environment: S must exceed 5 (equivalent to a relative humidity of 500%) to be nucleated, whereas for larger droplets, the required saturation ratio is considerably lower but still must exceed 1.0.

It would naturally be easier for a small number of water molecules (e.g., 20) in the vapor state to come together by chance. However, according to the values given by the Kelvin equation and shown in Table 6.2, the saturation ratio required for such a tiny droplet to stay together (i.e., nucleate) is extremely high (10.0), equivalent to a relative humidity of 1000%!

[2]This is why we do not see cloud droplets spontaneously nucleating in our classroom!

Table 6.2: The critical water droplet radius (r^*) at a given saturation ratio (S), assuming a temperature of 10 °C. Last column lists the corresponding number of water molecules each droplet contains.

S (e/e_s)	r* (μm)	# molecules
1	∞	∞
1.01	0.12	2.5×10^8
1.10	0.013	2.8×10^5
1.5	0.003	3,600
2	0.002	730
3	0.001	183
4	0.0009	91
5	0.0007	58
10	0.0005	20

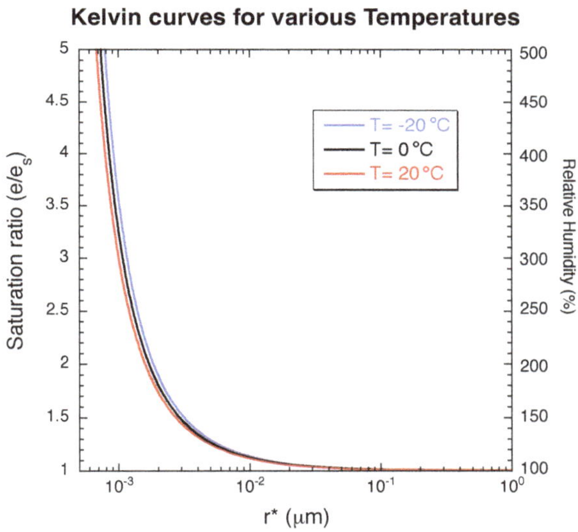

Fig. 6.5: The critical droplet radius versus the saturation ratio (left axis) or relative humidity (right axis) required for nucleation of pure water droplets, from Eq. (6.4). A unique curve exists for each temperature shown.

Such high values are never realized in Earth's atmosphere. In the troposphere, supersaturations within clouds tend to be low, usually reaching a maximum of ~1–2% ($S = 1.01$ to 1.02).

If we consider a more realistic supersaturation that could occur in Earth's atmosphere, such as 1% ($S = 1.01$), the number of molecules that would need to come together "by chance" is over 250 million. This is statistically highly unlikely. Therefore, we must conclude that *homogeneous nucleation of pure cloud droplets from water vapor is not likely in Earth's clouds.*

Instead, cloud droplets in Earth's atmosphere must form from water vapor through a different mechanism.

So, did we perform this entire analysis for nothing? Do not worry—we *will* find that we can use these principles associated with the Kelvin equation later. By discussing these relevant concepts, we will better understand how cloud droplets nucleate in reality.

We have also learned something important about the equilibrium state of a pure water droplet. Recall that the Kelvin equation describes the point at which the droplet is exactly in equilibrium with its environment. Thus, the value of e in (6.4) essentially defines a *new* equilibrium (saturation) vapor pressure for a pure water droplet of radius r. Solving (6.4) for e, and now instead denoting it as $e_{s,r}$, we find:

$$e_{s,r} = e_s \exp\left(\frac{2\sigma_{lv}}{nkTr^*}\right). \tag{6.5}$$

This equation shows that the smaller the droplet, the greater the vapor pressure required to maintain it. In essence, while a flat water surface requires a vapor pressure of e_s to be in equilibrium (i.e., a relative humidity of exactly 100 percent), a curved droplet needs a *higher* vapor pressure, $e_{s,r}$, i.e., a relative humidity greater than 100 percent, to be in equilibrium. As r^* decreases, the difference between $e_{s,r}$ and e_s grows. Consequently, larger droplets reach equilibrium more easily, nucleating at lower saturation ratios than smaller droplets. Compared to the saturation conditions discussed at the start of this chapter for a plane water surface, a curved droplet requires a higher vapor pressure surrounding it to be in equilibrium.

6.2.3 Homogeneous nucleation of ice crystals from water vapor

We can repeat an analysis similar to that in the previous section but for the *nucleation of ice crystals* directly from water vapor. Again, we can imagine a group of water molecules in the air coming together by chance at a temperature below 0 °C to form a temporary ice crystal.

For simplicity, we will account for the difference in the ice crystal area and volume compared to a sphere using correction factors β and α, respectively.[3] We then set up an energy equation like (6.3) but now replace the

---

[3]Consider a sphere of radius r inscribed *just within* the boundaries of an irregularly shaped ice crystal. The volume and surface area of the crystal can then be approximated by $V = \frac{4}{3}\pi r^3\alpha$, and $A = 4\pi r^2\beta$, respectively. Both α and β are > 1, but they approach values of exactly 1 as the crystal shape approaches that of a perfect sphere.

surface energy of the liquid–vapor interface with σ_{iv}, the surface energy of the ice–vapor interface, then use a value of n appropriate for hexagonal ice, replace e_s with e_{si} (the saturation vapor pressure over a flat ice surface) and finally denote the ice crystal radius by r_{iv}:

$$\Delta G = 4\pi r_{iv}^2 \beta \sigma_{iv} - \frac{4}{3}\pi r_{iv}^3 nkT\alpha \ln\left(\frac{e}{e_{si}}\right). \tag{6.6}$$

Here too, energy is required to build the ice surface, while energy is released due to the water molecules leaving a higher energy state (vapor) for a lower energy state (ice). Analogous to the droplet–vapor system, we find that for ΔG to decrease and thus for nucleation to be feasible, the air must first of all be supersaturated with respect to ice. We can find the critical radius of the ice crystal r_{iv}^* required for nucleation, obtaining an equation very similar to the Kelvin equation:

$$r_{iv}^* = \frac{2\beta\sigma_{iv}}{nkT\alpha \ln\left(\frac{e}{e_{si}}\right)}. \tag{6.7}$$

This equation can also be rewritten to show that a higher vapor pressure $e_{si,r}$ is required for an ice crystal to be in equilibrium compared to that of a flat surface of pure ice (e_{si}):

$$e_{si,r} = e_{si} \exp\left(\frac{2\beta\sigma_{iv}}{nkT\alpha r_{iv}^*}\right). \tag{6.8}$$

If we now plot the saturation ratio with respect to ice, $S_i \equiv e/e_{si}$, versus r_{iv}^* (Fig. 6.6), we observe that extremely high values of S_i are necessary to nucleate a tiny ice crystal. For the smallest crystals, which are statistically more likely to form by chance from water molecules in the vapor state, S_i exceeds five, corresponding to a relative humidity with respect to ice greater than 500%.

Maximum supersaturations with respect to ice in Earth's clouds are thought to be around 50% ($S_i \approx 1.5$), which is still not high enough to nucleate tiny ice crystals that could arise from chance collisions of water molecules in the vapor state. Thus, we must conclude that *homogeneous nucleation of ice crystals from water vapor is unlikely in Earth's clouds.*

6.2.4 Homogeneous freezing of water drops

We now examine the final system, involving the phase change from pure liquid water to pure ice, without considering water vapor. We start with a supercooled liquid droplet, where some water molecules in the liquid phase might momentarily align with the crystalline structure of ice.

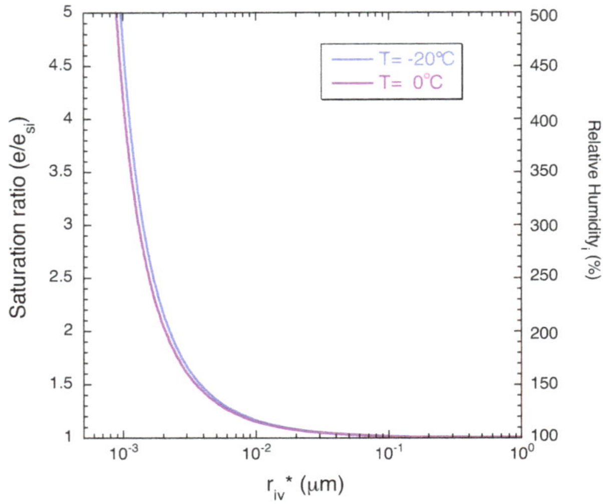

Fig. 6.6: The critical ice crystal radius for nucleation versus the saturation ratio with respect to ice (left axis) or relative humidity with respect to ice (right axis) of the surrounding air, from Eq. (6.7).

The question is whether this temporary ice crystal will persist and freeze the entire droplet or disintegrate and rejoin the liquid phase, leaving the droplet unfrozen.

Proceeding through the same energy analysis as in the other cases, we can determine the overall change in energy due to the formation of an ice crystal of a given size within the supercooled droplet. It again consists of two terms: one representing the energy required to build the crystal surface, and the other representing the energy released as the water molecules in the liquid phase convert to the ice (lower-energy) phase.

We again seek a solution where the overall energy of the system decreases due to the formation of an ice crystal within the droplet. This condition requires the temporary ice crystal to reach a critical size, denoted by r_{il}^*. For this system, the appropriate equation is

$$r_{il}^* = \frac{2\beta\sigma_{il}}{nkT\alpha \ln\left(e_s/e_{si}\right)}, \tag{6.9}$$

where the correction factors β and α are again used to account for the non-spherical ice crystal area and volume, respectively. Here, σ_{il} represents the surface energy of the ice-liquid interface, and e_s and e_{si} are the saturation vapor pressures over a surface of pure water and pure ice, respectively.

Because there is no water vapor in the system, the critical ice embryo radius has no dependence on the vapor pressure, e. Instead, e_s and e_{si} are the important factors, and they are both solely functions of temperature. By substituting different thermodynamic quantities, (6.9) can be manipulated and rewritten as

$$r_{il}^* = \frac{2\beta\sigma_{il}T_0}{n_i L_f \alpha T_s}, \tag{6.10}$$

where $T_0 = 273.15$ K, $T_s = T_0 - T$ is the amount of supercooling, and n_i is the molar density of ice (approximately 0.05089 cm^{-3}). This equation clearly demonstrates that the critical ice crystal size for nucleation (freezing of the droplet) is only a function of the amount of supercooling. That is, the further the droplet is cooled below 0 °C, as shown in Fig. 6.7 by plotting (6.10), the smaller the critical ice crystal within the droplet that is required to freeze the droplet entirely. Rather than the system needing to reach very high supersaturation to nucleate a new phase, as in the previous systems we have examined, in this case, the system must be supercooled to a very low temperature, far below 0 °C, to bring about the phase change we call **homogeneous freezing**.

Analogous to other systems presented in this chapter, the free energy barrier to freezing liquid water is much lower (i.e., much less supercooling of the droplet is required) for a very large initial ice crystal. In this case, a large number of water molecules within the droplet must organize by chance into an ice crystal, which is statistically unlikely. Thus, we would expect that the lowest temperatures shown in Fig. 6.7 would be required to nucleate ice homogeneously from pure water droplets.

For the smallest crystal sizes shown, can these temperatures be achieved in Earth's clouds? Our answer is an emphatic "yes"! *Thus, homogeneous freezing of droplets is possible in Earth's clouds.* Temperatures in the upper troposphere can easily be near or below −40 °C, and we typically observe that clouds at these temperatures, typically cirrus, *contain no liquid droplets*, i.e., are completely glaciated.

It is worth noting that the likelihood of ice crystal formation within a water drop increases with the number of water molecules present, i.e., with larger drop sizes. Statistically, larger drops are more likely to freeze homogeneously than smaller droplets, albeit not drastically so. Under controlled laboratory conditions, pure water drops as large as a few millimeters in diameter can freeze homogeneously; however, they still require substantial supercooling to temperatures as low as −35 °C.

Although homogeneous freezing of water droplets is possible within Earth's clouds at low temperatures, ice is often observed in clouds at much higher temperatures. Thus, homogeneous nucleation alone cannot explain the amount of ice observed. Some other mechanism for ice nucleation must be active, a topic that will be addressed in the next chapter.

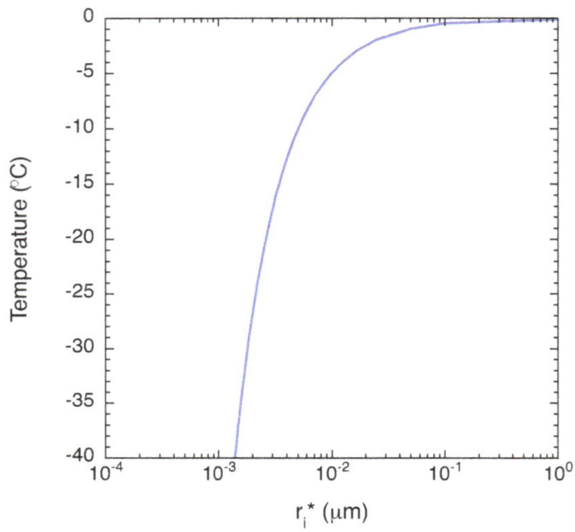

Fig. 6.7: Critical ice radius for nucleation within a supercooled droplet versus the temperature of the droplet, based on Eq. (6.10).

Questions for Review

1. What occurs at the molecular level when liquid water or ice reaches thermodynamic equilibrium with the water vapor in the air surrounding it, and how does the temperature control the point at which this equilibrium occurs?
2. What is meant by subsaturated air and supersaturated air?
3. What is the Gibbs free energy, and what is its role in determining whether a phase change will occur?
4. Why do clouds not form spontaneously whenever or wherever the air contains some water vapor?
5. How does the saturation vapor pressure over a flat surface of pure water differ from that over a pure water droplet?
6. What is the key difference between the saturation vapor pressures over liquid and over ice?
7. Why do not all (or sometimes any) of the liquid drops in clouds freeze when the cloud temperature first falls below 0 °C, and why do they *all* freeze when the cloud temperature falls below −40 °C?
8. What basic phase changes of pure water are possible in Earth's atmosphere without the assistance of any external agents?

Homework problems and other resources for students and instructors can be found at bit.ly/4dYBwB5, or scan this code:

Further Reading

For more on homogeneous nucleation [advanced]:

Fletcher, N. H., 1962: *The Physics of Rainclouds*. Cambridge University Press

CHAPTER 7

Heterogeneous Nucleation of Droplets and Ice

The previous chapter showed that homogeneous nucleation of water droplets or ice from water vapor is highly unlikely, because the supersaturations required are so extreme. Homogeneous freezing of droplets can occur, but only at extremely low temperatures approaching −40 °C. In this chapter, we learn how droplets and much of the ice within a cloud are actually nucleated.

In Earth's atmosphere, most hydrometeors arise from **heterogeneous nucleation**, where specific aerosol particles in the air, or within a droplet, significantly assist nucleation by decreasing its energy barrier. Aerosol particles that tend to nucleate cloud droplets are called **cloud condensation nuclei** (**CCN**); those that tend to nucleate ice crystals from vapor or freeze water drops are called **ice nucleating particles** (**INP**). CCN and INP can range in size from less than one micrometer to hundreds of micrometers in radius. The smaller particles are more numerous in the air than the larger ones. We will characterize aerosol particles in general, and CCN and INP in particular, more in Chapter 8.

For now, we simply recognize that aerosol particles can have a variety of chemical compositions. Some are **soluble** in water (i.e., they dissolve when mixed with liquid water, like salt), while others are **insoluble** (i.e., they retain their structure when mixed with liquid water, like sand). This property, among others, will affect their nucleating abilities.

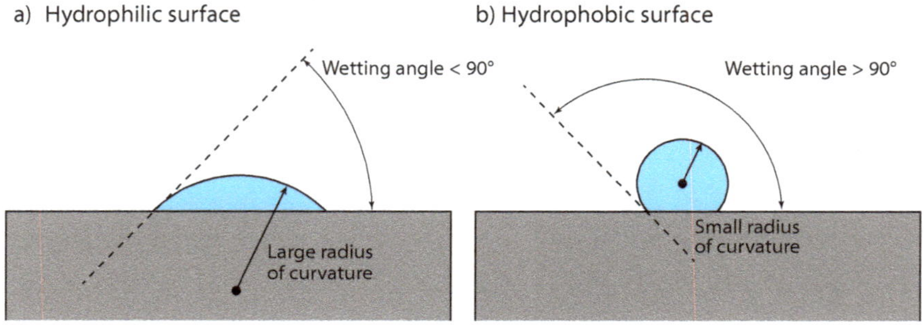

Fig. 7.1: Illustration of the behavior of a small water droplet on a) a hydrophilic surface and b) a hydrophobic surface. The total volume of water in each droplet is the same, but the radius of curvature, and thus the equilibrium vapor pressure $e_{s,r}$, depends strongly on the wetting angle.

7.1 Droplet Nucleation on Insoluble Particles

First, we consider aerosol particles (AP) in the atmosphere that are insoluble. These particles can be further subdivided into two categories: those with surfaces that are **hydrophilic** (i.e., wettable) and those with surfaces that are **hydrophobic** (i.e., water-repelling).

Hydrophilic substances allow a drop of water to readily spread out across their surface (Fig. 7.1a). Concrete is a good example of a hydrophilic surface: a raindrop falling on dry concrete will quickly flatten and spread out. Many naturally occurring rocks and minerals are also hydrophilic to varying degrees.

Hydrophobic surfaces, on the other hand, cause the water to "bead up" (Fig. 7.1b), similar to what is observed on a freshly waxed car or a non-stick frying pan. Most plastics and many other organic solids are at least moderately hydrophobic.

The degree to which a surface is hydrophilic or hydrophobic can be characterized by its *wetting angle* (also known as *contact angle*), which is generally much less than 90° for a hydrophilic surface and much larger for a hydrophobic surface, possibly approaching 180° in extreme cases. For droplets of equal total volume, we see that the radius of curvature of the droplet is strongly influenced by the wetting angle: that radius is much smaller for the hydrophobic surface than for the hydrophilic one.

The crucial point here is that the radius of curvature of a droplet on the surface of an aerosol particle can be substituted into the Kelvin equation (6.4) to obtain the equilibrium vapor pressure $e_{s,r}$.[1] Thus, we see that a wettable aerosol particle can behave as a much larger incipient droplet of

[1] You see, that entire derivation in Chapter 6 wasn't for naught!

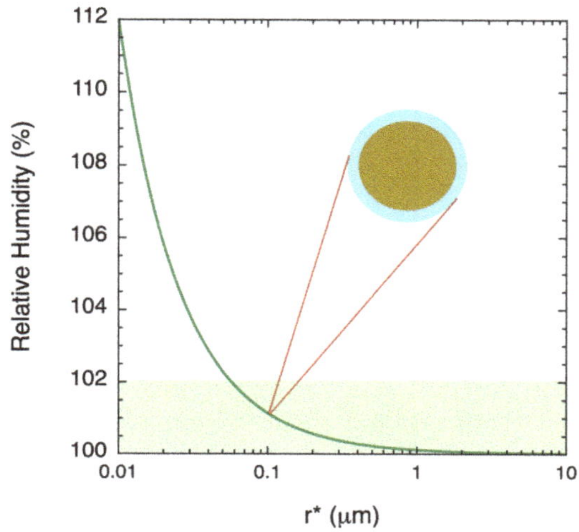

Fig. 7.2: A Kelvin curve depicting equilibrium conditions for insoluble, wettable AP in a supersaturated environment at a temperature of 10 °C. Inset circles represent an insoluble, wettable aerosol particle (brown) with a coating of liquid water on its surface (blue), which effectively makes the droplet large enough to be in equilibrium at typical supersaturation values found in Earth's clouds (shaded green rectangle, bottom).

pure water, giving it a considerable head start toward achieving its critical radius r^*.

Revisiting the principles of the Kelvin curve, we now address heterogeneous nucleation of water droplets from water vapor, *aided by insoluble, hydrophilic AP*. Recall that for homogeneous nucleation, it was argued that nucleation of cloud droplets from water vapor alone would be extremely unlikely because, for the supersaturations typically observed in clouds, an improbably large number of water molecules in the air would have to come together by chance to achieve the necessary critical radius, r^*.

Consider an insoluble, hydrophilic aerosol particle in the atmosphere. If the wetting angle is sufficiently small, the radius of curvature of an incipient droplet forming on its surface can approach that of the aerosol particle itself. Indeed, its entire surface could become wetted, resulting in the radius of curvature of the water surface equaling or exceeding that of the aerosol particle.

Based on the Kelvin curve plotted in Fig. 7.2, we see, for example, that condensation of water onto the surface of an insoluble, hydrophilic aerosol particle with a radius of 0.1 μm or larger could potentially nucleate a droplet at a supersaturation typically observed within many clouds. Thus, we conclude that some cloud droplets are nucleated heterogeneously with the assistance of these insoluble, hydrophilic AP *provided that those particles are large enough to help the droplet overcome the energy barrier.*

However, the number of AP in the atmosphere decreases with increasing size, so there is rarely a sufficient number of large, insoluble, hydrophilic AP to explain all the droplets we typically observe in clouds. Therefore, other AP must also nucleate cloud droplets.

7.2 Droplet Nucleation on Soluble Particles

Considering the bigger picture, the barrier to homogeneous nucleation of droplets lies in the requirement for very high supersaturation. But the typical maximum supersaturation with respect to water in Earth's clouds is less than a few percent. We must therefore consider whether there might be some AP that could reduce the extreme levels of supersaturation needed to nucleate droplets.

Such a population of AP does exist in Earth's atmosphere: *soluble AP*, also referred to as *hygroscopic* AP. Similar to table salt, they dissolve in liquid water to form a solution. For certain soluble AP, the critical supersaturation needed to nucleate a droplet can be significantly reduced.

7.2.1 The importance of soluble aerosol particles

The presence of a **solute**—i.e., some chemical component that will dissolve in water and create a **solution**—significantly decreases the saturation vapor pressure and, thus, the supersaturation required to nucleate a cloud droplet. The basis for this action is well known from **Raoult's Law**, which states that the saturation vapor pressure in the air above a liquid solution is *reduced* from that over pure water by an amount proportional to the strength of the solution:

$$\frac{e_s'}{e_s} = \frac{n_w}{n_s + n_w}, \tag{7.1}$$

where e_s' is the saturation vapor pressure over a plane (flat) surface of the solution (primed variables in this chapter will always refer to values for the solution), e_s is the saturation vapor pressure over a plane surface of pure water, n_w is the number of moles of water in the solution, and n_s is the number of moles of solute in the solution. Thus, the more concentrated the solution—i.e., the greater n_s relative to n_w—the lower the saturation vapor pressure over the solution.

Recalling the jars containing pure water and air from Fig. 6.1, if the pure water in the jars is replaced by a solution of water and some solute, the presence of the solute disrupts some of the hydrogen bonds between neighboring water molecules, hindering them from breaking free of the liquid surface as easily, and thus reducing the number of water molecules at the surface that are available to enter the vapor phase. The flux of molecules returning to the liquid phase from the vapor phase, on the other hand, is proportional to the vapor pressure. For the two to be in balance as required for saturation, the modified vapor pressure e_s' must be lower.

Most AP relevant to this discussion are *ionic*, that is, they at least partly dissociate into positively and negatively charged ions when the original substance is dissolved in water. For example, table salt (sodium chloride, NaCl) dissociates into Na^+ and Cl^- in water, resulting in two ions for every original molecule. Ammonium sulfate, $(NH_4)_2SO_4$, a common constituent of AP, dissociates into three ions, of which NH_4^+ and SO_4^{2-} are distinct. Table sugar, in contrast, is an example of a soluble particle that is non-ionic and does not dissociate in water. Consequently, for a given solute mass, it is less effective at lowering the saturation vapor pressure.

Importantly for this discussion, it is the number of moles of *ions*, where applicable, that must be used for n_s in (7.1). This is calculated by taking the mass m_s of the solute, dividing it by its molecular weight, M_s, and then multiplying by the *van 't Hoff factor*, i. For the solutes most relevant to cloud formation, such as ammonium sulfate and sodium chloride, experimental data suggest that $i \approx 2$.

We then have

$$n_s = \frac{i\, m_s}{M_s}\,. \tag{7.2}$$

The number of moles of water in the solution, n_w, can then be found as the remainder:

$$n_w = \frac{V\rho' - m_s}{M_w} \tag{7.3}$$

where V is the total volume of the solution, ρ' is its density, and M_w is the molecular weight of water. Substituting these expressions for n_s and n_w into (7.1) yields:

$$\frac{e_s'}{e_s} = \left[1 + \frac{i m_s M_w}{M_s(V\rho' - m_s)}\right]^{-1}. \tag{7.4}$$

Thus, for any soluble AP with a single chemical composition, we can find the amount by which the equilibrium saturation vapor pressure in the air over this solution will be reduced.

7.2.2 The Köhler equation

Equation (7.4) is relevant for a solution with a flat surface, but our goal is to understand the nucleation of spherical cloud droplets. We now derive an equation that specifies the saturation vapor pressure over a *solution droplet* of radius r. When the vapor pressure of the air surrounding the droplet equals the saturation vapor pressure required for equilibrium, the droplet will be nucleated.

i) Derivation

From the previous section, we know that a solution *decreases* the saturation vapor pressure over its surface compared with the saturation vapor pressure over a flat surface of pure water. On the other hand, the Kelvin equation reveals that the curved surface of a pure water droplet *increases* the saturation vapor pressure around it compared with the saturation vapor pressure e_s over a flat surface of pure water.

We now combine these two opposing effects. The ratio of the modified saturation vapor pressure $e'_{s,r}$ surrounding a solution droplet to that over a flat surface of pure water (e_s) can be written as the product of the right-hand sides of (6.4) and (7.4):

$$\frac{e'_{s,r}}{e_s} = \left[\exp \frac{2\sigma'}{n'kTr}\right] \left[1 + \frac{im_s M_w}{M_s(\frac{4}{3}\pi r^3 \rho' - m_s)}\right]^{-1}, \tag{7.5}$$

where the first term in brackets on the right-hand side is from the Kelvin equation, with primed variables now denoting values for a solution droplet.

Equation (7.5) is known as the **Köhler equation**, a fundamental equation in cloud microphysics.[2] Simply put, it relates the saturation vapor pressure surrounding a solution droplet, $e'_{s,r}$, to that above a flat surface of pure water (e_s) found using the familiar Clausius-Clapeyron equation. Note that when the mass of the solute $m_s = 0$, the Köhler equation reduces to the Kelvin equation for a droplet of pure water.

Similar to the Kelvin equation, the Köhler equation can be used in two ways: to find the saturation vapor pressure needed to nucleate or maintain a solution droplet of radius r or to find the minimum size (critical radius r^*) of solution droplets that can nucleate or exist at a given saturation ratio (e/e_s, where e is equivalent to $e'_{s,r}$).

Focusing on the right-hand side of (7.5), the first bracketed term represents the "curvature effect" of the droplet, which increases the saturation

[2]Aerosol scientists use a more generalized Köhler equation that also considers AP that do not dissociate into ions but still have an affinity for liquid water, such as some organic compounds. Instead of using the van 't Hoff factor, i, their derivation employs kappa (κ), a hygroscopicity parameter that changes with the chemical composition of the AP, including those containing multiple chemical elements (Petters and Kreidenweis, 2007).

vapor pressure of the air needed for the droplet to be in equilibrium, as discussed in Chapter 6. For hygroscopic AP, however, the second bracketed term, the "solute effect," decreases it (note that this term is raised to the -1 power). Both terms also depend on the droplet radius r. We now seek to understand the tradeoff between these two competing terms and identify the situations that might help to lower $e'_{s,r}$ relative to e_s to facilitate the nucleation of a cloud droplet.

To aid our analysis, some simplifications to (7.5) can be made:

1. $2\sigma/n'kTr$ is a small number (we denote it temporarily as x), allowing us to approximate $e^x \approx 1 + x$.
2. The solute term (call it y) is also a small number, so we can use the approximation $(1+y)^{-1} \approx 1 - y$.
3. The two new terms above would be multiplied, i.e., $(1+x)(1-y)$, and the cross-term xy is very small, so we neglect it and set it equal to zero.
4. In the denominator of the second bracketed term, $(4/3)\pi r^3 \rho'$ is much greater than m_s (indicating that the solution is sufficiently dilute), so we can neglect m_s there.

With these simplifications, the Köhler equation becomes

$$S = \frac{e'_{s,r}}{e_s} = 1 + \frac{a}{r} - \frac{b}{r^3}, \tag{7.6}$$

where

$$a = \frac{2\sigma'}{n'kT} \quad \text{and} \quad b = \frac{3im_s M_w}{4\pi\rho' M_s} \tag{7.7}$$

and recall $S \equiv e/e_s$ is the saturation ratio of the air immediately surrounding the droplet.

The dependence of S on the radius of a given solution droplet r from (7.6) can be plotted. These functions are called **Köhler curves**, and they specify the conditions under which hygroscopic AP can nucleate cloud droplets. Note that because both a and b are functions of the chemical composition and the mass of the solute in the droplet, different Köhler curves result from various combinations of these two aspects of the AP.

Figure 7.3 shows Köhler curves for soluble AP composed of sodium chloride or ammonium sulfate versus relative humidity rather than the saturation[3] ratio S used in (7.6). These two types of soluble AP are the most prevalent in the troposphere and are responsible for the majority of

---

[3]Recall that the relative humidity $RH = S \times 100\% = s + 1$, where S is the saturation ratio, and s is the supersaturation.

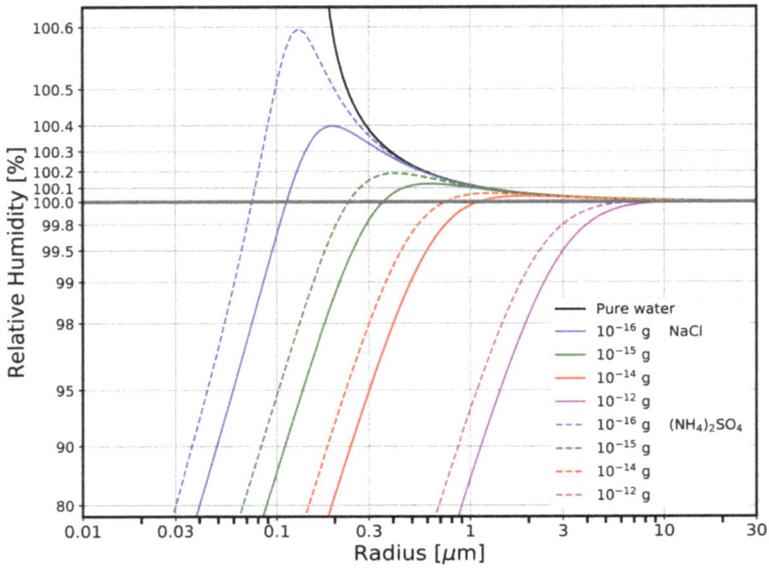

Fig. 7.3: Köhler curves for solution droplets nucleated on sodium chloride particles (colored solid curves) or on ammonium sulfate particles (colored dashed curves), with their respective masses as labeled. These calculations assume $i = 2$ for all particles and use a temperature of 20 °C. The Kelvin curve for pure water droplets is included for reference (black curve).

cloud droplets. Specifically, the Köhler curves illustrate the relative humidity needed for droplets containing one of these dissolved solutes to be at equilibrium with their environment and thus to exist, i.e., be *nucleated*, or to continue to exist if already nucleated. For any given initial radius r on the x-axis, if the environmental relative humidity exceeds the corresponding value given by that point on the curve, the droplet will grow by condensation and r will increase; if it is below that value, the droplet will lose water by evaporation and r will decrease.

ii) Haze droplets

The Köhler curves reveal that less restrictive conditions (lower RH, i.e., smaller supersaturation) are needed for the nucleation of solution droplets compared with similar-sized droplets containing only pure water that may have formed on insoluble AP; the Kelvin curve (black solid curve) plotted in Fig. 7.3 always lies above any Köhler curve. For smaller droplet sizes, the Kelvin curve extends far above the vertical axis shown in this plot, illustrating how very high the relative humidity must be to nucleate droplets of those sizes without the assistance of hygroscopic AP.

The Köhler curves show that these hygroscopic AP can nucleate droplets at relative humidity levels significantly below 100%, a phenomenon known as **deliquescence**.[4] The soluble AP represented by the Köhler curves can begin to attract water molecules from the vapor phase to their surfaces at relative humidity levels of approximately 80% or greater, thereby dissolving the solid particle into a solution droplet. Thus, even before the air is saturated (at 100% RH), these soluble AP can nucleate droplets, referred to as *haze droplets*, as discussed in Sec. 4.3.1.

You've likely encountered the meteorological phenomenon known as *haze*. It is characterized by murky air that reduces visibility measurably without being foggy or cloudy.

Because the Kelvin curve, which represents pure water droplets, *never* dips below 100% relative humidity, haze droplets can only nucleate on soluble (hygroscopic) AP. Furthermore, Fig. 7.3 shows that soluble AP of greater mass lead to the formation of larger haze droplets compared with those having less mass, given the same relative humidity. Their greater mass results in a more concentrated solution droplet, which promotes the addition of more water molecules from the vapor to facilitate droplet equilibrium.

iii) Trends in the peaks of the Köhler curves

All of the Köhler curves display peaks above the RH=100% level, indicating supersaturation conditions. The heights of the peaks are nearly indistinguishable from 100% for the largest masses of AP shown, but they exist nonetheless. Each Köhler curve illustrates that an increase in environmental vapor pressure, and thus a certain degree of supersaturation, is required to maintain droplet equilibrium until a specific droplet radius is reached at the peak. Beyond this peak, the supersaturation required for equilibrium actually *decreases*.

These changes result from the competing effects described by the Köhler equation (7.6). The a/r curvature term, related to the Kelvin equation, increases the equilibrium saturation vapor pressure over the droplet at smaller droplet sizes. However, its effect diminishes as the droplet size increases, because the droplet surface becomes less curved at larger sizes. Conversely, the b/r^3 solute term, related to Raoult's law, decreases the saturation vapor pressure over the droplet when the solution is concentrated, which occurs at smaller droplet sizes to the left of the peak in each curve.

---

[4]Deliquescence is the phenomenon responsible for the clumping of salt in your salt shaker on a very humid day.

The solute term "loses out" to the curvature term once the droplet has accumulated enough condensed water to significantly dilute its solution. This is why the Köhler curves converge toward the Kelvin curve at larger sizes.

Notice that the location of the peak in each Köhler curve is different: both the supersaturation and the droplet size at the peak vary. This variation arises from the amount of soluble mass in the original aerosol particle, which influences not only the concentration of the solution in the droplet but also its size at a given supersaturation.

For example, the peak in the solid red curve for the large salt (NaCl) mass in Fig. 7.3 occurs at a tiny supersaturation near 0.02%. However, less massive salt particles (solid green and blue curves) cannot achieve such concentrated solutions, so their peaks occur at higher supersaturation levels for equilibrium (approximately 0.12% and 0.4%, respectively). The ammonium sulfate particles [$(NH_4)_2SO_4$; dashed curves] exhibit similar trends to the salt particles: as their initial mass increases, the supersaturation required for equilibrium decreases.

However, for ammonium sulfate particles with the same original mass as the salt particles (comparing dashed and solid curves of the same color), the ammonium sulfate Köhler curves always require a higher supersaturation for equilibrium. This demonstrates the effect of the original chemical composition of the AP; some AP nucleate cloud droplets more easily—i.e., at lower supersaturation—than others. This follows from the inclusion of the molecular weights, m_s, in the Köhler equation by way of (7.7). Nevertheless, ammonium sulfate particles are usually more prevalent than salt particles in Earth's atmosphere, so they often dominate the nucleation of cloud droplets overall.

7.2.3 Particle evolution based on the Köhler curves

We can further use the Köhler curves to predict when nucleated droplets will grow larger by condensation. Suppose the air contains AP represented by the different Köhler curves shown in Fig. 7.3. If the ambient relative humidity is 90%, what sizes of haze droplets would be nucleated? Scanning across the different curves at a relative humidity of 90%, we see that the smallest droplets would form on the ammonium sulfate particles having the least mass (blue dashed curve), with radii slightly less than 0.04 μm. The largest haze droplets would nucleate on the most massive salt particles (solid magenta curve) with radii slightly greater than 1 μm.

Earlier, we stated that solution droplets occurring at a relative humidity less than 100% are called haze droplets. However, we also use this term to denote *any droplet to the left of the peak in its Köhler curve* (Fig. 7.4). These droplets are thermodynamically different from solution droplets larger than the size at the peak of their respective Köhler curves. Haze droplets

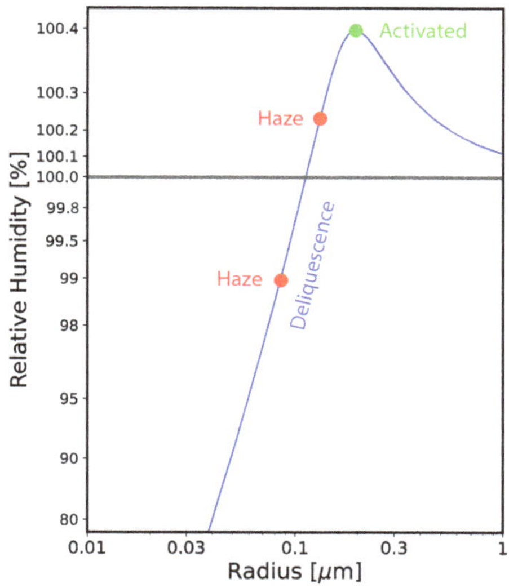

Fig. 7.4: A single Köhler curve, showing examples of different possible droplet equilibrium conditions. At a relative humidity less than 100%, a haze droplet has been nucleated by deliquescence and is in stable equilibrium. At a relative humidity exceeding 100%, a haze droplet is in stable equilibrium with supersaturated air because it is smaller than the critical radius in its Köhler curve. A droplet that has reached its critical radius (referred to as "activated") is now in unstable equilibrium; it will continue to grow as long as the relative humidity of its environment exceeds that for its current size on its Köhler curve.

are in a state of *stable equilibrium*—if more water molecules join the solution droplet than leave it, its size increases, and the Köhler curves predict that the supersaturation required for its equilibrium must also *increase*.

However, if the supersaturation of the air is not increasing, then the increase in droplet size due to condensation takes it out of equilibrium with the supersaturation of the surrounding air. The droplet will return to equilibrium through evaporation, reducing its size. In other words, *haze droplets must follow their respective Köhler curves and can only increase or decrease in size when the supersaturation of the surrounding air increases or decreases.* If the supersaturation of the surrounding air never increases, they will never grow larger.

7.2.4 Activation

A different situation arises for solution droplets with sizes at or to the right of the peaks of their respective Köhler curves. These droplets are thermodynamically *unstable*, as illustrated by the following scenario.

Consider a solution droplet exactly at the peak supersaturation of its Köhler curve (green dot in Fig. 7.4). If any additional water vapor condenses onto the droplet and its size increases even slightly (moving to the right along the x-axis), the supersaturation required for equilibrium will actually *decrease*, according to its Köhler curve. Even if the air remains at a constant supersaturation, the droplet will now be in air that is *more* supersaturated than required for equilibrium. This encourages additional droplet growth by condensation, attempting to bring it back to equilibrium with its environment. As the solution droplet continues to grow, its Köhler curve indicates that the supersaturation needed for equilibrium decreases further, taking it progressively further from equilibrium.

Soluble AP are considered **activated** when they nucleate droplets that grow beyond the peaks of their Köhler curves. These droplets are thermodynamically favored to grow to larger sizes, limited only by the available water vapor for condensation within the droplet population. As a primary goal in this book is to understand the chain of events in a cloud leading to precipitation, these droplets are of greatest interest to us.

Once their CCN are activated, these droplets continue to grow by condensation even if the supersaturation of the air decreases, as long as it does not fall below the value on its Köhler curve at the current droplet size. If entrainment of dry air decreases the relative humidity below that value, the droplet will evaporate until it becomes a haze droplet again, reaching the stable equilibrium size for the now lower relative humidity. If the relative humidity decreases further, below that required for deliquescence, the droplet will evaporate entirely.

7.2.5 Critical radius and critical saturation ratio

The **critical drop radius** r^* and the **critical saturation ratio** S^* define the position and height of the peak, respectively, in any given Köhler curve. These parameters are central to determining whether a particular size and composition of AP in a specific environment will activate to form cloud droplets. Specifically, the maximum supersaturation achieved in a rising parcel of air determines which AP get activated; any AP with critical saturation ratios higher than that maximum will remain, at best, haze droplets.

Expressions for the values of r^* and S^* can be obtained from (7.6) using standard calculus methods to give

$$r^* = \sqrt{\frac{3b}{a}} \tag{7.8}$$

and

$$S^* = 1 + \sqrt{\frac{4a^3}{27b}}, \tag{7.9}$$

where a and b are given by (7.7).

Based on (7.9), we can now make the following general statement: *the larger the mass of the solute in the droplet, the smaller the critical saturation ratio (or supersaturation) required for it to be activated and continue to grow in even a minimally supersaturated environment.*

7.2.6 Application to an adiabatic parcel

We can use the Köhler curves to predict the evolution of a population of haze and cloud droplets growing by condensation in adiabatic air parcels at the initial stage of saturation. Suppose we have an ascending and thus cooling unsaturated air parcel that contains some soluble AP. As the temperature of the air decreases during adiabatic ascent, the relative humidity increases. The soluble AP will deliquesce, and the larger ones will do so first, so that at a given relative humidity, the largest AP will have nucleated the first (and largest) haze droplets. (This can be seen by choosing a relative humidity such as 80% in Fig. 7.3.)

As the air parcel continues to ascend and cool, its relative humidity will eventually surpass 100%—i.e., it will become supersaturated, and the haze droplets will continue to follow their Köhler curves, growing by condensation to their equilibrium sizes for the new supersaturation value. As the cloudy parcel ascends further, the supersaturation continues to increase. The largest (most massive) soluble AP will be the first to experience the environmental relative humidity exceeding their respective critical saturation ratios S^* at the peaks in their Köhler curves. These AP will become activated first, and the droplets they nucleate can grow further by condensation, even if there is a slight reduction in S.

Meanwhile, as the cloudy air continues to ascend and cool, resulting in increasing supersaturation, it may surpass the peaks of the Köhler curves for some of the smaller AP, leading to their activation. If any insoluble, hydrophilic AP are present in the parcel, they may also nucleate droplets according to the Kelvin curve and the current supersaturation of the air parcel, but this will occur in far fewer numbers because they are less common than the soluble AP.

Eventually, either the cloudy parcel stops ascending and cooling, or enough of the cloud droplets have activated and are growing by condensation, thereby depleting water vapor and preventing further increases in supersaturation. At this point, the maximum supersaturation for that adiabatic parcel has been achieved, and the supersaturation of the air parcel then begins to decrease. Consequently, no additional droplets can be nucleated. The droplets that have already surpassed the peaks in their Köhler curves continue to grow by condensation, while any haze droplets shrink by evaporation as the supersaturation decreases.

Recall that at the start of this chapter, we sought a way for cloud droplets to form that requires lower supersaturations—that is, values less than the couple of percent that naturally occur in Earth's clouds. Now we know that *cloud droplets form in Earth's atmosphere via heterogeneous nucleation, primarily on soluble aerosol particles that require much lower supersaturations for nucleation*. Some of the largest insoluble, hydrophilic aerosol particles also nucleate some droplets, though far fewer.

Whether soluble or merely hydrophilic, we call the AP that nucleate cloud droplets *cloud condensation nuclei* (*CCN*). We will discuss more of their characteristics in Chapter 8. Then, in Chapter 9, we will discuss the droplet condensation and evaporation processes in more detail.

7.3 Heterogeneous Nucleation of Ice

In Earth's clouds, cloud droplets, drizzle, and raindrops can be supercooled to temperatures significantly below 0 °C before freezing. It is common to find clouds containing very little ice at temperatures warmer than −10 °C. In Chapter 6, we discussed that homogeneous freezing of liquid droplets occurs only for temperatures near −40 °C. Additionally, we noted that homogeneous nucleation of ice crystals from water vapor is highly unlikely due to the requirement for extremely high supersaturations with respect to ice. Nevertheless, it is rare to find clouds lacking ice when their minimum temperature falls much below −15 °C. Thus, the key to ice nucleation in clouds must involve heterogeneous nucleation.

The following discussion on the heterogeneous nucleation of ice will be more qualitative than quantitative. Many of the more quantitative aspects require concepts beyond the scope of this book and may yet be under debate within the scientific community. The primary reason for having fewer well established details of heterogeneous nucleation is that the chemistry, thermodynamics, and physics of ice are much more complex than those for liquid water droplets. Unlike water droplets, the molecules in ice must be arranged in a specific lattice structure, making nucleation highly sensitive to the molecular structure at the surface of any aerosol particle that could potentially assist it. Additionally, we will see that there are multiple *modes of ice nucleation*, each requiring different thermodynamic conditions to occur. As a result, there are no well-defined thermodynamic equations (like the Kelvin and Köhler equations for water droplets) that can be applied to all possible modes of heterogeneous ice nucleation.

Interestingly, the chemical components of AP that are most important for nucleating cloud droplets are mostly irrelevant for nucleating ice. Ice nucleates predominantly upon *insoluble* AP, because water molecules need a stable base upon which to build the lattice structure of ice. This restriction, coupled with the fact that the molecular structure of the AP must be similar to that of ice, makes the subset of AP that are *ice nucleating particles* (INP) very small compared with CCN. As a result, the number concentra-

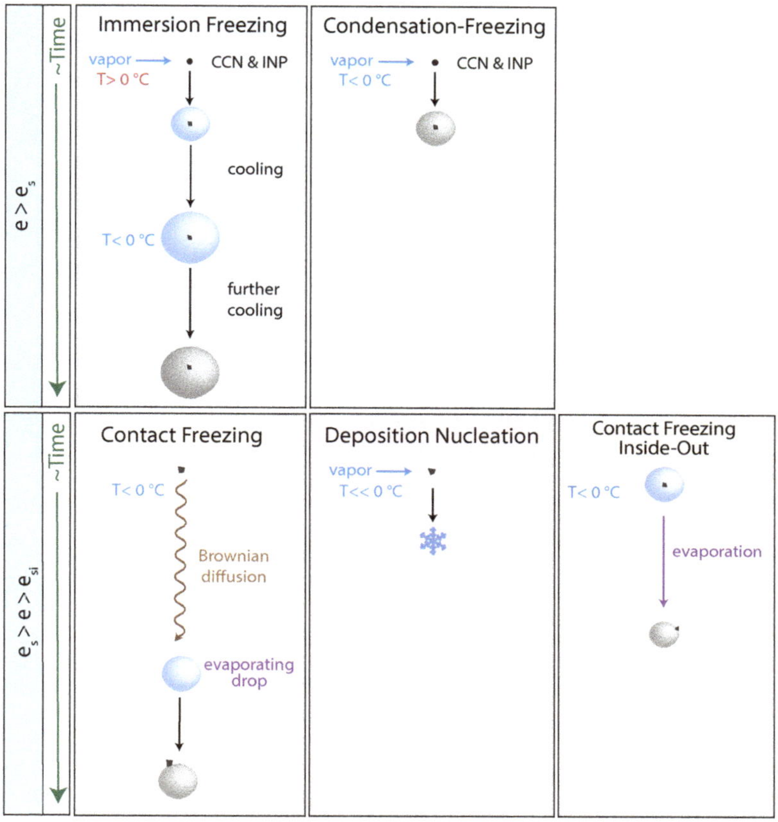

Fig. 7.5: Different modes of heterogeneous nucleation of ice. Liquid drops are shown as blue spheres; frozen drops are shown as gray spheres. Top row shows modes which require the air to be supersaturated with respect to liquid water. Bottom row shows modes which only require the air to be supersaturated with respect to ice. An approximate relative time scale is indicated by the depth of each process depicted within each enclosing rectangle. Mixed AP that contain elements of both CCN and INP are shown as black circles; INP are shown as irregular shapes.

tions of ice crystals found in clouds tend to be orders of magnitude smaller than those for cloud droplets, except for special cases discussed in Sec. 7.4.

There are currently five recognized **modes of heterogeneous nucleation of ice** in Earth's clouds, as summarized in Fig. 7.5. Each mode operates within a specific range of thermodynamic conditions and time scales that govern where they can be active, as shown in Fig. 7.6. Within a given mode, the exact temperature at which ice nucleation occurs depends on the chemical compositions of the INP and on their sizes and surface char-

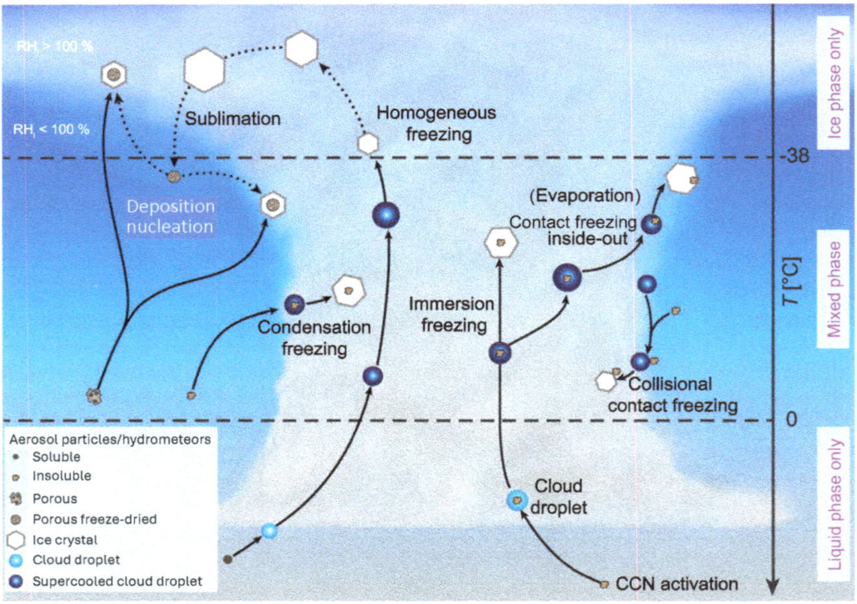

Fig. 7.6: Favored locations for all currently known mechanisms for primary ice nucleation in clouds. The designation of some aerosol particles as *porous* indicates their tendency to have cavities on their surfaces that favor deposition nucleation. *Adapted from Kanji et al. (2017). © American Meteorological Society. Used with permission.*

acteristics, which will be discussed in Chapter 8. The following sections introduce these five distinct modes of heterogeneous nucleation.

7.3.1 Nucleation of ice crystals from water vapor

Recall that the difficulty of homogeneous nucleation of ice from water vapor is due to the requirement for a large number of molecules to come together by chance to form an ice crystal large enough to survive at the low supersaturations common in Earth's clouds. However, if an ice nucleating particle is present, which has a molecular structure similar to that of ice, it provides a "pattern" for water molecules in the vapor state to build upon its surface and form an ice crystal at much lower supersaturation than is needed for homogeneous nucleation.

Deposition nucleation is the nucleation of ice directly from water vapor, without involving a liquid phase. Water molecules transition from the vapor phase and adhere to INP, growing according to the supersaturation with respect to ice in the surrounding air. This process is analogous to droplet nucleation upon insoluble particles. The INP must be sufficiently large for nucleation to occur within the range of ice supersaturation typical of Earth's clouds, as indicated by the homogeneous nucleation arguments used to derive (6.7) and the equilibrium curves shown in Fig. 6.6. Deposition nucleation is unlikely to occur on INP smaller than approximately 0.1 μm in radius. This mode of ice nucleation does not rely on the freezing of a liquid drop, allowing it to occur in parts of the cloud that are subsaturated with respect to liquid water; however, the surrounding air must still be supersaturated with respect to ice. It requires lower temperatures, and so it is more common in colder regions of the cloud compared with other modes of ice nucleation discussed in the following section.

7.3.2 Nucleation of ice from freezing of liquid drops

The other four modes of heterogeneous nucleation involve the freezing of water drops. Generally, these modes present lower energy barriers than deposition nucleation, allowing them to occur at higher temperatures.

In some of the modes of ice nucleation discussed in this section, it is important that the aerosol particles contain more than one chemical component, described as "internally mixed" AP. Internally mixed AP are quite common in Earth's atmosphere. In some cases, these AP can contain a soluble component that can act as a cloud condensation nucleus and an insoluble component that can act as an ice nucleating particle.

Immersion freezing occurs when the soluble component first nucleates a droplet at a temperature above 0 °C, immersing the insoluble component inside the droplet. Then, once it has sufficiently supercooled—and has grown larger while doing so—the drop freezes with the assistance of the insoluble component (if it is an ice nucleating particle). In this case, the cooling of the drops requires some length of time—for example, carried to greater heights within the cloud, as suggested by Fig. 7.6—to finally reach the lower temperatures required for the INP to nucleate ice. The specific temperature at which the ice nucleates, often referred to as the *activation temperature* of the INP, depends on their specific chemical composition, size, and surface characteristics.

INP effective for immersion freezing are believed to be more prevalent than those effective for deposition nucleation, as the size of INP needed to initiate ice from the liquid phase is smaller than that required from the vapor phase. Immersion freezing is also thought to occur at somewhat higher temperatures than deposition nucleation because of the greater ease in the phase transformation between liquid drops and ice, compared with water vapor and ice. Immersion freezing is considered a major mechanism for ice nucleation in convective clouds with stronger updrafts. The air must be supersaturated with respect to liquid water to maintain and grow drops until they freeze. Larger drops freeze more readily than smaller droplets.

Condensation-freezing is similar to immersion freezing, except droplet nucleation on the soluble component of the mixed aerosol particle occurs at a temperature near the activation temperature of the ice-nucleating component (i.e., *below 0 °C*), so the nucleation of the droplet is *followed almost immediately by its freezing*. The air must again be supersaturated with respect to liquid water to nucleate the droplet, and the solution comprising the droplet must be dilute enough to allow it to freeze—a solution droplet requires more supercooling to freeze.

This mode of ice nucleation, like immersion freezing, is arguably easier than deposition nucleation because a lower energy barrier (from liquid to ice) must be overcome, compared with the transition from vapor to solid in deposition nucleation. Consequently, it also does not require the air to be as cold as for deposition nucleation. Due to the requirement that the air be supersaturated with respect to liquid water to first nucleate the droplet, condensation-freezing is less likely in glaciated regions of the cloud.

Contact freezing, sometimes also referred to as *collisional contact freezing*, occurs when an ice nucleating particle is initially located *outside* a supercooled droplet and subsequently comes into contact with the droplet surface, freezing it. Contact freezing is considered relatively infrequent, because it relies on the low probability that INP are transported to the supercooled droplets by *Brownian motion*.[5] The movement of air molecules near a droplet growing by condensation tends to hinder the transport of particles to its surface. However, during droplet evaporation, such as at the edges of clouds or within downdrafts, the transport towards the droplet surface is more likely.

Contact freezing appears to require the least supercooling compared with all other modes of ice nucleation. In laboratory studies, it may require a temperature just a few degrees less than 0 °C, depending upon the composition of the INP. However, as it requires the presence of water droplets, the cloud must be supersaturated with respect to liquid water, at least initially.

Contact freezing inside-out may instead occur when an ice nucleating particle is already located inside a supercooled droplet, but is brought into contact with the droplet surface *from the inside* as the droplet experiences evaporation in a subsaturated environment. Like immersion freezing and condensation-freezing, in this case the droplets likely originated from mixed AP containing both soluble and insoluble components. Contact freezing inside-out can result in droplets freezing at a relatively higher

---

[5] Brownian motion refers to the random movement of particles suspended in air, resulting from fluctuations in air molecule collisions.

temperature, like contact freezing in general. This mode of ice nucleation may also help explain common observations of more ice being present in the evaporating regions at the edges of clouds.

7.3.3 *Dependence of ice nucleation mode on INP composition*

In summary, five different modes of heterogeneous nucleation of ice are thought to be possible in Earth's clouds, alongside the homogeneous mode of droplet freezing at temperatures near −40 °C. It can be quite challenging to identify which mode is active in a particular cloud, and it is likely that multiple modes are simultaneously active in different regions within deep mixed-phase clouds, as suggested by Fig. 7.6. Different chemical compositions of INP will nucleate ice at varying temperatures and supersaturations.

Furthermore, the same INP can nucleate ice under different thermodynamic conditions, depending on the active mode. For example, laboratory studies have indicated that for a given chemical composition of INP, condensation-freezing can occur at higher temperatures than if the same INP nucleate ice through deposition nucleation. Consequently, it is very challenging to predict the temperatures at which drops inside a cloud will freeze unless the homogeneous nucleation limit near −40 °C is reached.

Chapter 8 will examine the types of AP that are particularly effective in nucleating ice and their specific characteristics. We will find that significantly fewer AP in the troposphere qualify as INP compared with the much larger number of CCN. Consequently, we should expect the observed number concentrations of droplets in Earth's clouds to typically far exceed the observed number of ice crystals, which is indeed supported by data—as shown by the typical number concentrations of various hydrometeor types presented in Fig. 4.1.

7.4 Secondary Ice Production

At times, the measured number of ice crystals in a cloud greatly surpasses the measured number of INP in the atmosphere. Multiple mechanisms of **secondary ice production** have been hypothesized to explain these occurrences, whereby more than one frozen hydrometeor may be produced from a single ice nucleating particle. These processes are termed *secondary* because some ice must already exist in the cloud for the additional ice to be created. Several examples of possible secondary ice production observed in laboratory experiments are shown in Fig. 7.7.

Some mechanisms are as simple as *mechanical fracture*, in which broken fragments are produced after the collision of two ice crystals—one having significantly more mass than the other—or when a more fragile ice crystal

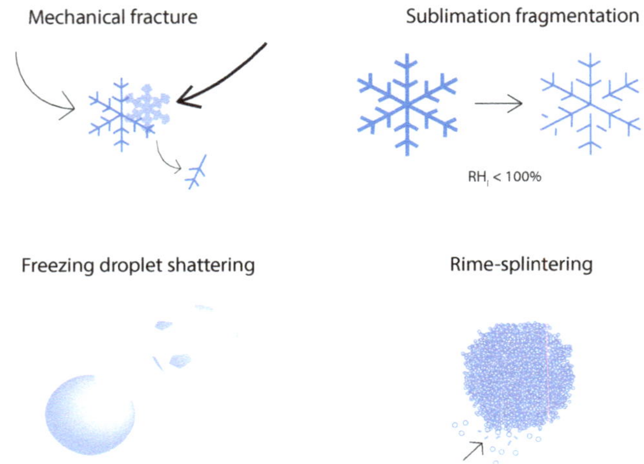

Fig. 7.7: Possible mechanisms of secondary ice production. *Top left:* a heavily rimed ice crystal has collided with a more fragile unrimed ice crystal, breaking off a branch of the unrimed crystal. *Top right:* a sublimating ice crystal has had some branches and sub-branches that became disconnected from the main crystal because of faster sublimation at those locations. *Bottom left:* a supercooled liquid droplet that shattered as it froze. *Bottom right:* as supercooled droplets (open circles) rime onto a graupel particle, they emit some ice splinters (marked with an arrow) as they freeze upon the graupel surface.

collides with a graupel particle. It has also been suggested that colliding graupel particles emit small bits of ice from their surfaces.

Another hypothesized mechanism of secondary ice production occurs when a sublimating ice crystal has a segment that breaks off, a process called *sublimation fragmentation*. This may happen when a thinner area on the crystal sublimates entirely before the surrounding regions.

Freezing drop shattering occurs when isolated pockets of liquid water are trapped within a solid ice shell as the droplet freezes. When these liquid pockets freeze and expand, the internal pressure within the ice shell increases, causing it to burst and create multiple fragments.

A more complex and sometimes controversial mechanism is *rime-splintering*. This process is thought to occur within a narrow temperature range between approximately −9 °C and −3 °C, when a large frozen hydrometeor rimes supercooled droplets having radii greater than 24 μm. Some laboratory experiments suggest that under these conditions, the droplets emit *splinters* of ice as they freeze onto the frozen hydrometeor surface.

All these secondary ice production mechanisms can produce up to several orders of magnitude more ice than would be predicted by the number of INP alone, given enough time. Any ice fragments can continue to grow on their own by deposition in a supersaturated environment or, if large enough, by riming. They can also freeze larger water drops on contact.

Because quantitative information about the productivity of these mechanisms outside the laboratory is difficult to obtain, current research often utilizes numerical modeling to understand which mechanisms may be active under different cloud conditions. Researchers compare these numerical modeling results with aircraft observations within real clouds to build confidence in our understanding of the significance of these hypothesized mechanisms for secondary ice production.

Questions for Review

1. What is the difference between soluble and insoluble aerosol particles and between hydrophilic and hydrophobic surfaces?
2. What role does the Kelvin equation play in understanding droplet nucleation on different surfaces?
3. Why do soluble aerosol particles make droplet nucleation easier?
4. What does the Köhler equation describe, and how is it applied to explain the nucleation of haze droplets and, possibly, activation of a CCN within a droplet that favors its growth to larger sizes?
5. Describe the progression of activation of CCN within the updraft of an ascending adiabatic cloud parcel.
6. What are the five known modes of heterogeneous ice nucleation, what type of aerosol particle is required for each mode, and which modes require a droplet to be nucleated first?
7. Why are supercooled drops common in Earth's clouds?
8. Why does the number of frozen hydrometeors observed in a cloud sometimes greatly exceed the number of ice nucleating particles in its environment?

Homework problems and other resources for students and instructors can be found at bit.ly/4dYBwB5, or scan this code:

Further Reading

For more information on secondary ice production [advanced]:

Field, P. R., and Coauthors, 2017: Secondary ice production: Current state of the science and recommendations for the future. *Meteorological Monographs*, **58**, 7.1–7.20

Korolev, A., and T. Leisner, 2020: Review of experimental studies of secondary ice production. *Atmospheric Chemistry and Physics*, **20 (20)**, 11,767–11,797

CHAPTER 8

Atmospheric Aerosol Particles

In the last two chapters, we explored the different ways that cloud droplets and ice can—or cannot!—be nucleated to create Earth's clouds. Aerosol particles of certain chemical compositions are most helpful: soluble particles for droplet nucleation, and insoluble particles for ice nucleation. In this chapter, we first consider the population of aerosol particles suspended in Earth's atmosphere as a whole, including their sources and sinks. We then focus on the prevailing types needed for droplet and ice nucleation.

8.1 Terminology, Sizes, and Prevalence

Technically, an **aerosol** is a mixture of air with small (solid and/or liquid) particles within it. The small particles are called **aerosol particles** (AP), but they are sometimes also loosely referred to as just "aerosol(s)."

The names[1] for typical size ranges of AP of concern for clouds are shown in Fig. 8.1. In this book, we will use the approximate particle radius to define these different categories, but some scientists may instead use the

[1]Atmospheric aerosol scientists use different names that encompass and extend these size ranges based on the approximate particle diameter D and associated behavior: *nucleation mode* ($D < 0.01$ μm), *Aitken mode* (0.01 μm $< D <$ 0.1 μm), *accumulation mode* (0.1 μm $< D <$ 1 μm), and *coarse mode* ($D > 1$ μm) aerosol particles.

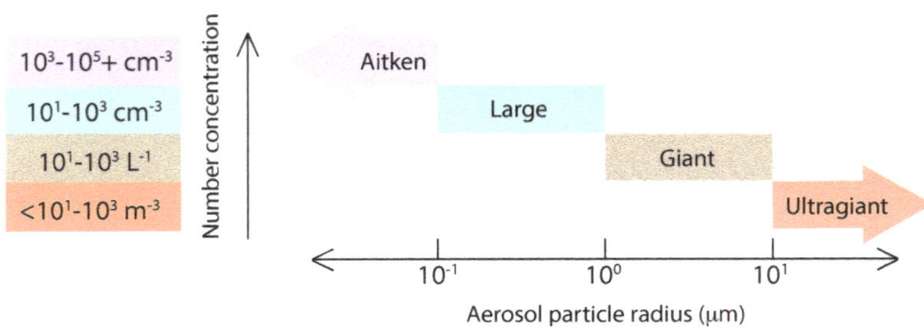

Fig. 8.1: Size ranges and typical number concentrations of AP that significantly influence cloud formation in Earth's atmosphere.

particle diameter or maximum dimension. These size categories include **Aitken** particles, with radii less than 0.1 μm; **large** particles, with radii ranging from 0.1 to 1 μm; **giant** particles, with radii greater than 1 μm; and **ultragiant** particles, with radii of 10 μm or larger.

Many AP are not spherical, as shown in Figs. 8.2 and 8.3. They can consist of organic material, sulfates, soot, metals, minerals, or combinations of these materials, referred to as *internally mixed*. They are usually too small to be visible to the human eye alone and must be magnified to see their shapes. Ultragiant particles (Fig. 8.3) are the only AP visible to the human eye. They include the tiny reflective particles you notice floating inside your home or office when illuminated by a beam of sunlight entering a window.

Aerosol particles are abundant in the troposphere and are present in much greater numbers than the droplets or ice crystals found in clouds. Aitken particles are much more plentiful than larger AP, with giant and ultragiant particles being far fewer. Typically, these larger particles are so scarce that there is significant uncertainty in their measurement.

It should come as no surprise that pollutants from anthropogenic activities are major contributors to the overall number of atmospheric AP. Consequently, a high concentration of AP is found in the air within and around urban centers, while significantly fewer are present in the cleaner air over the oceans.

Because AP originate primarily near the Earth's surface, their concentrations significantly decrease with altitude in the troposphere. This is particularly true for giant and ultragiant particles (Fig. 8.4), which have the largest individual masses among all AP and, consequently, the highest **settling velocities** (i.e., fall speeds). These particles cannot remain in the troposphere for long unless there is upward motion in the atmosphere to support them. Their concentrations can vary considerably on any given day and in any given area.

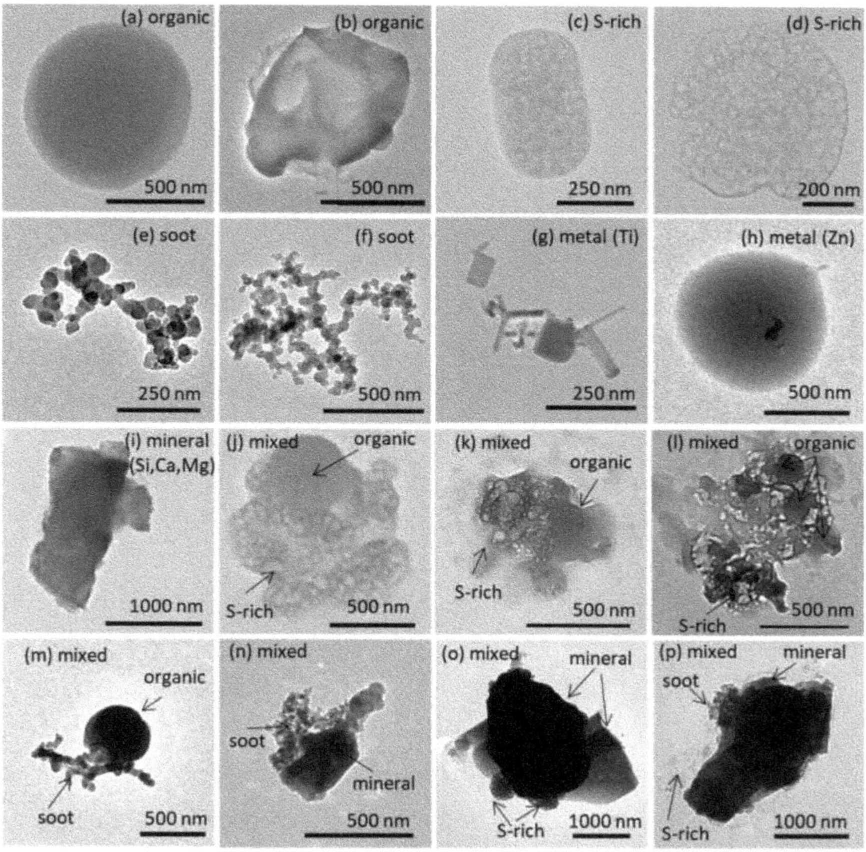

Fig. 8.2: Images of Aitken, large, and giant AP, as seen using transmission electron microscopy (TEM). Particle compositions and reference scales are noted in each frame. Note that 1000 nm = 1 μm. *From Wang et al. (2021). (CC-BY)*

8.2 Sources and Sinks

8.2.1 Primary sources

There are numerous sources of AP in our atmosphere, which we will now categorize. In some geographical regions, more than one of these sources may be an important contributor.

Biological sources of AP include pollen and spores naturally occurring from vegetation on the Earth's surface, as well as bacteria, algae, protozoa, fungi, and viruses. Particles from *naturally occurring* forest fire smoke are also included in this category.

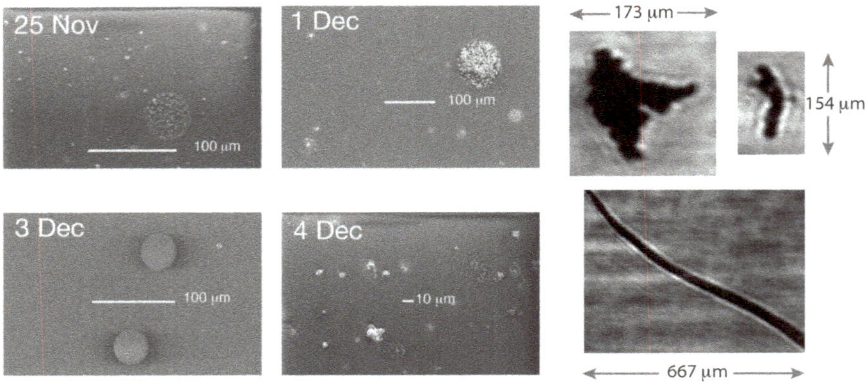

Fig. 8.3: Examples of ultragiant AP. Soluble particles appear as flat circles (top left, top center and bottom center images), having been solution droplets that dried on the slide. Some insoluble particles are spherical (lower left), but most are not (all panels). *Adapted from Lasher-Trapp and Stachnik (2007) and Lasher-Trapp et al. (2008). Both sources © American Meteorological Society. Used with permission.*

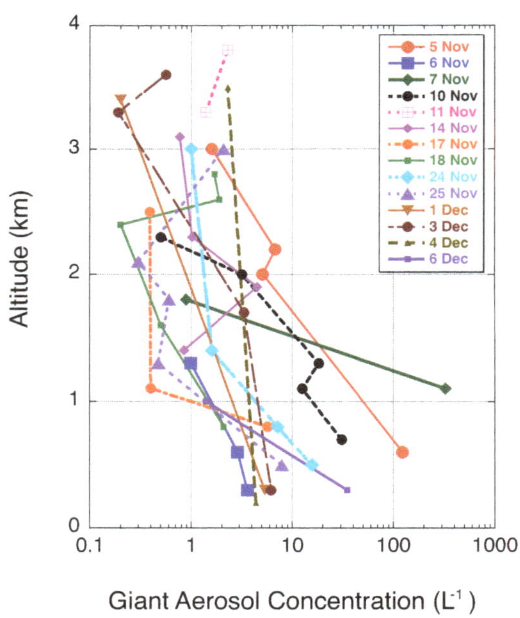

Fig. 8.4: Observed number concentrations of giant AP versus height over the same continental area in the U.S. Midwest over successive days. *From Lasher-Trapp and Stachnik (2007). © American Meteorological Society. Used with permission.*

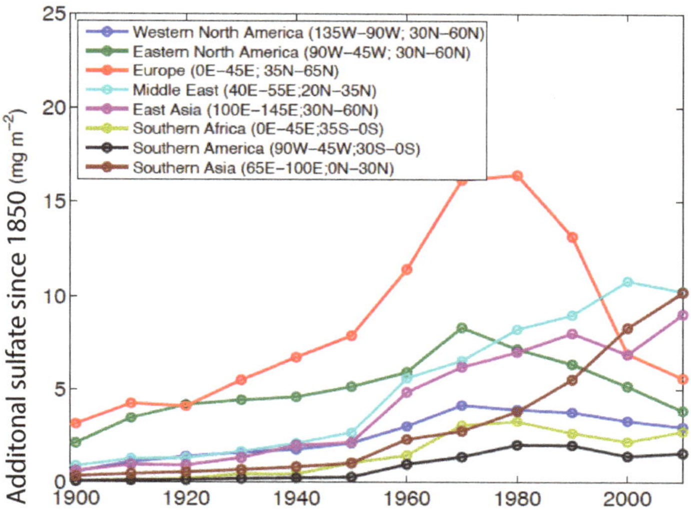

Fig. 8.5: Amounts of sulfate (a subset of anthropogenic AP) over various regions of the world from 1900 to 2010, expressed as a difference from their relative values in 1850. *Adapted from Skeie et al. (2011). (CC-BY-SA)*

Solid earth sources of AP include minerals and soil particles transported from the Earth's surface to the atmosphere by winds and turbulence, such as Saharan dust, which can encircle the globe, as well as particles from volcanic eruptions.

Oceans are also a source of AP, as sea spray and bubble bursting on the ocean surface release tiny droplets of seawater into the atmosphere. These droplets can evaporate, leaving solid salt particles suspended in the air. At sufficiently high relative humidity, these particles may also remain in their liquid state as haze droplets.

There are also many *anthropogenic activities* that contribute to the population of AP. Estimates suggest that the combined mass of AP from these activities is approximately 20% of that contributed by natural sources.

As illustrated by the trends for Europe and Eastern and Western North America in Fig. 8.5, the amount of anthropogenic AP in the atmosphere has been declining in recent decades. This decline is attributed to steps taken by the governments of industrialized nations to reduce pollutants due to health hazards. However, an increase in aerosol emissions from economically developing nations, as they engage in more industrial activities, has also been predicted. Their impact is becoming evident, as shown by the trends for East Asia, South Asia, and the Middle East.

Some of these activities produce AP of the same composition as natural sources, but human activities increase the overall amounts. For example, constructing roads or tilling the land promotes additional lofting of soil particles into the atmosphere. When humans set forest fires, either deliberately or inadvertently, the resulting smoke and ash plumes can add substantially to those from naturally occurring forest fires, such as those initiated by lightning strikes.

Many AP result from the combustion of fuel for transportation or industrial processes. Black carbon in Earth's atmosphere, primarily a byproduct of burning fossil fuels or biomass burning, has continued to increase globally since pre-industrial times.

8.2.2 Gas-to-particle conversion

Not all AP originate from the direct injection of particulate matter into the atmosphere. Instead, many AP, especially the smallest Aitken particles, are formed via a process known as **gas-to-particle conversion**. This process occurs when certain gas-phase chemicals, termed *precursors*, undergo chemical reactions in the air that result in the condensation of the reaction products into droplets or solid particles, known as *secondary aerosol particles*. This conversion is often facilitated by ultraviolet radiation from the sun and/or aqueous chemical reactions when the gases are absorbed into cloud droplets.

Precursor gases include both natural and anthropogenic emissions. Some of these are listed below:

i) Natural

- *Biogenic Volatile Organic Compounds (BVOCs)*: Emitted by vegetation, these include compounds such as isoprene and monoterpenes, which can subsequently form secondary organic aerosol (SOA) through oxidation reactions.
- *Oceanic Emissions*: A gas called dimethyl sulfide (DMS) is emitted from oceanic plankton and contributes to sulfate aerosol formation.
- *Soil Emissions*: Nitrogen oxides (NO_x) released from soil can contribute to nitrate aerosol.
- *Volcanic Activity*: Eruptions emit sulfur dioxide (SO_2) and other gases that can convert to sulfate aerosol.
- *Forest Fires and Biomass Burning*: These emit organic compounds and SO_2 that can form organic and sulfate aerosol.
- *Lightning*: Produces NO_x gases, contributing to nitrate aerosol formation.

ii) Anthropogenic

- *Fossil Fuel Combustion*: Vehicles and industrial processes emit NO_x and SO_2, contributing to nitrate and sulfate aerosol formation.
- *Industrial Processes*: Factories often emit volatile organic compounds (VOCs), NO_x, and SO_2.
- *Agriculture*: Fertilizers and animal waste emit ammonia (NH_3), which can form ammonium aerosol.
- *Waste Treatment*: Landfills and waste treatment can emit CH_4, VOCs, and other compounds.
- *Residential Cooking and Heating*: The burning of natural gas, propane, or fuel oil emits compounds that can form organic aerosol.

8.2.3 The role of sinks

Given the many sources of AP, one might wonder why they do not continuously build up over time to greater and greater concentrations in the atmosphere. Of course, humankind's efforts to decrease anthropogenic sources help, but these efforts, by themselves, can only reduce the *rate of increase*.

More importantly, there are natural *sinks* (removal mechanisms) for AP. The population of particles at any given time and location is not static but reflects a dynamic balance between sources and sinks. When sources collectively exceed sinks, the concentration of AP increases; conversely, when sinks exceed sources, the concentration decreases.

We can write this semi-mathematically as follows:

$$\frac{\mathrm{d}N}{\mathrm{d}t} = \sum_i \text{Source}_i - \sum_i \text{Sink}_i \tag{8.1}$$

In words, the *rate of change* of the concentration N of some AP of interest—which might be a subset of all AP, such as just Aitken particles—is equal to the sum of the rates of contributions from all sources minus the sum of the rates due to all sinks.[2]

Key to interpreting the above simple balance equation is the observation that the sources of AP generally do not depend on the existing N,

[2]Equation (8.1) is quite general and can be applied to almost any physical component of any natural system, from the amount of water vapor in the atmosphere to the number of moose in the state of Alaska!

whereas sinks are often proportional to N.[3] That is, if you double the concentration of AP, you usually also double their rate of removal.

If we define O as the sum of all sources, then we can write

$$\frac{dN}{dt} = O - \nu N, \tag{8.2}$$

where the constant ν determines the combined magnitude of all sinks for a given concentration N. Note that ν has units of inverse time, so its reciprocal $\tau \equiv 1/\nu$ can be interpreted as a **residence time**; that is, as an *average time* that a particular constituent of interest remains in the atmosphere before being removed.

Putting it all together, we see that if the sources O remain constant for a period of time, the concentration N will increase or decrease until sources and sinks are in balance and N reaches a steady state. That steady-state value is obtained by setting $dN/dt = 0$ and solving for N:

$$N_{\text{steady}} = \frac{O}{\nu} = \tau O. \tag{8.3}$$

Thus, large sources and/or a long residence time can give rise to a large concentration of AP, whereas small sources and/or a short residence time can give rise to a small concentration. If humans were to succeed in reducing global emissions of aerosol pollutants by, say, half, (8.3) tells us that the steady-state concentration would also be reduced by half.

Crucially, the above analysis can be applied separately to any subpopulation of AP. In general, different size classes and compositions have distinct sources, sinks, and residence times. These characteristics can also depend strongly on location, season, and other factors.

8.2.4 Size-dependent sinks

For the smallest AP of our concern, the Aitken particles, their primary sink within a few tens of meters above the ground is due to the effects of Brownian motion, which diffuses these tiny particles toward the ground and other surfaces such as trees, buildings, etc., to which they adhere and are thus removed from the air. At greater heights above the ground, the dominant sink of Aitken particles is **coagulation**, where they collide with each other (again due to Brownian motion) and stick together, forming *large* AP. These larger particles are more susceptible to other sinks because of their

---

[3]There are exceptions. For example, depletion of Aitken particles by coagulation is more nearly proportional to the *square* of the concentration.

increased size. Aitken particles can also collide with a cloud droplet or a frozen hydrometeor and become embedded within it, thus removing them from the air.

The main sink for the *large* category of AP, as well as some of the larger Aitken particles, is their *nucleation of cloud droplets*, acting as CCN if they have a soluble component, which is further discussed in the next section. Insoluble particles at the larger end of the *large* category of AP may also act as INP, as discussed in Sec. 8.4. Some giant and ultragiant AP can also nucleate cloud droplets or ice, but this is not the predominant process removing them from the atmosphere.

Instead, the giant and ultragiant categories of AP are primarily removed from the atmosphere by two other processes: **precipitation scavenging** and **sedimentation**. Precipitation scavenging refers to the process by which precipitation collides with and collects these AP as it falls toward the ground. Sedimentation refers to the fallout of these AP toward the ground due to their larger fall velocities, which result from their significantly greater mass relative to the smaller categories of AP.[4]

As previously discussed, the residence time of each AP size category indicates how long the particles typically remain in the atmosphere. When comparing all AP sizes, the residence time for the *large* category is the longest. These particles are big enough that coagulation is not very efficient, yet small enough that they do not have significant fall velocities.

At the opposite ends of the AP spectrum, both Aitken and giant/ultragiant particles have at least one effective sink, resulting in much shorter residence times. A summary of the sources, sinks, and typical residence times of AP resulting from these processes is given in Fig. 8.6.

8.3 Cloud Condensation Nuclei (CCN)

Not all AP serve as *cloud condensation nuclei* (CCN); that is, not all are able to nucleate droplets at supersaturations typical of Earth's clouds. As we have seen in Chapter 7, the "best" CCN, according to the Köhler equation—i.e., those that will nucleate droplets under the least amount of supersaturation—are *soluble* and *large*. Some insoluble but hydrophilic large, giant, and ultragiant AP can also serve as CCN, as shown by the Kelvin equation, but they are not a major source of CCN.

---

[4]Aerosol scientists include the combined processes of removal of the Aitken particles by Brownian motion and the sedimentation of giant and ultragiant particles as a single sink called "dry deposition" because they do not involve hydrometeors.

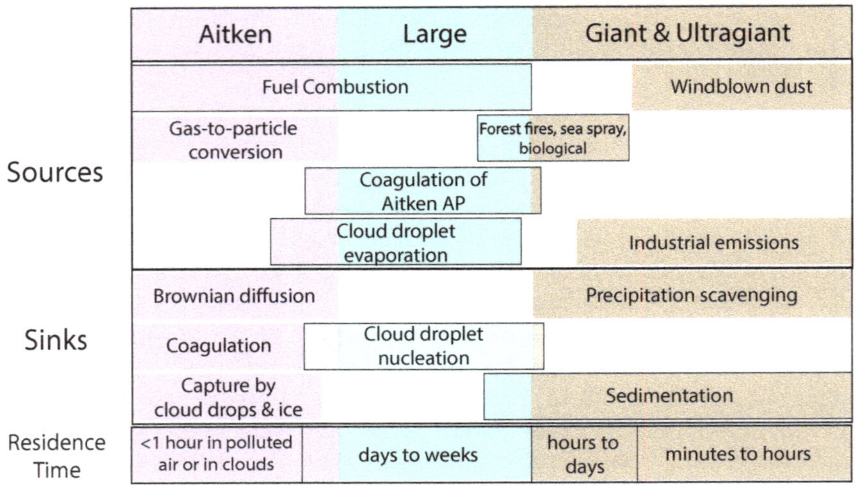

Fig. 8.6: Sources, sinks, and residence times of aerosol particles of concern to cloud microphysics, as described in the main text. Processes outlined in black boxes span more than one size category of AP. Lighter shading for cloud droplet nucleation signifies that only the subset consisting of soluble AP would experience this sink. *After Wallace and Hobbs (2006).*

However, we also know from our earlier discussion in this chapter that the largest AP are far fewer in number relative to the smaller sizes of AP in the troposphere. Indeed, the few giant and ultragiant CCN typically found in the air cannot alone explain the number of droplets observed in clouds. Thus, the dominant sizes of CCN for formation of cloud droplets are the Aitken (primarily towards the bigger end of that size range) and "large" categories.

CCN can be measured using *thermal diffusion chambers*, which can either be based on the ground ormounted on research aircraft. The general principle behind these instruments is that air is brought into the chamber, where it flows between parallel, wetted plates set at different temperatures. The temperature difference between the plates creates a specific supersaturation value within the chamber, typically ranging from a few tenths of a percent to slightly greater than 1%. As the CCN nucleate droplets in the chamber, their number is recorded by an optical particle counter (Sec. 5.2.3).

Like AP in general, CCN also tend to decrease in number with increasing height in the troposphere, because their main source is near the ground. Vertical profiles can show variability depending upon the local meteorology, however. Shown in Fig. 8.7 are some vertical profiles of CCN measured by a research aircraft for a supersaturation of 1%. In Fig. 8.7a, we see the typical decrease in the number of CCN with height up to the

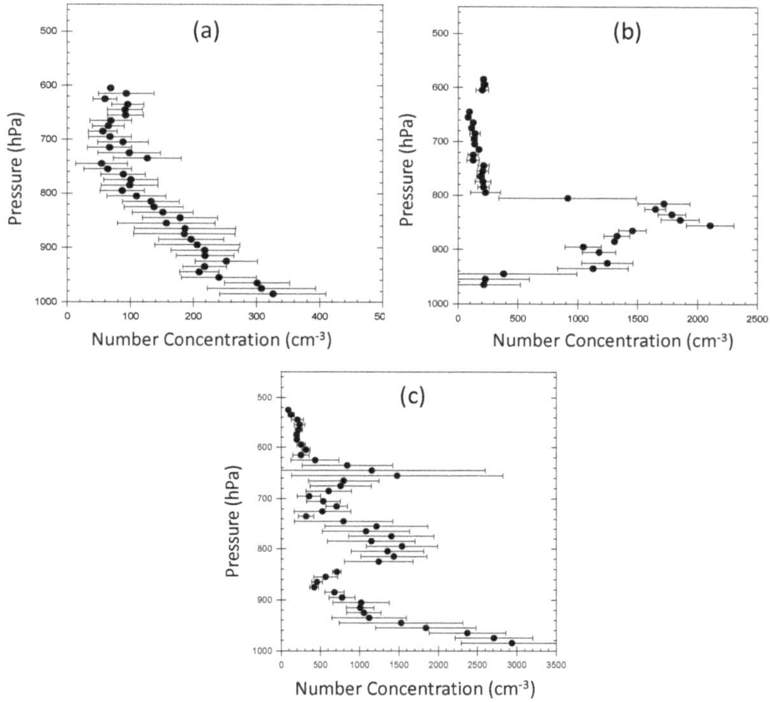

Fig. 8.7: Examples of vertical profiles of CCN data collected during the AIRS II Field Campaign over (a) Lake Erie, (b) near Cleveland, OH, and (c) directly over Cleveland, OH. All CCN measurements are representative of air at 1% supersaturation. Error bars are shown on each data point based on the variability of the number concentrations measured at each altitude. *Data courtesy of J. Hudson.*

maximum altitude at which data were collected. This profile was taken over one of the North American Great Lakes, so the maximum CCN number concentrations are only ~300-400 cm^{-3} near the ground.

A highly polluted air mass over a major urban area (Cleveland, Ohio) is evident in Fig. 8.7c, where the number concentrations of CCN near the ground often exceed 3,000 cm^{-3}. This profile also generally shows the number concentration of CCN decreasing with height, but on this day it exhibits substantial variability, possibly due to cloud layers, which can be a source of CCN when cloud droplets evaporate.

On the other hand, CCN data collected on a different day near Cleveland, Ohio (Fig. 8.7b) show number concentrations *increasing* with height up to about the 850 hPa level and then suddenly decreasing to very small values above 800 hPa. Variable vertical profiles like this can result from several possible meteorological factors. A temperature inversion (indicating a layer of stable air) was located near 800 hPa on this day. This inversion prevented the vertical mixing of CCN upwards, effectively cutting off that air from the main source of CCN near the ground, and thus greatly reducing the measured CCN number concentrations at and above that level. Additionally, a precipitating stratiform cloud was located below 800 hPa at this time. Precipitation can *scavenge* (collect and remove) CCN and other AP as it falls toward the ground, reducing the CCN number concentrations to much smaller values.

The most common chemical composition of CCN is sulfate, either as sulfuric acid or ammonium sulfate, with a major source being combustion products, as discussed earlier in this chapter. Sea salt particles can also be important, particularly over the oceans or near coastlines, but they are typically greater than 1 μm in radius (thus being giant and ultragiant CCN) and are therefore far less numerous.

However, these are idealizations. Most CCN (and AP) have a mixed composition, containing multiple chemical components, and sometimes even both soluble and insoluble substances. In such cases, the nucleation of cloud droplets and their possible subsequent activation are determined by the characteristics of the soluble part of the AP, while their ice nucleating ability is determined by the characteristics of the insoluble part of the AP.

8.3.1 Activity spectrum

Extensive observations of CCN in the atmosphere, collected using thermal diffusion chambers, have shown that their ability to activate cloud droplets can be expressed as a power law called an **activity spectrum**:

$$N_{\mathrm{CCN}} = Cs^k \tag{8.4}$$

where N_{CCN} is the number of CCN activated into cloud droplets, in units of number per cm^3; s is the supersaturation, expressed in percent; and C (in units of number per cm^3) and k (unitless) are constants that depend on the amounts and chemical compositions of the CCN in that air mass.

In maritime air, C often ranges between 30 and 300 cm^{-3}, and k tends to range from 0.3 to 1.0. In continental air, C is larger, ranging from 300 to 3,000 cm^{-3}, and k has a wider range, from 0.2 to 2.0. For a supersaturation of 1%, the number of CCN activated equals C, and the slope of the curve when N versus s is plotted on a log scale is k (Fig. 8.8). Because we know from our earlier study of the Köhler equation that larger CCN tend to be activated at lower supersaturations—meaning they have lower peaks in their Köhler curves—it should come as no surprise that the curves all tilt upward for higher s. This indicates that fewer CCN are activated at low supersaturation, whereas more CCN, sometimes by orders of magnitude, are activated at higher supersaturation.

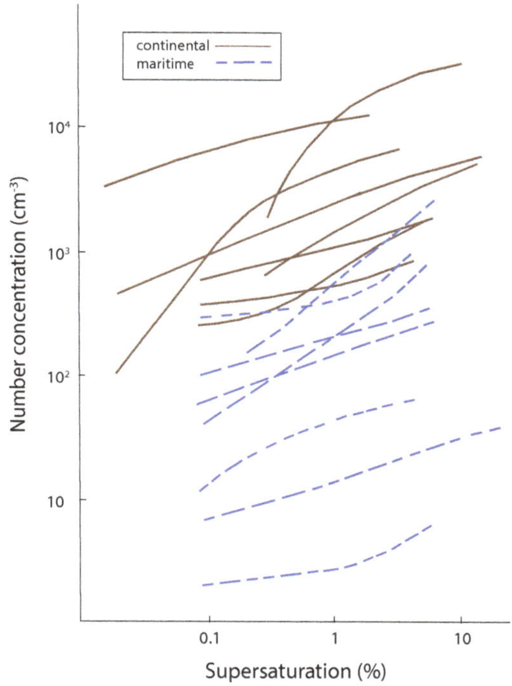

Fig. 8.8: Examples of the number of activated CCN versus supersaturation for data collected in the vicinity of Sydney, Australia, when the prevailing winds were from the continent (brown, solid lines) or from the ocean (blue, dashed lines). *After Twomey (1959).*

However, recall that in real data such as these, there can also be variability in the chemical composition of different sizes of CCN. These variations affect the critical supersaturations needed for activation among the population of CCN. Also, the maximum supersaturations achieved in clouds typically fall between about 1% and 2% in cumuliform clouds with stronger updrafts, but can be much less than 1% in stratiform or cirriform clouds with weaker updrafts. For both reasons, it is evident that the number of cloud droplets activated in different air masses, or by different cloud types within these air masses, can vary by orders of magnitude.

In addition, Fig. 8.8 shows that CCN follow the trends of all AP in the atmosphere: many more are present in polluted air masses over land (solid brown lines) than in cleaner air masses over the oceans (dashed blue lines). Remote land masses lacking major population centers, such as those near the poles, contain fewer CCN than other land masses, but still typically more than over the oceans. Generally, we should expect that *clouds over land will contain more cloud droplets than over the ocean*, sometimes by an

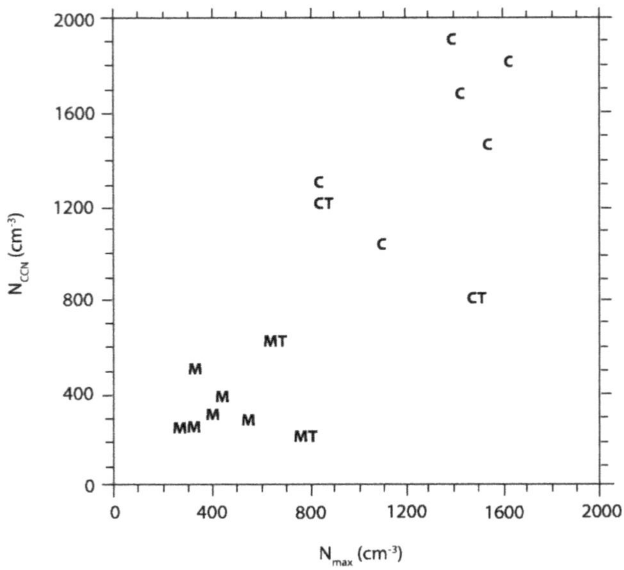

Fig. 8.9: The relationship between measured CCN number concentrations and measured maximum cloud droplet number concentrations for small Florida cumuli influenced by maritime (M), continental (C), or transitioning (MT or CT) air masses. *Adapted from Hudson and Yum (2001). © American Meteorological Society. Used with permission.*

order of magnitude or more, due to the more numerous CCN present in the air over land. Indeed, this is found to be the case when we collect observations of cloud droplets in maritime versus continental clouds.

Near coastlines influenced by varying air masses, the concentration of CCN can differ significantly based on the direction of the prevailing winds. This variation illustrates the difference in CCN between these air masses. As shown in Fig. 8.9, when clouds over the Florida coast develop within a continental air mass with higher CCN concentrations, or fewer CCN from a maritime air mass, the droplet number concentrations in the clouds increase or decrease accordingly. Transition air masses with both continental and maritime influences exhibit CCN and droplet number concentrations that fall between these two extremes.

Because there are more CCN in the air over land than over the oceans, and thus higher droplet concentrations in continental clouds than in maritime clouds, the sizes of the cloud droplets in each type of cloud are affected. Specifically, because the total water vapor available for condensation must be distributed among all nucleated droplets, *maritime clouds generally have fewer but larger droplets relative to continental clouds*. We will see in later chapters that this tendency has important ramifications for the ability of these different types of clouds to produce precipitation.

8.3.2 Updraft-supersaturation-CCN relations

In Chapter 7, we saw that soluble AP have critical supersaturations that must be reached to activate and form cloud droplets, according to their Köhler curves. Insoluble, hydrophilic AP also require critical supersaturations to be attained, based on the Kelvin curve. Using some assumptions, along with consideration of the relationship between supersaturation and CCN activation, we can conceptualize how many CCN in a population of AP could be activated within a cloud updraft of a given magnitude.

In an adiabatic parcel, the time rate of change in the saturation ratio S can be expressed as:

$$\frac{dS}{dt} = Q_1 w - Q_2 \frac{dq_c}{dt} \tag{8.5}$$

where w is the updraft speed, q_c is the cloud water mixing ratio (Sec. 2.3.10), and Q_1 and Q_2 are positive coefficients that depend on the air parcel temperature, pressure, and other thermodynamic quantities.[5]

The first term on the right-hand side represents the tendency of the updraft to *increase* the saturation ratio (and thus the supersaturation) by cooling due to adiabatic expansion of the parcel. The second term represents the rate of *reduction* in saturation ratio due to the conversion of vapor to cloud water during condensation. Thus, at or just above cloud base—where a significant updraft exists but few droplets have yet been nucleated—the first term dominates, and the saturation ratio (and supersaturation) increases most rapidly.

However, once the overall condensation rate across a population of nucleated drops becomes large enough, the second term begins to dominate, and the saturation ratio (supersaturation) decreases. Consequently, at some intermediate location of the parcel above cloud base, a maximum in the saturation ratio (supersaturation) is attained.

Recall from Chapter 7 that once the maximum supersaturation has been attained in a cloud, any CCN not yet activated will remain as haze droplets. These haze droplets will shrink in size (by evaporation) as the supersaturation decreases from its maximum value. Typically, within 50 m above the cloud base, the maximum supersaturation is reached in the updraft, and the maximum cloud droplet number concentration in an ascending, adiabatic cloudy air parcel is established.

The entrainment process, which modifies a cloudy parcel from its adiabatic characteristics, can lead to the activation of newly entrained CCN while also causing the complete evaporation of some pre-existing cloud

[5] Specifically, $Q_1 = \frac{g}{R_d T}\left(\frac{\varepsilon L_v}{C_p T} - 1\right)$ m^{-1} and $Q_2 = \rho\left(\frac{R_d T}{\varepsilon e_s} + \frac{\varepsilon L_v^2}{p T C_p}\right)$, where $\varepsilon = 0.622$.

droplets. The net effect of entrainment is typically a reduction in the maximum droplet number concentrations observed in the cloud.

Now that we have explored the basic factors controlling the maximum saturation ratio (or maximum supersaturation) in a cloud, we can revisit the activity spectrum power law given by (8.4) and examine the simplified, practical relationships derived from it. These relationships help conceptualize the impact of an observed CCN spectrum on the maximum number of cloud droplets that can form within a cloud.[6]

Starting with the activity spectrum given by (8.4), the maximum supersaturation s_{max} that can be achieved in an adiabatic parcel with an updraft of maximum speed w_{max} can be approximated as

$$s_{max} \approx 3.6 \left[\frac{0.0016\, w_{max}^{3/2}}{C} \right]^{1/(k+2)} \tag{8.6}$$

where s_{max} is in units of percent, w_{max} is in units of cm s^{-1}, and C (units of number per cm^3) and k (unitless) have the values appropriate for the given air mass. Note, however, that k must have a value between 0.4 and 1.0 due to the mathematical approximations made in producing this simplified equation.

The corresponding maximum droplet number concentration formed in this same updraft can be approximated as

$$N_{max} \approx 0.88\, C^{2/(k+2)} \left[0.07\, w_{max}^{3/2} \right]^{k/(k+2)} \tag{8.7}$$

where N_{max} represents the number of droplets nucleated per cm^3 at a supersaturation of s_{max}, and C and k correspond to the values used in (8.6).

These formulas show that the maximum updraft speed plays a central role in determining the maximum supersaturation achievable in a cloud, and thus the maximum number of CCN that nucleate cloud droplets. The updraft speed affects the rate of air cooling, with stronger updrafts producing higher supersaturations and activating more CCN. However, the maximum supersaturation also depends on the properties of the CCN ingested by the cloud and the number of cloud droplets they can activate at the lower supersaturations encountered on the way to the maximum value.

---

[6]These simplified expressions are sufficient for our purposes here, but more complex and accurate calculations are required for use in numerical models and research.

8.4 Ice Nucleating Particles (INP)

Ice nucleating particles (INP) present greater challenges for observation and quantification in the atmosphere compared to CCN. These challenges arise from both their scarcity relative to CCN and the multiple ice nucleation modes, each requiring specific conditions to be met. Moreover, these conditions can vary even if the chemical composition of the INP remains the same. Effective INP—those that can nucleate ice at minimal supersaturation or minimal supercooling—exhibit certain key characteristics:

1. They must have a molecular structure similar to that of ice;
2. They are generally water insoluble (a stable substrate is needed);
3. They are generally *large* AP (0.1–1 μm radius);
4. They have inhomogeneities on their surfaces (e.g., cracks, crevices, cavities) that serve as preferred sites for ice formation.

Given these constraints, only about one aerosol particle in every *10 million*, on average, is a good ice nucleating particle at temperatures above -20 °C. Thus, *INP are far fewer in number than CCN*, often by four or more orders of magnitude. Like CCN, INP can also vary by orders of magnitude in number concentration across different geographical regions (Fig. 8.10). They have not been studied as extensively as CCN, due largely to the very specialized (and expensive) equipment and analytical methods required.

Numerous methods exist to quantify the number of INP in the atmosphere, as no single method can address all the different modes of heterogeneous nucleation. Because INP are so few in number, measurements must be taken over long periods to collect enough particles for statistically meaningful results.

For deposition nucleation, an instrument called a *continuous flow diffusion chamber* (CFDC) can be used in a laboratory or mounted on aircraft. Air containing potential INP is introduced into the chamber, where it is exposed to two ice-coated plates maintained at different temperatures to achieve a specific supersaturation with respect to ice. An optical particle counter is then used to count the ice crystals produced within the chamber at the given temperature and supersaturation.

So-called *offline methods* use AP filters, designed to capture AP of specific sizes based on the spacing of the threads within the filter. These filters are exposed to air either at ground level or outside a research aircraft. The filters are later analyzed under controlled laboratory conditions. To measure INP acting via the deposition mode, the filter is placed in a chamber set to a particular temperature and supersaturation with respect to ice. An optical particle counter is used to detect ice formation when it occurs.

For detection of INP acting in the condensation-freezing mode, the experiment is conducted similarly but requires a higher supersaturation,

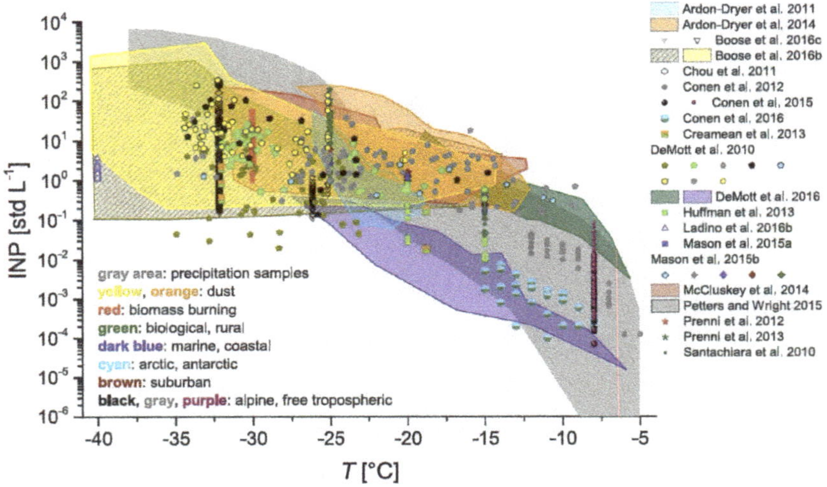

Fig. 8.10: Measurements of INP collected over various regions of the world by numerous researchers, including all modes of ice nucleation. Colored symbols or shading designate source region or type of particles; two-colored symbols designate multiple sources. Hatched areas and open symbols designate INP active for supersaturations less than those for liquid water. *From Kanji et al. (2017). © American Meteorological Society. Used with permission.*

exceeding that necessary for liquid drops to initiate the condensation-freezing process. To measure INP acting in the immersion-freezing mode, the collected aerosol particles on the filter are immersed in water droplets in the laboratory, which are then cooled to progressively lower temperatures until freezing occurs.

The number of INP versus their **threshold temperature**—that is, the highest temperature at which they can nucleate ice given their physical and chemical properties—is shown in Figs. 8.10 and 8.11. Both data summaries show that, of all INP, there are many orders of magnitude fewer that have higher threshold temperatures (near −5 °C) compared to those with much lower threshold temperatures (near −40 °C). It follows that, for any given temperature, and depending on the type of INP present, the number of ice particles nucleated can also vary by many orders of magnitude.

8.4.1 Sources of INP

At present, the highest temperature at which natural INP are known to induce ice formation is attributed to the bacteria *Pseudomonas syringae* (usually shortened to *P. syringae*) found on decaying leaves (Fig. 8.11). Recent studies over the Earth's oceans, however, indicate that some marine phytoplankton can also nucleate ice at similar temperatures (Fig. 8.10, dark

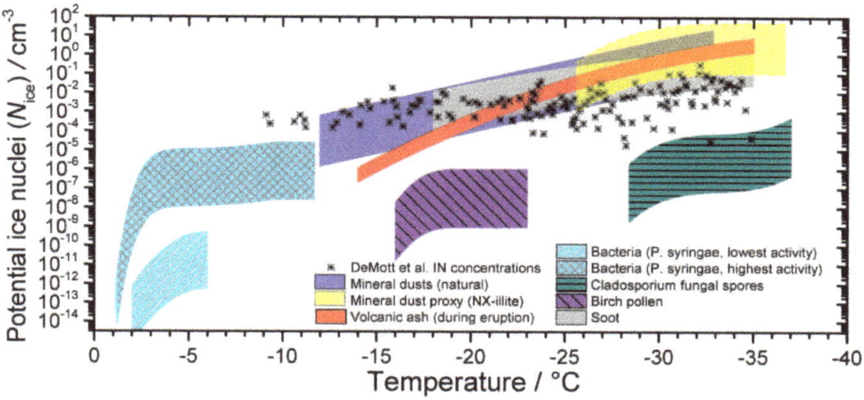

Fig. 8.11: Observations of INP of various compositions, acting only in the immersion mode, versus the temperature ranges over which they freeze liquid drops. *From Murray et al. (2012). Used with permission from The Royal Society of Chemistry.*

blue shading). Nevertheless, as depicted in both figures, these INP are rare.

More numerous INP consist of mineral dusts (e.g., kaolinite), volcanic ash, birch pollen, and soot. However, these nucleate ice at temperatures less than -10 °C. Some fungal spores and dusts nucleate ice only at very low temperatures approaching the homogeneous freezing limit of -40 °C.

The types of INP shown in both figures are not the only INP in our atmosphere, but they have been observed the most and thus are believed to be the most common. Not shown in these figures is a human-made chemical, *silver iodide*, which has a molecular structure so similar to ice that it is able to nucleate ice at temperatures as high as -4 °C. It was used in many early cloud seeding experiments in the 1970s.

In modern practice, ski resorts in mountainous regions frequently augment snowfall by spraying water mixed with an additive derived from dead bacteria, such as *P. syringae*. This additive allows the water droplets to freeze with minimal supercooling.

8.4.2 INP relations

As the number of observations of INP has continued to grow over decades of research, different relationships have been proposed to predict the number of ice crystals that INP can nucleate at different temperatures or supersaturations.

The earliest and simplest relationship, proposed by Fletcher (1962) based on measurements at that time, is referred to as the **Fletcher curve**:

$$N_i = A \exp\left[\beta(273.15 - T)\right], \tag{8.8}$$

where N_i is the total number of ice crystals per liter nucleated at an air temperature T (in kelvin). The constant A often has a value of 10^{-5} per liter, though it can vary by several orders of magnitude. The parameter β varies from ~0.4 to 0.8 K^{-1} in different air masses, with a global average of 0.6 K^{-1}. The coefficients A and β encapsulate the number of INP and the physical and chemical properties that influence their ice nucleating abilities, respectively. This relationship is expected to hold over the temperature range of 243.15 K to 263.1 K (−30 to −10 °C) and was intended to combine all modes of ice nucleation into a single relationship.

A pair of equations, colloquially referred to as the **Meyers scheme**, was later proposed based on additional measurements of INP and, in addition to temperature, takes into account the ambient supersaturation with respect to ice. The first equation represents deposition nucleation, condensation-freezing, and immersion freezing combined and is given by

$$N_i = \exp\left\{a + b\left[100\left(S_i - 1\right)\right]\right\}, \tag{8.9}$$

where N_i is the total number of ice crystals per liter produced at a saturation ratio of S_i, with $a = -0.639$ and $b = 0.1296$ and no nucleation occurring at temperatures greater than −5 °C.

The second equation is for contact nucleation only and is given by

$$N_{ic} = \exp\left[c + d\left(273.15 - T_c\right)\right], \tag{8.10}$$

where N_{ic} is the number of ice crystals per liter created by contact of INP with supercooled droplets having temperature T_c in kelvin, $c = -2.80$, and $d = 0.262$. This equation is not applicable for droplet temperatures greater than 271.15 K (−2 °C).

Equations (8.9) and (8.10) have been used extensively in numerical cloud models to predict ice since the early 1990s, but they are based on limited observations and do not appear universal to INP around the world.

As more and more data on INP are collected globally, it becomes evident that none of these equations, or others proposed in the past, are universal. As shown in Fig. 8.12, significant discrepancies in various equations used to predict INP are especially prominent at higher temperatures. Additionally, some data sets (each represented by a given marker) do not appear to be well correlated with temperature.

8.4.3 Importance of larger AP for ice nucleation

In contrast to observations of INP, total number concentrations of AP are routinely measured worldwide, primarily from ground-based locations. Recently, some researchers have identified a relationship between INP and larger AP in certain air masses. When the number of INP active at a particular temperature is plotted against *all* AP greater than 0.5 μm in diameter

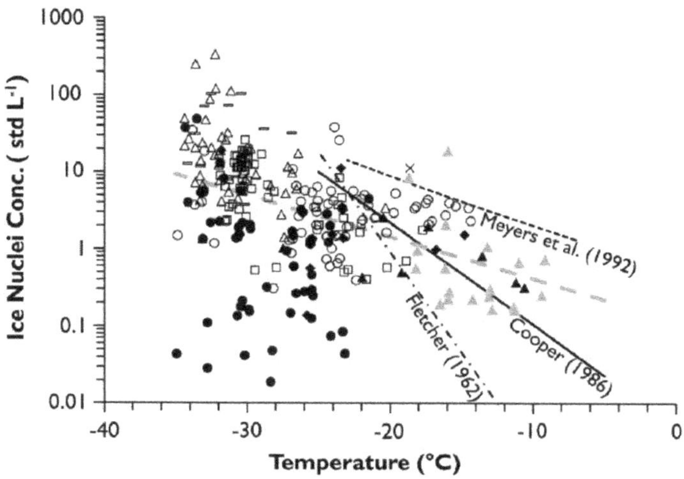

Fig. 8.12: INP number concentration (at supersaturation equal to or exceeding that for liquid water) versus temperature, for temperatures between −9 and −35 °C. Symbols denote different observations collected worldwide: Colorado, USA (gray triangles); Ontario, Canada (open circles); Eastern Canada and Great Lakes region (x's); Amazon basin, Brazil (squares); Colorado and Wyoming, USA (solid circles); mountains in Western Colorado, USA (diamonds); North Slope of Alaska, USA (black triangles); and Pacific Ocean basin (open triangles). Various proposed relationships to predict the number of INP nucleating ice versus temperature are plotted as lines and labeled. A single line fit to these data is shown as an unlabeled gray dashed line. *From DeMott et al. (2010). Used with permission from Proceedings of the National Academy of Sciences.*

in those same air masses, under conditions just saturated or supersaturated with respect to liquid water, clearer relationships emerge (Fig. 8.13).

This focus on AP with diameters 0.5 μm and larger reveals that these particles—specifically, the high end of the *large* range—primarily dominate ice nucleation at high supersaturations and over the temperature range of −9 to −35 °C.[7] The so-called **DeMott relationship** for INP in a continental air mass, based on the total amount of AP larger than 0.5 μm in diameter in that air mass, is given by:

$$N_{\mathrm{INP}} = \left[N_{\mathrm{AP,0.5}}{}^{c(273.16-T)+d} \right] a(273.16 - T)^{b} \tag{8.11}$$

where N_{INP} is the number concentration (per liter) of INP at an air temperature T (in kelvin), $N_{\mathrm{AP,0.5}}$ is the number concentration (per cm^3) of

---

[7]The data shown in Fig. 8.13 exclude marine environments, where large sea salt particles can dominate the larger AP population but have no role in ice nucleation.

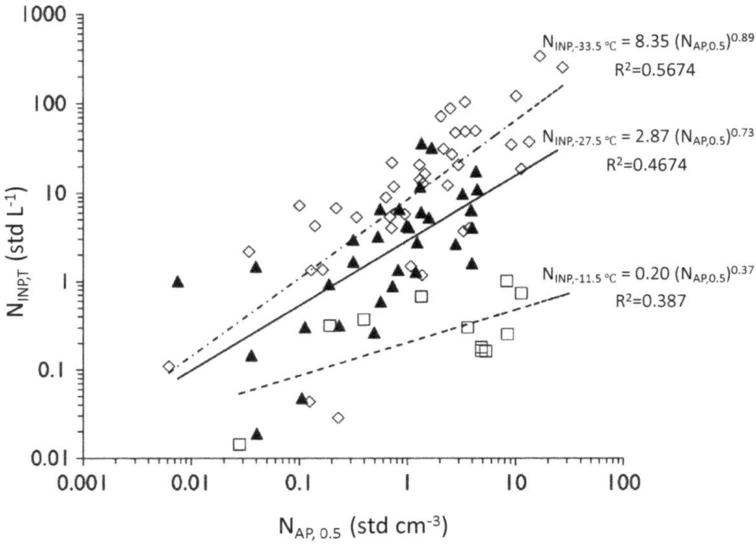

Fig. 8.13: Measured number concentrations of INP ($N_{INP,T}$) from the same data shown in Fig. 8.12 versus measured AP number concentrations $N_{AP,0.5}$ having diameters exceeding 0.5 μm. Measurements are shown by symbols. Predicted values of INP using Eq. (8.11) are plotted as lines for three different 3 °C temperature intervals: top dashed line centered at −33.5 °C (measurements are open diamonds), middle solid line centered at −27.5 °C (measurements are solid triangles), and bottom dashed line centered at −11.5 °C (measurements are open squares). Squared correlation coefficients (R^2) between the predicted and observed $N_{INP,T}$ are also listed. *Adapted from DeMott et al. (2010). Used with permission from PNAS.*

AP with diameters larger than 0.5 μm, and the constants are given by $a = 5.94 \times 10^{-5}$, $b = 3.33$, $c = 2.64 \times 10^{-2}$, and $d = 3.3 \times 10^{-3}$.

There is still some scatter in the data compared to the lines predicted by this equation, as shown in Fig. 8.13, which is likely due to differences in chemical composition across the set of AP shown. The predictive skill of the equation decreases with increasing air temperature. Equation (8.11) does not hold for INP with threshold temperatures greater than −9 °C, nor does it hold for air masses over the oceans or coastal regions partly influenced by marine air masses. Research is ongoing to characterize the INP in maritime air masses.

More recently, advanced relationships based on various modes of ice nucleation have been proposed and introduced into cloud, weather, and climate models. However, improving their accuracy requires significantly more extensive observations of INP across diverse environments. As of this writing, researchers continue to seek a simplified, unified model that can accurately predict ice nucleation across all global regions, threshold temperatures, and nucleation modes.

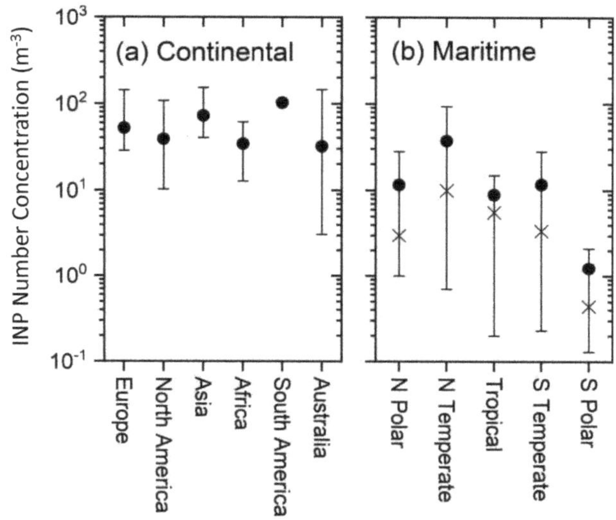

Fig. 8.14: (a) Averaged INP number concentrations at a threshold temperature of -15 °C from ground measurements taken over various continental regions. (b) INP number concentrations at -15 °C averaged over latitude, from ship measurements over various oceanic regions (dots). Vertical bars indicate the range of the measurements excluding the 10% highest and 10% lowest values. Crosses in (b) mark the modes of the number concentration measurements. *From Welti et al. (2020). (CC-BY)*

8.4.4 *Influence of the number of INP from different air masses*

Because there are more AP over land than over ocean, we would expect there to also be more INP in air masses over land than over ocean. Measurements of INP confirm that this is often the case (Fig. 8.14). Recent measurements aboard ships over the world's oceans show that there are typically a few orders of magnitude fewer INP in those air masses than over land, but with greater variability, as shown in Fig. 8.14b.

The increased variability over the oceans is likely due to seasonal influences such as the transport of Saharan dust and the enhanced metabolism of microscopic ocean organisms during particular seasons. The variability in INP over the continents shown in Fig. 8.14a may be underestimated, as the data presented are all for a single season, and seasonal variability can exist in INP over the land as well.

Although maritime air masses contain fewer INP compared to continental air masses, it cannot be generally concluded that maritime clouds will have fewer frozen hydrometeors than continental clouds. Based on the frequency of measured ice particle concentrations greatly exceeding typical INP concentrations, secondary ice production mechanisms (dis-

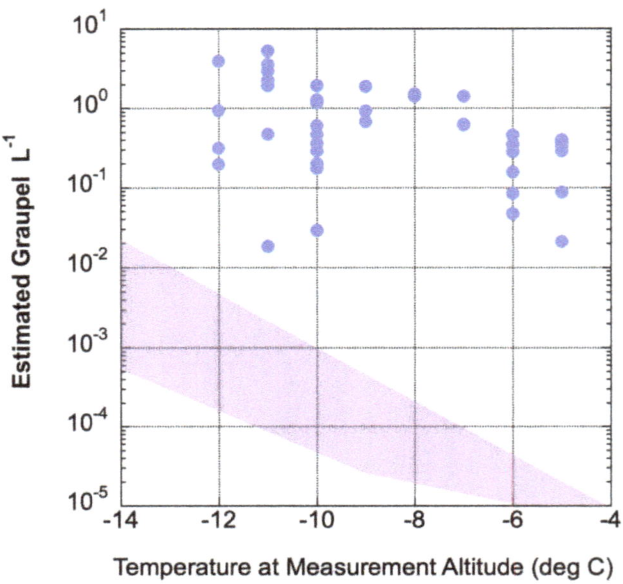

Fig. 8.15: Number concentrations of graupel observed in the updrafts of cumulus clouds over the Caribbean Sea for three different days (blue dots) versus the temperature at which they were measured within the clouds. Because the graupel particles were measured within the updrafts and had fall speeds slower than the updrafts, it is assumed that they must have originated from ice nucleated at temperatures warmer than those at which they were observed. The measured range of number concentration (including uncertainty) of INP acting in the immersion mode for two of the days is shown as purple shading. *Adapted from Lasher-Trapp et al. (2016). © American Meteorological Society. Used with permission.*

cussed in Sec. 7.4) may be more prominent in maritime clouds than in continental clouds.

This possibility is particularly evident at higher temperatures (Fig. 8.15), especially above −10 °C, and is even more noteworthy since the number of INP having threshold temperatures in this range in maritime air masses is typically much lower than in continental air masses. It has been hypothesized that the increased activity in secondary ice production may be related to the larger size of cloud droplets in maritime clouds, which can enhance riming and droplet freezing, and thus potentially rime-splintering and freezing drop shattering.

Given the difficulties in obtaining direct measurements of INP, the various modes of ice nucleation, and potential secondary ice production mechanisms, a simple relationship analogous to (8.4) for differentiating the number of frozen hydrometeors observed in continental versus maritime clouds has not yet been found. It is not clear that such a relationship is possible.

Questions for Review

1. How does the number of AP vary with increasing size in Earth's atmosphere?
2. What are the major sources and sinks of these AP?
3. How can the residence time of a given kind of AP be found mathematically, given its sources and sinks?
4. What characteristics of AP allow them to serve as CCN, how are they measured, and how is the number of CCN activated in a cloud with increasing supersaturation expressed mathematically?
5. Why does cloud droplet number concentration vary significantly between maritime and continental clouds, and what implications does this have for the sizes of cloud droplets produced within those clouds?
6. What factors predominantly determine the maximum supersaturation achieved in an ascending adiabatic cloud parcel, and why is it achieved at some intermediate height above the cloud base?
7. What characteristics make an AP an effective INP, and how are different kinds of INP measured?
8. How has the number of INP active at different temperatures been expressed mathematically, both in the past and more recently?

Homework problems and other resources for students and instructors can be found at bit.ly/4dYBwB5, or scan this code:

Further Reading

For more information on INP [advanced]:

Kanji, Z. A., L. A. Ladino, H. Wex, Y. Yvonne Boose, M. Burkert-Kohn, D. J. Cziczo, and M. Krämer, 2017: Overview of ice nucleating particles. *Meteorological Monographs*, **58**, 1.1–1.33

Cziczo, D. J., L. Ladino, Y. Boose, Z. A. Kanji, P. Kupiszewski, S. Lance, S. Mertes, and H. Wex, 2017: Measurements of ice nucleating particles and ice residuals. *Meteorological Monographs*, 8.1–8.13

CHAPTER 9

Diffusional Growth of Droplets and Ice

We now begin to study the chain of microphysical processes that occur after droplet or ice nucleation, ultimately producing hydrometeors large enough to fall from a cloud and reach the ground as precipitation. These processes are summarized in Fig. 9.1.[1]

We have already referred to the **diffusional growth** of cloud droplets and ice particles in Chapters 6 and 7. The nucleation of droplets or ice from the vapor state requires diffusional growth, which involves water molecules in the vapor state moving via molecular diffusion towards the droplet or ice crystal and incorporating into it. **Diffusion** results from the random motions of individual molecules within gases or liquids, with the net effect of transporting molecules and/or energy from regions of higher concentration to regions of lower concentration.

We use different terms depending on the direction of the diffusion of water molecules toward or away from a droplet or ice crystal. The diffusion of water molecules leaving the vapor phase and joining the liquid phase is called **condensation**; when they leave the vapor phase and join the ice phase, it is called **deposition**. These processes are listed in the top row of Fig. 9.1. If the vapor pressure of the air surrounding the droplet or ice crystal exceeds the level needed for equilibrium, the droplet or ice crystal will increase in size through these processes.

[1]To refresh your big-picture view of how the various microphysical processes work together to form precipitation, you may find it helpful to refer back to this figure as you progress through future chapters.

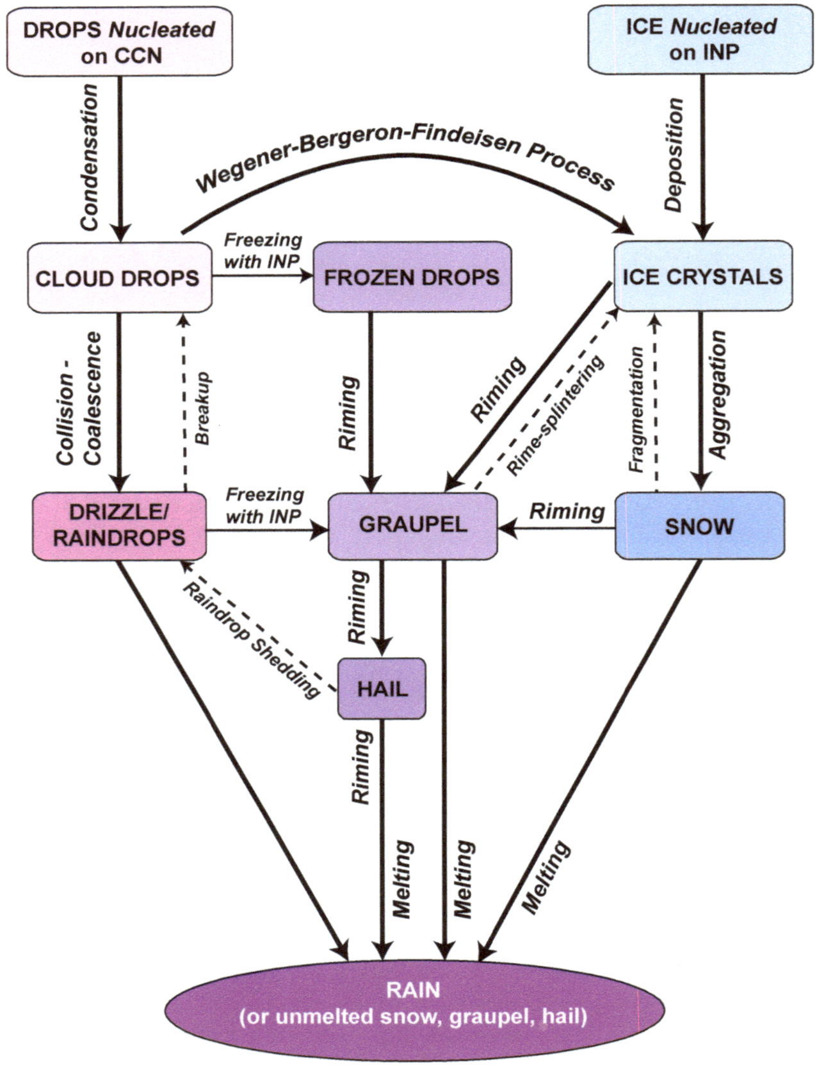

Fig. 9.1: Summary of the microphysical processes involved in the chains of events that can produce precipitation at the Earth's surface. Hydrometeors are labeled within boxes; microphysical processes that promote precipitation are labeled alongside black arrows. Processes that can produce additional small hydrometeors are denoted by dashed arrows. *After Lamb (2001).*

Conversely, if the vapor pressure of the air falls below that needed for equilibrium of an existing water droplet or ice crystal, then the water droplet or ice crystal will decrease in size. When some of the water molecules diffuse away from the droplet surface and enter the vapor phase, it is called **evaporation**; when some of the water molecules diffuse away from the ice surface and enter the vapor phase, it is called **sublimation**. We now study in detail how these processes occur.

First, we note that certain complications important for the nucleation of droplets become much less influential once the newly formed droplets exceed a certain size, as indicated by the radius at which the Köhler curve flattens on the right side of its peak. Solution droplets become sufficiently diluted to be indistinguishable from pure water, and their radii become large enough that curvature is no longer important.

Second, we note that the different modes of ice nucleation discussed in Chapter 7 are no longer relevant. Once an ice crystal or frozen drop exists, it will grow or shrink by diffusion in the same manner.

However, there are new issues to consider:

- the *gradient* of water vapor density around the droplet or ice crystal, which drives the diffusion of water vapor at a specific rate and direction;
- the latent heating or cooling that results from the phase change of water molecules;
- the typically non-spherical shapes of ice crystals undergoing deposition or sublimation; and
- the variable surface affinities for water molecules joining or leaving the droplet or ice crystal surface.

Central to the processes of condensation and deposition are the diffusion of water vapor from the ambient environment *toward* the droplet or ice crystal and the diffusion of latent heat released during condensation *away* from the droplet or ice crystal into its environment. Although water vapor and heat are different physical entities, the molecular (diffusive) transfer of both quantities through air is governed by the *diffusion equation:*[2]

$$F_\phi = -D_\phi \frac{\mathrm{d}\phi}{\mathrm{d}x}, \tag{9.1}$$

[2]This equation is variously known as *Fick's first law of diffusion* for mass transfer and *Fourier's law* for heat transfer. Mathematically, these laws are equivalent. The broad applicability of the diffusion equation cannot be overstated; it is relevant to everything from the loss of heat through wall insulation in houses to the spread of pollutants in underground water reservoirs.

where ϕ is the property (e.g., heat) or substance (e.g., vapor) being transferred by molecular motions, and F_ϕ is the *flux* of ϕ in the x-direction, having units of ϕ per unit area per unit time. D_ϕ is the appropriate *diffusion constant*, taking into account not only what ϕ represents but also the properties of the medium through which ϕ is diffusing.

In words, (9.1) states that when a gradient in some quantity ϕ (such as temperature or vapor) is present, there will be a flux of ϕ with a magnitude proportional to both the diffusion constant D_ϕ and the local gradient $d\phi/dx$. The minus sign indicates that the flux is directed from higher to lower values of ϕ, i.e., opposite to the gradient of ϕ.[3]

When diffusion is at work, gradients can persist only if another process acts to maintain them. In the vicinity of hydrometeors, water vapor gradients are sustained by continuous condensation or evaporation of liquid water at the droplet surface, or continuous deposition or sublimation of ice at the crystal surface. Temperature gradients are maintained by latent heat release or uptake.

9.1 Condensation and Evaporation of Droplets

The theory behind the diffusional growth of water droplets has been understood for over a century. The derivation presented in the next section is based on an original formulation by James Maxwell in 1890, with modifications and simplifications by B. J. Mason in 1971. Although subsequent studies have refined these equations to account for surface effects that can sometimes be significant, the basic equations remain unchanged.

9.1.1 Diffusion equation for a water droplet, neglecting heat diffusion

Consider a water droplet of radius R (Fig. 9.2; blue circle) surrounded by air that contains water molecules in the vapor phase. The water vapor density and temperature in the infinitely thin layer of air in direct contact with the surface of the droplet are given by $\rho_{v,R}$ and T_R, respectively. At a sufficiently large distance from the droplet, the water vapor density and temperature are those of the unperturbed environment, as indicated by $\rho_{v,\infty}$ and T_∞ at the right edge of Fig. 9.2.

---

[3]For example, if a bowl of thick soup is reheated in a microwave oven, the temperature may initially be very uneven: very hot around the edges and still cold at the center. Setting the bowl aside for a few minutes will allow thermal diffusion (heat conduction) to move heat from hotter regions to cooler regions, so that eventually the temperature of the soup is nearly the same throughout the bowl.

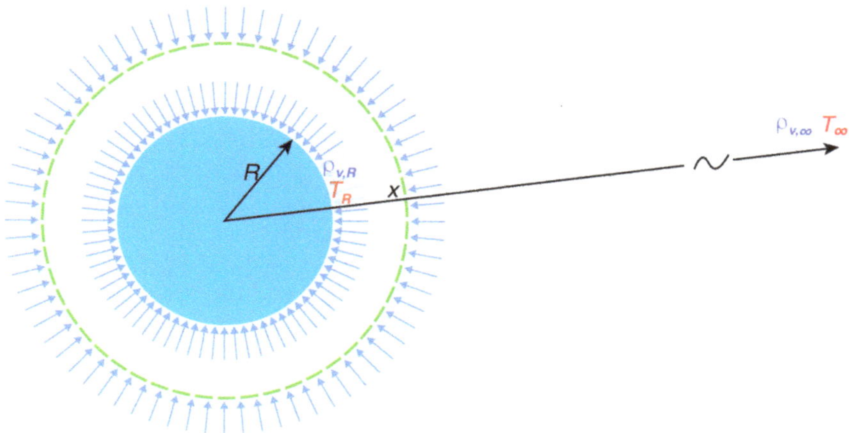

Fig. 9.2: Schematic illustrating the flux of water vapor toward a droplet of radius R (blue circle) growing by condensation, i.e., for the case that $\rho_{v,\infty} > \rho_{v,R}$. The dashed green circle depicts an imaginary sphere with arbitrary radius $x > R$. Each blue arrow represents an equal mass flux of water vapor, with the combined flux over each sphere (total number of arrows) being independent of x, in keeping with the steady-state assumption.

To derive the basic diffusional growth equation for a water droplet, we start with some simplifying assumptions:

1. The spherical droplet of radius R is isolated from any nearby droplets and is suspended (not falling) in the environment.
2. The distributions of temperature and water vapor in the air are *radially symmetric*. In other words, they exhibit the same functional dependence on the radial distance x from the droplet center in all directions extending away from the droplet.
3. The air in direct contact with the droplet has the same temperature T_R as the droplet and a water vapor density $\rho_{v,R}$ equal to that required for the droplet to be in equilibrium, i.e., proportional to the saturation vapor pressure of the air immediately around the droplet.
4. The distributions of temperature and water vapor in the air are temporarily in a *steady state*. This means that the ambient temperature T_∞ and water vapor density $\rho_{v,\infty}$ remain constant, as does the water vapor density near the droplet $\rho_{v,R}$.
5. An energy balance exists at the surface of the droplet, such that any latent heat released during the phase change is balanced by the diffusion of heat away from the droplet, resulting in the droplet temperature, T_R, being (temporarily) constant in time.

For the droplet to grow by condensation, $\rho_{v,\infty}$ must be greater than $\rho_{v,R}$. This establishes a gradient in water vapor density, causing vapor to diffuse from the environment towards the droplet. However, this gradient is not constant along the path toward the droplet.

To see why, we need to consider the total mass flux of vapor at various distances from the droplet. Recall that, according to (9.1), the flux of water vapor F_v, measured in units of mass of water vapor per unit time *per unit area* normal to the gradient, is proportional to the local density gradient $d\rho_v/dx$, where x is the distance from the center of the droplet. However, the *total* mass flux is equal to the product of F_v and the total surface area $4\pi x^2$ of an imaginary sphere with radius x, illustrated for one value of x as the green dashed circle in Fig. 9.2.

In steady state, *the total flux through the imaginary sphere must be independent of its radius* x. Otherwise, mass would either accumulate or disappear at some intermediate distance along the path, changing the density profile. The short blue arrows represent the flux of vapor across the spheres at both R and x. Note that while the arrows are more widely spaced at x, indicating a lower *flux per unit area*, the total number of arrows is the same for both spheres, indicating equal *total mass fluxes*.

For the case when $x = R$, the total flux is the flux that reaches the surface of the droplet and must therefore equal the condensation rate, which we can now express as

$$\frac{\mathrm{d}M}{\mathrm{d}t} = -4\pi x^2 F_v(x), \tag{9.2}$$

where M is the mass of the droplet. The minus sign appears because growth occurs when F_v is directed inward and therefore negative. Substituting the diffusion equation (9.1) for F_v, we can write

$$\frac{\mathrm{d}M}{\mathrm{d}t} = 4\pi x^2 D_v \frac{\mathrm{d}\rho_v}{\mathrm{d}x}, \tag{9.3}$$

where we have used the **diffusivity of water vapor in air**, D_v, given by

$$D_v = 0.211 \left(\frac{T}{273.15}\right)^{1.94} \left(\frac{1013.25}{p}\right), \tag{9.4}$$

where T is in units of kelvin, p is in units of hPa, and D_v has units of $\mathrm{cm^2\ s^{-1}}$. At 20 °C and 1000 hPa, $D_v \approx 0.2452\ \mathrm{cm^2\ s^{-1}}$.

We do not yet know the value of $\mathrm{d}M/\mathrm{d}t$, but we do know that it is independent of x; we can therefore treat it as a temporarily unknown constant Y and rewrite (9.3) as

$$Y = 4\pi x^2 D_v \frac{d\rho_v}{\mathrm{d}x}. \tag{9.5}$$

This is now a first-order linear differential equation in $\rho_v(x)$ that we can solve via the standard technique of separation of variables,

$$d\rho_v = \frac{Y}{4\pi D_v} x^{-2}\, dx, \tag{9.6}$$

followed by integrating both sides from R to an arbitrary x:

$$\int_{\rho_{v,R}}^{\rho_v(x)} d\rho_v = \frac{Y}{4\pi D_v} \int_R^x x'^{-2}\, dx', \tag{9.7}$$

where we now use the dummy variable x' inside the integral to distinguish it from the upper limit of integration. Performing the integration, we now get

$$\rho_v(x) - \rho_{v,R} = \frac{Y}{4\pi D_v} \left[\frac{1}{R} - \frac{1}{x}\right] \tag{9.8}$$

which can be rearranged to give

$$\rho_v(x) = \rho_{v,R} + \frac{Y}{4\pi D_v} \left[\frac{1}{R} - \frac{1}{x}\right]. \tag{9.9}$$

Taking the limit $x \to \infty$, we find that

$$\rho_{v,\infty} = \rho_{v,R} + \frac{Y}{4\pi D_v} \frac{1}{R} \tag{9.10}$$

so

$$\frac{Y}{4\pi D_v} \frac{1}{R} = \rho_{v,\infty} - \rho_{v,R}. \tag{9.11}$$

Before solving for the mass growth rate, we can examine the vapor density field surrounding the droplet by first solving this equation for $\rho_{v,R}$ and substituting it into (9.9) to obtain a final expression for $\rho_v(x)$:

$$\rho_v(x) = \rho_{v,\infty} - \frac{Y}{4\pi D_v x}. \tag{9.12}$$

This relationship is depicted in Fig. 9.3 as the blue curve. It demonstrates a simple $1/x$ dependence transitioning from $\rho_{v,R}$ at the surface of the droplet to the ambient density $\rho_{v,\infty}$ at a sufficiently large distance from the droplet.

Finally, recalling that Y is a shorthand for the condensation rate dM/dt, we can revisit (9.11) and rearrange it to obtain

$$\frac{dM}{dt} = 4\pi R D_v (\rho_{v,\infty} - \rho_{v,R}). \tag{9.13}$$

This equation indicates that the mass rate of condensation (or evaporation, if $\rho_{v,\infty} < \rho_{v,R}$) is proportional to the droplet radius R, the diffusion coefficient D_v, and the difference between the environmental vapor density $\rho_{v,\infty}$ and the equilibrium vapor density at the droplet surface $\rho_{v,R}$.

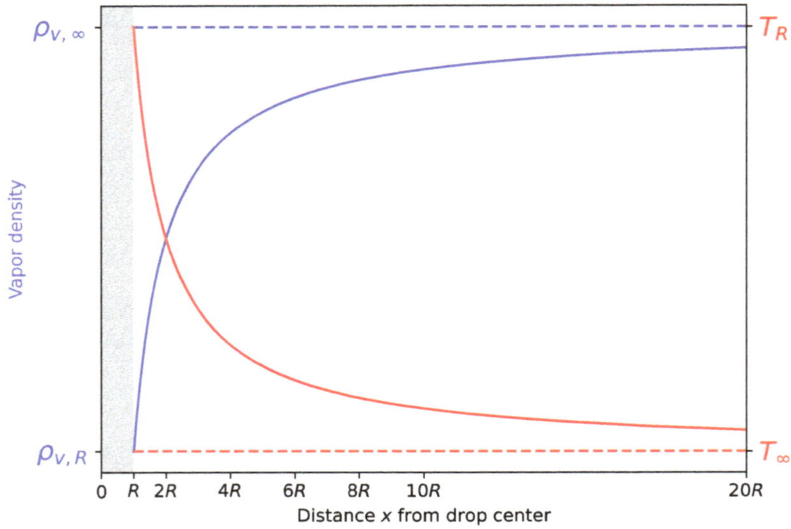

Fig. 9.3: Vapor density and temperature profiles in the vicinity of a spherical droplet during steady-state diffusional growth by condensation, following equation (9.12) and its equivalent for temperature (9.23). Shaded gray area indicates the interior of the droplet. Blue curves and labels are for vapor density ρ_v; red curves and labels are for temperature T.

One might conclude from (9.13) that as the droplet becomes larger (R increases), the rate of growth also increases, allowing a droplet to grow exponentially to, say, the size of a raindrop. But this is misleading, because two different measures of droplet size appear in this equation, and they are not independent! Specifically, the droplet mass M equals the density of pure water (ρ_l) times the volume of the droplet and is therefore related to R:

$$M = \rho_l \frac{4}{3}\pi R^3. \tag{9.14}$$

Now we can eliminate M from (9.13) by noting that the left-hand side can be rewritten as

$$\frac{\mathrm{d}M}{\mathrm{d}t} = \frac{\mathrm{d}}{\mathrm{d}t}\left[\rho_l \frac{4}{3}\pi R^3\right] = 4\pi\rho_l R^2 \frac{\mathrm{d}R}{\mathrm{d}t}. \tag{9.15}$$

We then have

$$4\pi\rho_l R^2 \frac{\mathrm{d}R}{\mathrm{d}t} = 4\pi R D_v(\rho_{v,\infty} - \rho_{v,R}) \tag{9.16}$$

which simplifies to our desired expression for the rate of change of the droplet *radius* by diffusion:

$$\frac{\mathrm{d}R}{\mathrm{d}t} = \frac{1}{\rho_l R} D_v(\rho_{v,\infty} - \rho_{v,R}). \tag{9.17}$$

We now see that as the radius increases, the rate of radius growth actually slows down! Conversely, as the droplet radius decreases under conditions of evaporation, i.e., if $\rho_{v,\infty} < \rho_{v,R}$, the rate of decrease becomes larger.

Let's now use this equation to find an expression for the functional dependence of R on time, $R(t)$. Starting from (9.17) and again separating variables and integrating, we have

$$\int_{R_0}^{R(t)} R\, dR = \frac{1}{\rho_l} D_v (\rho_{v,\infty} - \rho_{v,R}) \int_0^t dt', \tag{9.18}$$

which evaluates to

$$\frac{1}{2}[R^2(t) - R_0^2] = \frac{1}{\rho_l} D_v (\rho_{v,\infty} - \rho_{v,R}) t, \tag{9.19}$$

or

$$R(t) = \sqrt{\frac{2D_v}{\rho_l}(\rho_{v,\infty} - \rho_{v,R})t + R_0^2}. \tag{9.20}$$

Here, R_0 is the initial droplet size at time $t = 0$.

From the above result, we observe that the growth rate expressed in terms of radius exhibits a *square-root* dependence on time. Consequently, we expect the growth rate to slow significantly as the droplet enlarges, or, conversely, the evaporation rate to markedly increase as the droplet diminishes.

Now suppose the air surrounding a droplet has a temperature of 10 °C and a pressure of 700 hPa. Assuming vapor saturation near the droplet surface, (6.1) and (2.6) yield a vapor density of $\rho_{v,R} = 9.4 \times 10^{-6}$ g cm^{-3}. Equation (9.4) gives a vapor diffusivity $D_v = 0.328$ cm^2 s^{-1}. If we retain all of our original assumptions, including steady-state conditions, we can plot the droplet radius as a function of time for two different initial radii R_0 growing under these conditions, as shown by the set of red curves in Fig. 9.4.

As expected from (9.20), small droplets grow quite quickly at first. For instance, for an initial radius of 1 μm, the droplet doubles in size to 2 μm in just about half a second! In comparison, for a slightly larger initial radius of 5 μm (dashed red line), the droplet requires about 10 seconds to double its size to 10 μm.

On the other hand, even with such a large vapor density difference, more than 6 hours would be required for a small raindrop with a radius of 0.5 mm to form solely through diffusional growth. A raindrop with a radius of 1 mm would require more than a day to form this way. Most clouds do not last nearly that long, yet many clouds produce rainfall during their lifetimes, sometimes in as little as 15 minutes.

If we repeat these calculations for a moderate vapor density difference between the droplet surface and its environment, specifically 0.1%, we find

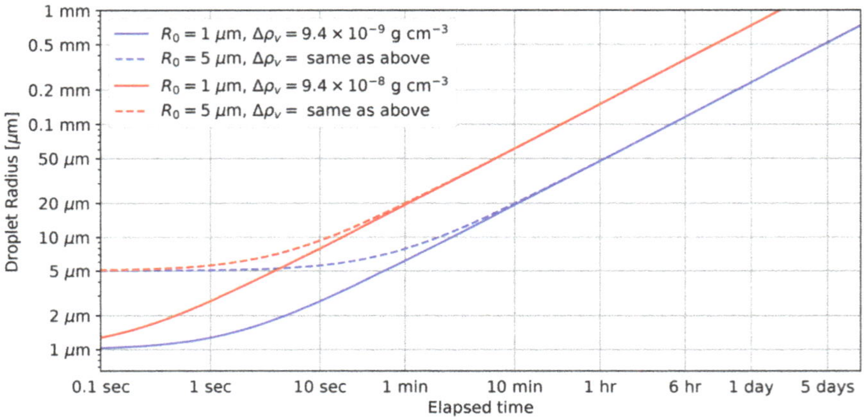

Fig. 9.4: Growth of water droplets by diffusion (condensation), assuming the indicated values for $\Delta\rho_v = \rho_{v,\infty} - \rho_{v,R}$ and D_v, using (9.20), which ignores latent heat effects. Results are shown for two different initial radii R_0.

$\Delta\rho_v = 9.4 \times 10^{-9}$ g cm^{-3}. Consequently, droplet growth by diffusion is only one-tenth as fast (blue curves), requiring nearly 5 hours to produce drizzle drops with radii of 0.1 mm, and more than 5 days to produce small raindrops.

Based on this admittedly simplified analysis, we must conclude that *some other microphysical process besides diffusional growth from water vapor must be involved in the production of raindrops*. In later sections, we will further refine our understanding of the growth rate of droplets by diffusion, taking into account, for example, the effects of latent heating. However, those refinements will generally be found to further slow the growth rates shown in Fig. 9.4. Thus, they do not solve the problem of achieving raindrop sizes via condensation alone in any reasonable amount of time.

9.1.2 *Narrowing of droplet size distributions in time by condensation*

For the two initial sizes of the droplets considered in each set of calculations shown in Fig. 9.4, we observe that after a few minutes of growth, the initial radius R_0 becomes nearly inconsequential. This result is anticipated due to the $1/R$ dependence of the growth rate $\mathrm{d}R/\mathrm{d}t$ in (9.17), or the square-root dependence of droplet size on time when growing by condensation (9.20). Initially smaller droplets quickly catch up in size to the initially larger droplets growing in the same environment.

More generally, if a population of small droplets with widely varying initial sizes grows by condensation under similar conditions, their sizes will tend to converge until they become approximately the same size. Thus, a broad initial droplet size distribution will tend to become much

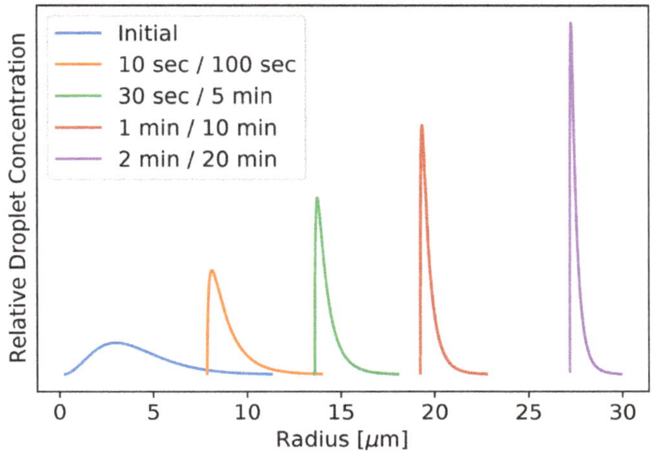

Fig. 9.5: Evolution of a hypothetical initial cloud droplet size distribution under the same conditions of diffusional growth as in Fig. 9.4. For each curve, the indicated elapsed times are for $\Delta\rho_v$ values of 9.4×10^{-8} g cm^{-3} and 9.4×10^{-9} g cm^{-3}, approximately corresponding to supersaturations of 1% and 0.1%, respectively.

narrower over time when growth occurs by condensation alone, especially when they all experience identical thermodynamic conditions, such as in an adiabatic parcel. This behavior is illustrated in Fig. 9.5, showing two different extremes in vapor density differences.

9.1.3 Droplet diffusional growth equation in terms of saturation ratio

Now, the vapor density gradient term used in our diffusional growth equation is not a quantity easily measured in the atmosphere, nor is it easily related to our discussions of droplet nucleation and activation from Chapter 7. The vapor density gradient surrounding the droplet can be related to the more familiar supersaturation, most easily using the saturation ratio $S = \frac{e}{e_s}$. The vapor pressure of the environment is proportional to the vapor density of the environment (see Eq. 2.6), whereas the vapor pressure at an infinitesimal distance from the droplet surface is equal to the saturation vapor pressure e_s.[4]

[4]Even if this were not initially true, enough evaporation would take place in an infinitesimally short time interval to eliminate the otherwise infinite gradient in vapor pressure at the droplet surface.

Mathematical manipulation of (9.17) yields the following:

$$\frac{\mathrm{d}R}{\mathrm{d}t} = \frac{1}{R} D_v \frac{\rho_{v,\infty}}{\rho_l} \left(S - 1 - \frac{a}{R} + \frac{b}{R^3} \right) \tag{9.21}$$

where we have incorporated the effects of droplet curvature (a/R) and solute strength ($-b/R^3$) from the Köhler equation. This allows (9.21) to be used for describing the diffusional growth of very small haze droplets.

For cloud droplets with radii of a micrometer or more, the curvature and solute effects are minimal and can be neglected. Thus, for most cloud droplets, we can simplify (9.21) to:

$$\frac{\mathrm{d}R}{\mathrm{d}t} = \frac{1}{R} D_v \frac{\rho_{v,\infty}}{\rho_l} \left(S - 1 \right) \tag{9.22}$$

It is now clear that a droplet will grow by condensation ($\mathrm{d}R/\mathrm{d}t$ is positive) when the saturation ratio is greater than one (i.e., the air is supersaturated), and will shrink by evaporation ($\mathrm{d}R/\mathrm{d}t$ is negative) when the saturation ratio is less than one (i.e., the air is subsaturated).

9.1.4 *Equation for diffusional transfer of latent heat around a droplet*

Up to this point, we have neglected the fact that the release of latent heat during condensation or the uptake of latent heat during evaporation changes the temperature of the air immediately surrounding the droplet. When latent heating (positive or negative) occurs, our steady-state assumption requires a continuous transfer of heat away from or toward the droplet. The mechanism by which this occurs is *thermal diffusion*, more commonly known as *heat conduction*.

Heat conduction occurs whenever temperature gradients are present. At locations of higher temperature, air molecules have higher mean kinetic energy. Through molecular collisions, the warmer molecules transfer some of their energy to nearby cooler molecules. The larger the temperature difference (or gradient), the more rapid the transfer of heat energy by this mechanism.

Equation (9.22) shows that the rate of condensation or evaporation depends on the saturation ratio S. Recall that S depends not only on the amount of water vapor in the air but also on the air temperature. Therefore, any temperature gradient between the droplet and the environment will affect the saturation ratio. A more precise treatment of droplet condensation must take this difference into account.

Because the same generic diffusion equation (9.1) governs both vapor diffusion and thermal conduction, an equation analogous to (9.13) can be derived for the diffusion of heat toward/from the droplet:

$$\frac{\mathrm{d}Q}{\mathrm{d}t} = 4\pi RK(T_R - T_\infty) \tag{9.23}$$

where $\mathrm{d}Q/\mathrm{d}t$ is the heat transfer rate near the droplet surface, K is the **thermal conductivity** (e.g., $K \approx 0.0259$ W m^{-1}K^{-1} at 20 °C and 1000 hPa), T_R is the temperature of the air immediately over the droplet surface, and T_∞ is the air temperature of the environment.

By utilizing the fact that $\mathrm{d}Q = C_l \rho_l \left(\frac{4}{3}\pi R^3\right) \mathrm{d}T$, where C_l is the specific heat capacity of liquid water, this differential equation can be integrated to find an expression for $T(x)$ analogous to (9.12) for water vapor density. However, in this case, we have $T_\infty < T_R$ when latent heat release due to condensation occurs. In Fig. 9.3, the corresponding temperature profile is depicted as a red curve. Note that *the gradient in temperature is always opposite to that of the vapor density.*

9.1.5 Diffusion equation for a water droplet considering heat diffusion

In steady-state conditions, the diffusion of heat away from or toward the droplet must equal the latent heating or cooling near the droplet surface, so

$$L_v \frac{\mathrm{d}M}{\mathrm{d}t} = \frac{\mathrm{d}Q}{\mathrm{d}t}. \tag{9.24}$$

Substituting the right-hand sides of (9.13) and (9.23) into the above yields:

$$L_v \left[4\pi R\, D_v(\rho_{v,\infty} - \rho_{v,R})\right] = 4\pi R\, K(T_R - T_\infty). \tag{9.25}$$

With some manipulation using the Clausius-Clapeyron equation and the equation of state for water vapor, we can find the **diffusional growth equation for a water droplet** that considers the effects of both the temperature gradient *and* the vapor density gradient:

$$\frac{\mathrm{d}M}{\mathrm{d}t} = \frac{4\pi R\,(S-1)}{\frac{L_v^2}{R_v T^2 K} + \frac{R_v T}{D_v e_s}} \tag{9.26}$$

where we have replaced the air temperature of the environment, T_∞, by simply T, and e_s and S are the corresponding saturation vapor pressure and saturation ratio of the environment, respectively.

This equation can also be rewritten in terms of the droplet radius:

$$R\frac{\mathrm{d}R}{\mathrm{d}t} = \frac{(S-1)}{\frac{\rho_l L_v^2}{R_v T^2 K} + \frac{\rho_l R_v T}{D_v e_s}}. \tag{9.27}$$

For small droplets (less than $\sim$1 μm radius), the a/R and $-b/R^3$ terms can again be added to the numerator on the right-hand side, to account for curvature and solute effects as in (9.21).

For simplicity, let's represent the denominator of (9.26) as $1/G$, and thus rewrite (9.26) as

$$\frac{dM}{dt} = 4\pi RG(S-1). \tag{9.28}$$

The factor G modifies the droplet growth rate: it includes both the diffusivity in our original equation (9.13) and latent heating effects. The air immediately over the droplet surface is warmer during condensation, resulting in a decrease in the value of G, effectively decreasing the saturation ratio experienced by the droplet and leading to a slower rate of growth.

On the other hand, when $S < 1$ and the droplet undergoes evaporation, the air near the droplet surface becomes cooler due to latent heat uptake. In this case, the factor G increases, representing the increase in the saturation ratio experienced by the droplet due to the cooling, which slows the rate of evaporation.

With the latent heating correction, the growth rates depicted in Figs. 9.4 and 9.5 are *reduced by roughly a factor of two*. We see once again that it would require an impossibly long time for condensation alone to achieve the size of a typical raindrop.

9.1.6 Droplet surface kinetic effects

Another challenge in accurately predicting the change in the mass or size of a droplet due to diffusion involves giving special consideration to the processes occurring just above the droplet surface, within a distance of approximately one molecular mean free path. These molecular processes are known as **surface kinetic effects**.

A small temperature discontinuity can occur between the droplet and the surrounding air. To account for this effect, a *thermal accommodation coefficient* is added to the equations we have derived. Additionally, some water vapor molecules may temporarily adhere to the droplet but then rebound within the thin layer over the droplet surface. This phenomenon is considered by incorporating a *condensation coefficient* into the equations used in this chapter. Both effects further decelerate droplet growth or evaporation, particularly within the first few minutes.

9.1.7 Droplet ventilation effects

One final complication requires our attention. Up to this point, we have considered the droplet to be stationary, i.e., not falling relative to the surrounding air. For droplets less than 10 μm in radius, this assumption holds very well, but by the time droplets have attained a 40 μm radius or more, it

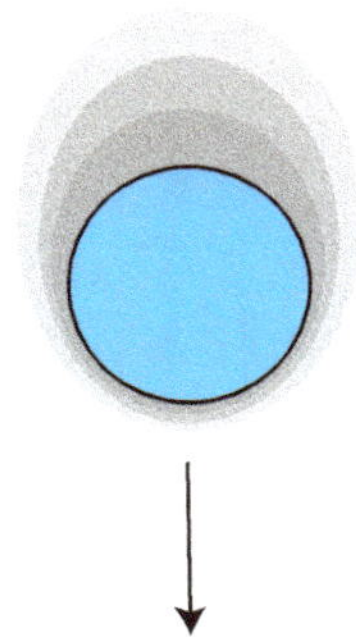

Fig. 9.6: Example of ventilation effect on a raindrop falling in unsaturated air. The vapor density gradient (darker gray representing higher ρ_v, and lighter gray representing lower ρ_v) is magnified on the bottom of the falling raindrop and thus speeds up its evaporation.

does not. Then we must consider **ventilation effects**, i.e., the distortion of the vapor and temperature fields around a hydrometeor as it falls through the air.

These effects are particularly important when calculating the evaporation rate of raindrops as they fall outside the cloud.[5] As shown in Fig. 9.6, as the raindrop falls in unsaturated air, the vapor gradient (and thus the temperature gradient) is no longer isotropic (symmetrical in all directions). Thus, on the lower side of the falling raindrop, a much larger (negative, in this case) vapor density gradient is encountered, which accelerates the evaporation of the raindrop.

These effects can also be included in (9.26) or (9.27) by multiplying the mass or radius rate of decrease by a factor scaled to the drop Reynolds number, which depends on the raindrop size, its fall speed, and the kinematic viscosity of air. For example, for raindrops with a radius of 1.2 mm, this factor is approximately 10 and increases for larger raindrops, significantly accelerating their evaporation.

9.1.8 Diffusional growth of a droplet population in an adiabatic parcel

With the consideration of latent heating, surface kinetic effects, and ventilation effects, the diffusional growth equation can no longer be solved analytically. We therefore solve (9.26) or (9.27) using numerical methods

[5]In this chapter, we mainly focus on the diffusional growth of cloud droplets and less so on raindrops. Raindrops can grow by diffusion as well, but we will see in Chapter 10 that their growth by collision-coalescence is much more important. However, raindrops can still shrink in size quite effectively by evaporation.

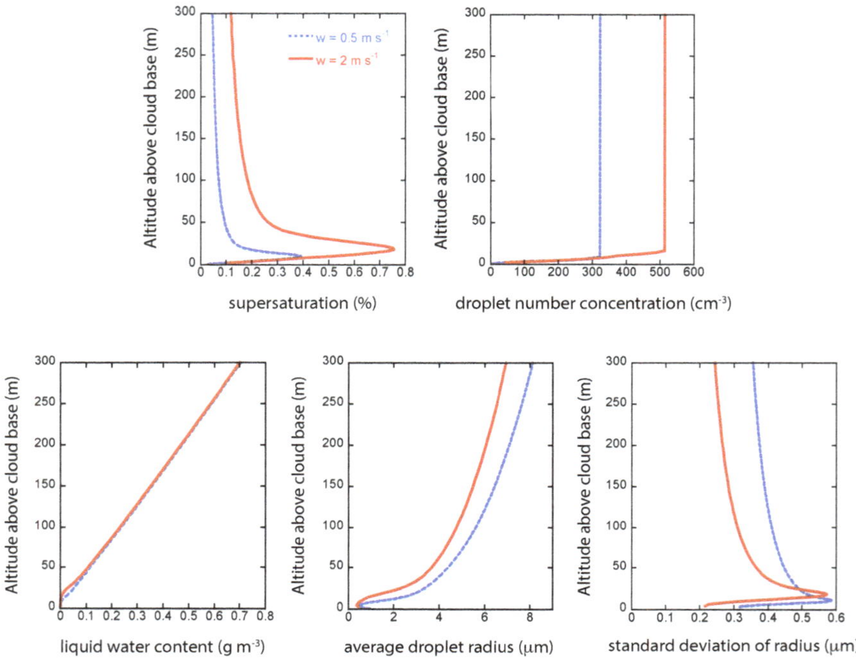

Fig. 9.7: Droplet nucleation and growth within an adiabatic cloud parcel ascending at a constant updraft speed of 0.5 m s^{-1} (blue dashed lines) or 2 m s^{-1} (red solid lines). Supersaturation, droplet number concentration, liquid water content, average droplet radius, and standard deviation of droplet radius shown versus parcel altitude. *After Rogers and Yau (1989).*

to calculate the growth of a population of droplets within a cloud. The results of these calculations for a simple case of two single adiabatic parcels, starting at the base of a cloud and ascending at constant updraft speeds, are shown in Fig. 9.7. Both adiabatic parcels start with the same CCN activity spectrum.

In such a simple model with a constant updraft speed, the parcel height is proportional to time, but the time required for a parcel to reach a particular height depends on its updraft speed. Calculations are shown for two different constant updraft speeds: 0.5 m s^{-1} and 2 m s^{-1}. Thus, the parcel with the slower updraft speed will require four times as much time to attain any given height compared to the parcel with the faster updraft speed.

In either case, as the air ascends and cools, the supersaturation initially increases (top left panel of Fig. 9.7), and more cloud droplets form on activated CCN (top right panel). The supersaturation increases to greater values for the stronger updraft, consistent with (8.5), and a higher droplet number concentration results.

Soon, however, a point is reached in both parcels where enough droplets are growing by condensation to deplete water vapor at a rate sufficient to offset any increase in supersaturation within the updraft. Thus, the supersaturation stops increasing and begins to decrease. This point is reached at a lower altitude for the parcel with the weaker updraft, because the parcel does not cool as quickly and thus cannot produce as large a supersaturation.

As the parcel continues to ascend and cool, it remains supersaturated, but at values less than the maximum that occurred lower in the cloud. Consequently, no additional CCN can be activated. *At this point, the maximum cloud droplet number concentration in the cloudy adiabatic parcel is established*, as approximated by (8.7) and shown in the upper right panel.

Because the supersaturation remains greater than zero above this altitude, the cloud droplets continue to grow by condensation. Consequently, the overall liquid water content of both parcels increases, as shown in the lower left panel. At lower altitudes, the liquid water content (resulting from the droplet mass growth rates) is slightly greater for the slow updraft case, especially near 20 m in this example. This is due to the larger droplet sizes (bottom center panel) and the longer growth time (four times greater) up to any given altitude. Over time, however, the liquid water content of the fast updraft case becomes essentially the same as that of the slow updraft case, both approximating the adiabatic liquid water content.

As previously discussed in Sec. 9.1.2 and shown in Fig. 9.5, smaller droplets grow more rapidly than larger droplets in the same environment, tending to "catch up" in size. Consequently, an initially broad droplet size distribution narrows over time. The spread in a size distribution can be quantified using the standard deviation of the droplet radii. In the lower right panel of Fig. 9.7, the standard deviations of the droplet radii of both parcels initially increase as additional droplets are nucleated. Once the maximum total droplet number concentrations are attained, the standard deviations decrease, i.e., the droplet size distributions narrow due to additional growth by condensation. The cloud parcel with the slower updraft both contains larger droplets and has a much smaller supersatuation, slowing the narrowing of the size distribution relative to the parcel with the faster updraft. We will see in Chapter 10 that the tendency of the droplet population to narrow in response to prolonged diffusional growth has significant implications for rain production.

9.1.9 Droplet condensation in maritime versus continental clouds

In a real cloud, a group of droplets grows by condensation at a rate proportional to the local supersaturation. The available water vapor for growth must be divided among the growing droplets. Consequently, the number of droplets in a cloud influences their growth.

Let's now reconsider the top right and bottom center panels of Fig. 9.7 under a different scenario. Suppose we have two adiabatic parcels *ascending at the same rate* (i.e., within updrafts of equivalent speeds). Assume that one parcel contains fewer CCN (denoted by the dashed blue line), while the other contains more CCN (denoted by the solid red line). In this scenario, we would expect the droplet number concentrations to resemble the upper right panel, where fewer droplets are nucleated in the air parcel with fewer CCN, and more droplets are nucleated in the air parcel with more CCN.

If the available water vapor for growth by condensation must be divided among all the droplets, then, given otherwise similar conditions in the two air parcels, *droplets growing by condensation must be smaller in the air parcel containing more droplets than in the air parcel containing fewer droplets.* This trend is illustrated in the average drop radius in the bottom center panel[6].

As discussed in Chapters 7 and 8, there are more CCN in continental air masses than in maritime air masses. As a result, there are more droplets growing by condensation in a continental cloud compared to a maritime cloud. These trends are evident in observations of continental versus maritime clouds. However, the difference in the number of droplets would be even greater than suggested in Fig. 9.7, as would be the differences in their average sizes. Such differences are also predicted by the diffusional growth equation used in numerical models when initialized with different amounts of CCN. We will see that this difference in droplet sizes due to condensation between continental and maritime clouds has an important impact on how easily these clouds create raindrops, at least when ice is not present.

9.2 Deposition and Sublimation of Ice

As noted previously, microphysical processes related to ice introduce additional complexities compared to those for liquid droplets. For liquid droplets, the fluid nature and surface tension result in a predictable spherical geometry, ensuring uniform radial growth. Conversely, ice crystals are typically non-spherical.

The different shapes, or *habits*, of ice crystals necessitate adaptations to the diffusional growth equation derived in Sec. 9.1. The molecular structure of ice, where each water molecule is bonded to four others, forms an

---

[6]Note that the difference in supersaturation between the two air parcels would no longer appear as those shown in Fig. 9.7. If the updraft speeds are identical, then the supersaturation in the air parcel containing more droplets would be less than that in the air parcel containing fewer droplets.

open lattice. This typically results in a crystal habit shaped like a right hexagonal prism. Occasionally, growth is preferentially in one direction of the prism over the other. In this section, we will examine how these factors are considered in the diffusional growth of ice.

9.2.1 *Ice crystal habits*

The **Magono and Lee Classification**, the earliest comprehensive attempt to classify natural ice crystals, was first published in 1966 and is still widely used today. It was based on observations in Japan. However, since then, additional habits have been observed in other regions, particularly in polar areas, prompting some authors to extend the original classification to include these new observations. This updated classification scheme now encompasses 121 distinct habits, yet infinite variations exist. These ice crystals and the conditions conducive to their formation have also been reproduced in laboratory studies.[7]

Some general habits are widely observed. Three of these were noted in Chapter 4: *columns* have a greater length than width, *plates* have a much greater width than height, and *dendrites* have similar dimensions to plates but possess fern-like branches. An example of each is shown in Fig. 9.8 along with other commonly observed habits that are variations on these three general types.

For example, columns can be *solid*, *hollow*, or somewhere in between (*skeletal*). Extremely long, narrow columns with pointed edges are called *needles*. *Sheaths* are hollow columns with open ends, similar to the sheath of a sword. *Bullets* are wider columns with one tapered end. *Bullet rosettes* consist of multiple bullets attached at their tapered ends.

Plates too can be *solid* or *skeletal*. Some plates are divided into sectors, conveniently called *sector plates*. *Stellars* are an extreme example where the sectors are so narrow that they appear as six single branches. If those branches have smaller branches emerging from them, they are then referred to as *dendrites*.

9.2.2 *Ice crystal geometry*

These different habits result from vapor deposition on ice crystals being rarely equal in all directions. Thus, it is useful to have specific names for the various directions in which crystal growth can occur. A right hexagonal prism has *eight* different faces: top and bottom faces and six sides of the hexagon, as shown in Fig. 9.9. The top and bottom faces are called the

[7] A fun website from Professor Ken Libbrecht, who grows and photographs ice crystals in his laboratory, can be found at snowcrystals.com.

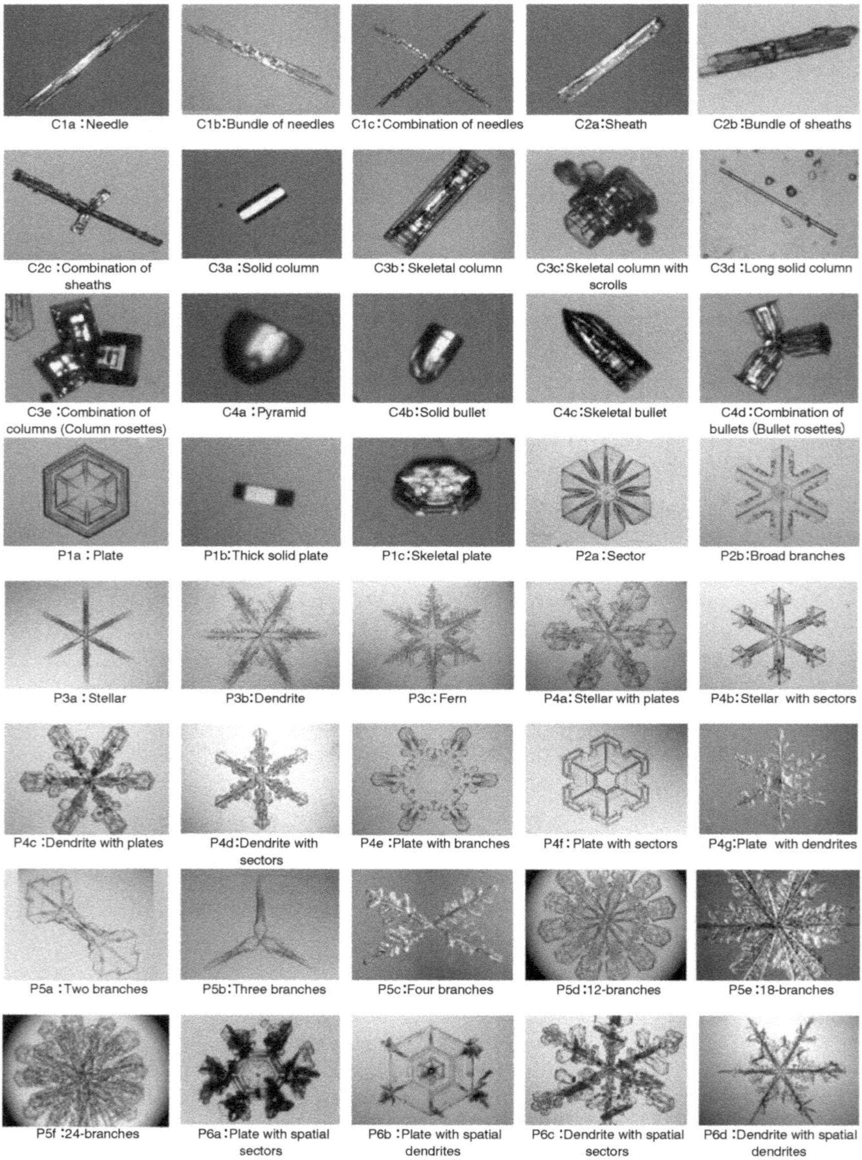

Fig. 9.8: Examples from the original Magono and Lee (1966) ice crystal classification scheme, as well as some newer habits classified by later studies. *From Kikuchi et al. (2013). ©2013, with permission from Elsevier.*

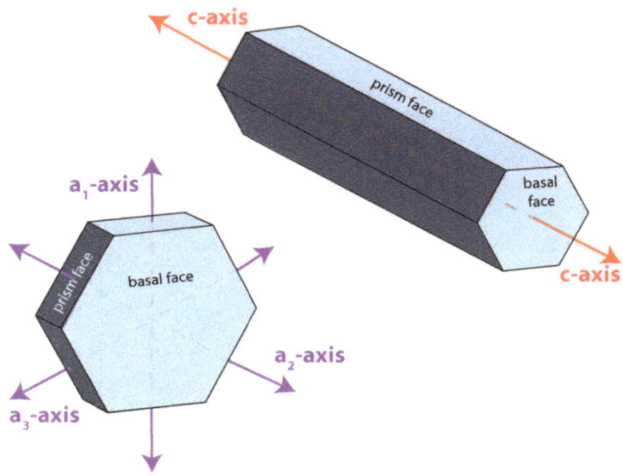

Fig. 9.9: A plate crystal (lower left) and a column crystal (upper right), showing the orientation of the c-axis through the basal faces and the three a-axes through the prism faces. Each crystal has a-axes and a c-axis, but the axes are shown on separate crystals for clarity.

basal faces; these faces have a hexagonal shape. The "**c-axis**" is the axis that passes through (perpendicular to) the basal faces. The six rectangular faces on the sides of the prism are called the **prism faces**. Through each pair of opposite prism faces, an "**a-axis**" intersects them both, resulting in three unique a-axes.

When we speak of growth along the c-axis, water molecules are added to the basal faces, elongating the crystal into a column. Conversely, if growth occurs along the a-axes, water molecules are added to the prism faces, widening the crystal into a plate. Water molecules preferentially begin a new layer of ice at the corners of a face, subsequently filling in the rest of the layer across the face, before initiating another layer at the corners.

In cases where the supersaturation is extremely high, however, water molecules join the crystal at the corners so quickly that the rest of the layer across the face is not filled in before a new layer is started at the corners. Needles, sector plates, and dendritic crystals (Fig. 9.10) are examples of these extreme cases of growth along either the c-axis (needles), or along the a-axes (sector plates and dendrites).

In reality, growth occurs along every axis, but usually at different rates. It is the relative rate of growth along the *c* and *a* axes that ultimately determines the final crystal habit. Moreover, the preferred direction of growth can change as the crystal passes through different thermodynamic conditions in the cloud, leading to a potentially infinite range of variations of the basic habits.

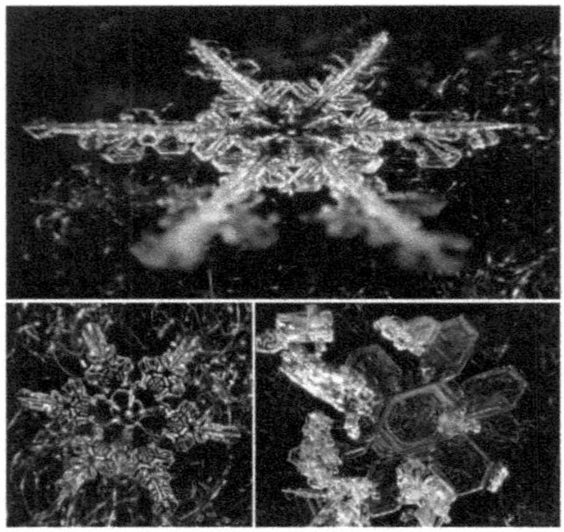

Fig. 9.10: Ice crystals photographed during a natural snowfall: dendrite (top, nearly a side view; bottom left, view from above) and a sector plate (bottom right, with some extra crystals attached). Magnification of photos shows fibers of cloth upon which the crystals fell. These crystals have a very small dimension along their c-axes, resulting in extremely thin crystals, but much larger dimensions along their a-axes. *Photos courtesy of P. Carrington-Hoekstra.*

9.2.3 Deposition growth types

During its time in a cloud, an ice crystal can be advected into, or fall through, various temperature and humidity conditions, each of which affects the crystal growth at that particular moment. Thus, *each crystal habit reflects the history of the thermodynamic conditions during its growth.*

Most habits result from six general types of deposition growth, as shown in Fig. 9.11:

tabular: producing plate-like characteristics

columnar: producing column-like characteristics

branched: dividing the plate into sectors of growth

side-branched: further dividing sectors into branches, like dendrites

tabular polycrystalline: producing a crystal with multiple plates

columnar polycrystalline: producing a crystal with multiple columns

Combinations of these growth types for a single crystal produce more complex habit types.

Laboratory experiments have shown that the growth rate of crystals along the c-axis relative to the a-axes varies significantly with temperature and supersaturation, with temperature being the primary influence on crystal habit. Natural observations have also supported this dependence.

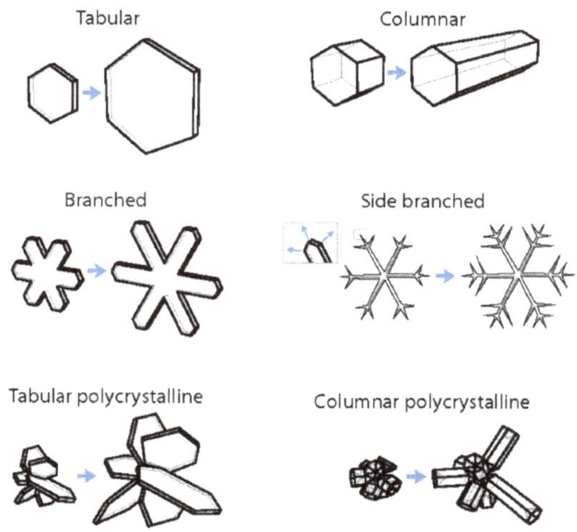

Fig. 9.11: Summary of the six main deposition growth types in the troposphere. *From Hueholt et al. (2022). © American Meteorological Society. Used with permission.*

We now examine the range of temperature and supersaturation conditions in tropospheric clouds that produce the six different types of deposition growth, resulting in specific categories of ice crystal habits. Fig. 9.12 summarizes the most current information.

Across the range of temperatures experienced by clouds in the troposphere (vertical scale to the left), the typical crystal habits change as different deposition growth types dominate. Starting at the highest temperatures, between approximately 0 and −4 °C, *tabular growth* tends to dominate, creating *plates*.

At lower temperatures between approximately −4 °C and −9 °C, an abrupt transition to *columnar growth* occurs, producing *columns*. At even lower temperatures, from −9 °C down to −40 °C, *tabular growth* again produces *plates*, but in the colder two-thirds of this range (below −20 °C), *tabular polycrystalline growth* produces *crystals with multiple plates*. Finally, at temperatures lower than −40 °C, *columnar polycrystalline growth* dominates, often creating crystals referred to as *bullet rosettes*.

At some temperatures, the supersaturation of the crystal environment (moving roughly from the lower left toward the upper right in Fig. 9.12) also has a notable influence. If the air is not supersaturated with respect to ice (gray shaded region), only sublimation of ice crystals can occur. At the lowest supersaturations with respect to ice, more "solid" crystals form that are nearly equal in width and height, with no single deposition growth type dominating (brown shaded region). As the supersaturation increases and surpasses the level of supersaturation with respect to liquid water (to

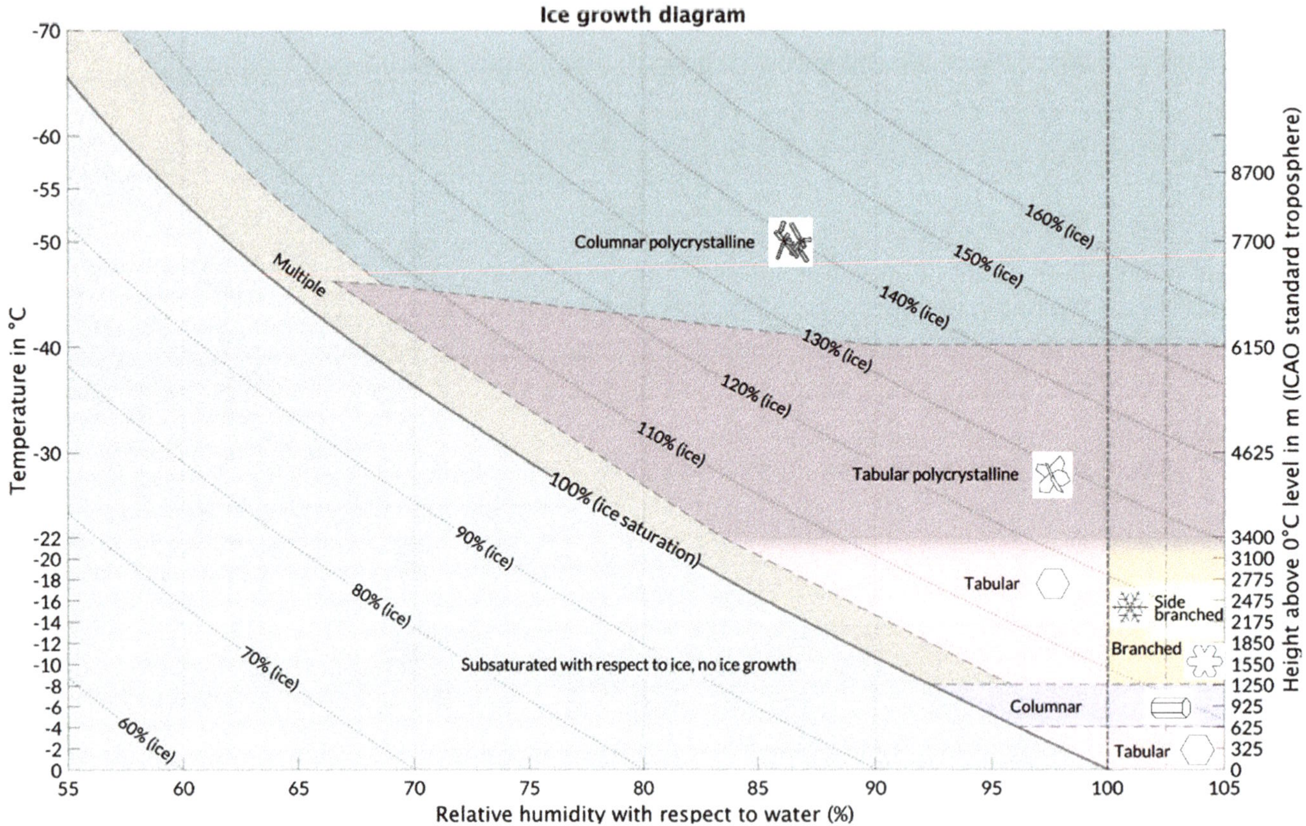

Fig. 9.12: Deposition growth types versus temperature and relative humidity with respect to water. Dashed contours of relative humidity with respect to ice are overlaid. Typical heights above the 0 °C isotherm for a standard atmosphere are shown on the right vertical axis. Dashed lines around different growth types indicate uncertainty in the boundaries. Darkened region above the roughly −20 °C level indicates uncertainty in the transition zone between ice growth forms. *Adapted from Hueholt et al. (2022). © American Meteorological Society. Used with permission.*

the right of the vertical dashed line), the growth modes result in crystals that are more "embellished" or "ornate."

For example, when the supersaturation exceeds that for liquid water, columnar growth starts to produce *sheaths* or *needles* in the temperature range of −4 °C to −9 °C. At lower temperatures between −9 °C and −20 °C, tabular growth converts to *branched growth*, creating *sector plates* or *stellars*. At a supersaturation exceeding that needed for liquid water within the temperature range of −12 °C to −18 °C, *side-branched growth* produces *dendrites*.

9.2.4 The electrostatic analogy

The different habits discussed in the previous section result from the diffusion of water vapor towards the crystal surface, where it subsequently deposits. The diffusional growth of ice crystals adheres to many of the same principles used when deriving the growth equation for the diffusional growth of water droplets.

Thus, we can start from an equation equivalent to (9.30), including latent heating effects by substituting the latent heat of sublimation L_s for L_v, and substituting the saturation vapor pressure with respect to ice e_{si} for e_s. As we would expect, the mass growth rate of the ice crystal then increases for a higher saturation ratio (higher supersaturation) with respect to ice.

However, *ice crystal habits are generally not spherical*, necessitating an alternative method to express the crystal habit other than the radius R. Crystal growth does not occur radially in equal amounts in all directions, as would be the case for a spherical drop growing by condensation, unless a small frozen drop is growing by vapor diffusion. In fact, more ornate crystal habits, such as needles, sheaths, sector plates, and dendrites, which occur at higher S_i, grow more quickly than other habits.

To represent different ice crystal habits when quantifying diffusional growth, we make use of the **electrostatic analogy**. In electrostatics, the particular shape of a charged metal object influences the gradient in electric potential surrounding it, and thus the movement of charge. The size and shape of the metal object can be represented by a quantity called the **capacitance shape parameter**, C_0, which has units of length. By mathematical analogy,[8] we now apply the capacitance shape parameter C_0 to represent the effect of the ice crystal shape and size upon the gradient in

[8] For an in-depth look at the mathematical similarities between electrostatics and diffusion that allow us to apply this analogy, see Pruppacher and Klett (2010).

vapor density surrounding it, which will thus influence the deposition or sublimation it experiences as a result of that vapor gradient. All ice crystal habits, even the polycrystals, are approximated by these shapes in most present-day numerical models, as shown in Fig. 9.13 and listed in Table 9.1.

Figure 9.14 illustrates that considering the deviation in shape of an ice crystal from a perfect sphere actually increases its growth rate compared to a perfectly spherical ice particle. At the center of the plot, where the aspect ratio $c/a = 1$, are the capacitance shape factor values for a perfect sphere. For any given volume of the sphere (following the curve of any single color), these values (which exactly equal the radius of the sphere) are the lowest. As the spheroid representing the crystal of the same volume becomes either more oblate (flattened—dashed lines) or more prolate (elongated—solid lines), the capacitance increases significantly compared to the value for a sphere.[9] Thus, if we had not used the electrostatic analogy and just assumed all ice crystals are perfect spheres, we would underestimate their rate of growth (or sublimation), potentially by very large amounts. As the volume of the crystal increases (moving from the blue curve towards the purple curve), ignoring the non-sphericity of the crystal becomes an even worse assumption.

For thin plate-like crystals, assuming that they are spherical would erroneously overestimate the rate of mass change through deposition or sublimation. The capacitance shape parameter for a thin plate is $2a/\pi$, approximately 0.64 times the radius of a spherical ice particle of equal mass. Therefore, using the half-width along an a-axis as the representative dimension would cause the ice crystal's mass to increase or decrease too rapidly during deposition or sublimation. However, the *width* $2a$ of the disk (plate/dendrite) would increase or decrease faster than that of a perfectly spherical ice particle, as most of its growth or reduction occurs within one plane, along the a-axes.

9.2.5 *Diffusional growth equation for any ice crystal habit*

Finally, we can adapt (9.28) to represent the **diffusional growth equation for an ice crystal** instead of a water droplet by replacing R with C_0, the capacitance shape parameter, and S with S_i:

$$\frac{\mathrm{d}M}{\mathrm{d}t} = 4\pi C_0 G_i (S_i - 1), \tag{9.29}$$

[9]One notices that the curves shown in Fig. 9.14 are not symmetric for a given spheroid volume. This is because the growth/decrease along the c-axis represents growth in one dimension, while the growth/decrease along the a-axes represents growth in two dimensions, which necessarily involves more molecules changing phase and thus is slower.

Table 9.1: Capacitance shape parameter (C_0) formulas for different shapes, and the ice crystal habits they are used to represent. In the case of oblate and prolate spheroids, a and c are the half-lengths along those respective axes (like a radius in each direction). The eccentricity for the oblate spheroid is given by $\epsilon_0 = \sqrt{1-(c/a)^2}$, and the eccentricity for the prolate spheroid is then $\epsilon_p = \sqrt{1-(a/c)^2}$. These relationships are depicted graphically in Fig. 9.14.

C_0	Shape	Habit
a	Sphere of radius a	Sphere
$\frac{2a}{\pi}$	Thin disc of radius a	Very thin plate
$\frac{a\epsilon_0}{\sin^{-1}\epsilon_0}$	Oblate spheroid	Thick plate
$\frac{2a\epsilon_p}{\ln\left(\frac{1+\epsilon_p}{1-\epsilon_p}\right)}$	Prolate spheroid	Column, needle

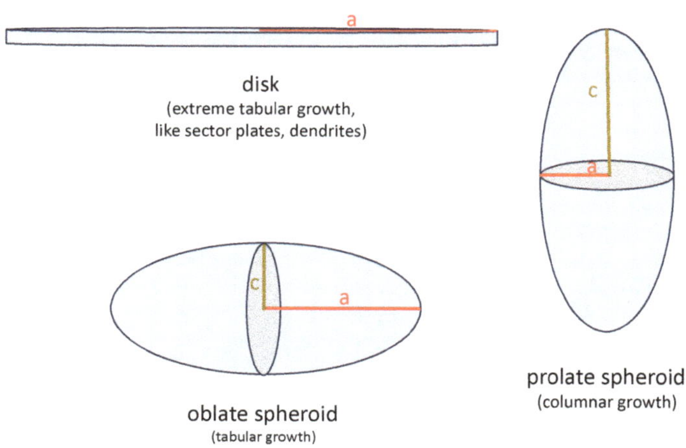

Fig. 9.13: Shapes typically used for the capacitance shape parameter C_0 quantified in Table 9.1, where a is the half-length along the a-axes of the ice crystal, and c is the half-length along the c-axis of the ice crystal. The vertical dimension of the disk is exaggerated here, and would, in actuality, be seen from the side as a single horizontal line of water molecules.

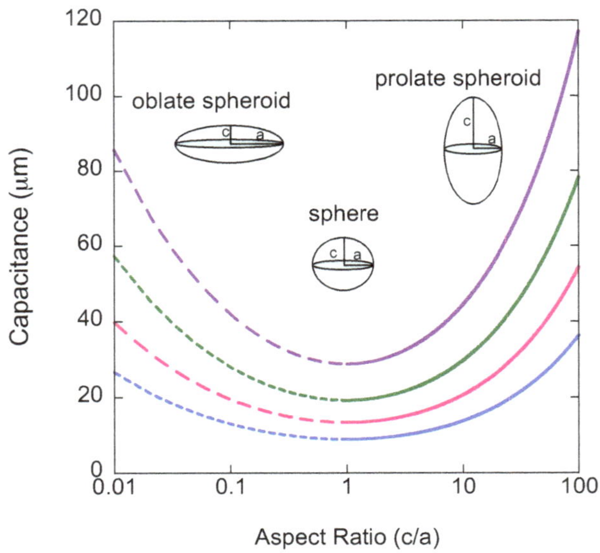

Fig. 9.14: Capacitance shape parameters from Table 9.1, plotted here as a function of c/a, where c represents the half-length of the crystal along the c-axis, and a represents the half-length along an a-axis. Dashed and solid lines of the same color represent values for oblate and prolate spheroids, respectively, for the same spheroid volume. Spheroid volumes are 3000 μm^3 (blue), 10^4 μm^3 (pink), 3×10^4 μm^3 (green) and 10^4 μm^3 (purple). For reference, a thin disk having the same volume as the blue curve would have an aspect ratio $\sim 9.5 \times 10^{-13}$ and capacitance shape factor of $\sim 2.4 \times 10^4$ μm far beyond the upper left corner of the plot, although for a single thin plate or dendrite to have this large a volume would be unrealistic.

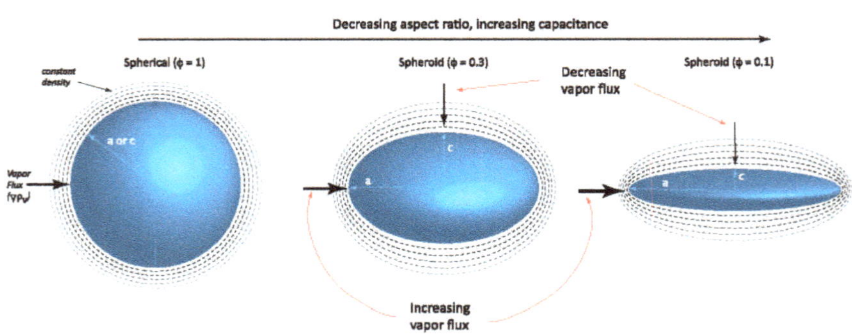

Fig. 9.15: The vapor density gradient around three ice ellipsoids of increasing non-sphericity, as given by their aspect ratio ϕ (ratio of height $2c$ to width $2a$). *From Harrington et al. (2013).* *© American Meteorological Society. Used with permission.*

where the value of C_0 is given in Table 9.1 for the appropriate ice habit. Here, G_i is analogous to G in (9.28), but with L_v and e_s replaced by L_s and e_{si}. In practice, numerical models utilize (9.29), assuming as necessary a mass-diameter relationship like (4.12) to determine a in the formulas for C_0.

Recall that for water droplets, a stronger vapor density gradient corresponds to a higher value of S; similarly, a higher value of S_i indicates a stronger vapor density gradient. So, how does the crystal habit affect the vapor density gradient and thus S_i?

Ice crystals with very small or very large aspect ratios increase or decrease their mass much more quickly than spherical ice particles because of the feedback that deposition or sublimation has on the surrounding vapor field. As shown in Fig. 9.15, a stationary spherical drop (or frozen drop) has an isotropic (equal in all directions) vapor field surrounding it, so the flux of water vapor toward or away from the drop is uniform over its entire surface, and this in turn keeps the vapor field isotropic. On the other hand, when ice crystals have one dimension ($2a$ or $2c$) that is greater than the other, the *vapor density gradient increases over the narrowest parts of the crystal* (near the sharpest edges with the smallest radius of curvature). This then *increases the rate of the vapor flux toward or away from the crystal,* enhancing deposition (or sublimation) in the already preferred direction, making those crystal edges even sharper, and thus the vapor density gradients even larger, resulting in a positive feedback at those locations. Such behavior, especially under conditions of high supersaturation, increases the tendency of the water molecules to join the pointed edges of the crystal face before filling out the entire layer on the face, resulting in needles, sector plates, and especially dendrites.

9.2.6 Ice crystal surface kinetic effects

Equation (9.29) is not yet complete, as it tends to overestimate the mass growth rate by as much as a factor of two. *Surface kinetic effects* were previously discussed in connection with vapor diffusion for water droplets, but their role for ice crystals is more complicated. For water droplets, once a water molecule joined it, it did not matter where it was located, as it simply increased the overall radius due to the fluidity of the drop. However, for an ice crystal, the location on the solid crystal surface where the molecule joins the lattice determines preferential regions of growth along the a-axes or c-axis, thus determining the crystal habit. These preferences feed back into the overall growth rate because of the influence of the habit on the vapor density gradient, as shown in Fig. 9.15.

Depositional growth of an ice crystal face occurs by layers or steps and involves a sequence of three events. First, water vapor molecules in the surrounding air diffuse towards the ice crystal face and weakly attach to it.

Then, these molecules randomly migrate across the crystal face. Finally, if one of these wandering molecules on the crystal face encounters one of the edges along a step, there is a chance for it to form a strong bond with other water molecules, joining the ice lattice as a permanent member. If it does not become incorporated into the lattice, the water molecule eventually detaches from the face and re-enters the vapor phase.

Once a new layer has begun in the lattice, an additional layer can develop on top of it, even if the bottom layer has not been completely filled. The molecules preferentially join the corners or edges of each layer in the ice lattice because less surface energy is required.

These surface kinetic effects are included in the diffusional growth equation for ice by specifying a *deposition coefficient*, which is the ratio of the number of water molecules incorporated into the ice lattice to the total number of water molecules on the crystal face. It is a function of both temperature and air supersaturation (with respect to ice) near the crystal face.

For temperatures where tabular growth dominates, the deposition coefficient on the prism faces is higher than on the basal faces. Conversely, for temperatures where columnar growth dominates, the deposition coefficient on the basal faces is higher than on the prism faces. *Why* molecular bonding is preferred on particular faces of the crystal over a specific temperature range remains unknown.

The effect of supersaturation is also important. For high supersaturation, diffusion acts rapidly, especially at the corners of the crystal faces, causing the vapor density to become non-uniform across the faces. This results in even faster growth at the corners, leading to the branched or side-branched structures that form sector plates, stellars, and dendrites observed at higher supersaturation.

Because of the differences in the preference for growth along one set of axes, the current state-of-the-art is to model the growth of a crystal using (9.29), but as a *system of two equations*: one for growth along the c-axis, and the other for growth along the a-axes (including the deposition coefficients applicable to each axis). When incorporated into numerical models, the dominance of one growth mode over another in different regions of the cloud with varying temperature and supersaturation conditions can be more accurately represented. These models also allow ice crystals to evolve over time into different habits, depending on the specific regions within the cloud that favor another type of deposition growth (Fig. 9.11).

Such calculations require significant computing power and have generally been limited to research applications. However, they are now starting to become available in numerical models used for weather prediction. In models where these calculations are not practical, a single equation like (9.29) is still used. The capacitance shape factor C_0 might even be assumed constant throughout the cloud, so that the assumed habit type cannot change even if the crystal moves into different temperature or supersaturation conditions.

9.2.7 *Ice crystal ventilation effects*

One final modification to (9.29) is again important for ice crystals, as it was for liquid drops: *ventilation effects* when the ice crystal is falling relative to the air. Ice crystals tend to fall with their largest dimension parallel to the ground. Thus, tabular crystals and branched crystals will fall with a basal face pointing downward, and columnar crystals will fall with a prism face pointing downward. As in the case of water droplets, ventilation intensifies the gradients occurring on the lower side of the crystal as it falls, especially for larger crystals that can fall faster (e.g., being several hundred micrometers wide in their maximum dimension). If large ice crystals are falling in subsaturated air, ventilation will speed their rate of sublimation. However, if the large ice crystals are falling in supersaturated air (i.e., still within the cloud), asymmetric ventilation on the face parallel to the ground may enhance their growth.

9.3 Diffusional Growth of Larger Hydrometeors

We will see in Chapters 10 and 11 that much faster growth rates for drizzle, raindrops, aggregates, graupel particles, and hailstones result from microphysical processes other than diffusional growth. Nonetheless, once any of these liquid or frozen hydrometeors are formed, they can still grow by condensation/deposition in supersaturated regions of the cloud or shrink by evaporation/sublimation in unsaturated regions inside or outside the cloud.

These diffusion processes acting on precipitation-sized hydrometeors must still be included in numerical models for accuracy and consideration of latent heating and cooling effects. For simplicity, these hydrometeors are often approximated as pure spheres, as higher accuracy is generally not warranted due to the relatively slow effects of diffusion compared to other growth processes.

9.4 Diffusion of Water Vapor for a Droplet/Ice Mixture

Up to this point, we have discussed water vapor diffusion for droplets and ice crystals separately. We now address how diffusion evolves when both supercooled droplets and ice crystals coexist in close proximity within the same region of a cloud.

Table 9.2: Consequences of different values of the environmental vapor pressure e relative to the saturation vapor pressure for water, e_s, and the saturation vapor pressure for ice, e_{si}.

Condition	Liquid droplets	Ice particles	Vapor
$e < e_{si} < e_s$	Evaporate ($S < 1$)	Sublimate ($S_i < 1$)	e increases
$e = e_{si} < e_s$	Evaporate ($S < 1$)	Equilibrium ($S_i = 1$)	e increases
$e_{si} < e < e_s$	Evaporate ($S < 1$)	Grow ($S_i > 1$)	e nearly steady
$e_{si} < e = e_s$	Equilibrium ($S = 1$)	Grow ($S_i > 1$)	e decreases
$e_{si} < e_s < e$	Grow ($S > 1$)	Grow ($S_i > 1$)	e decreases

Recall from Chapter 6 and Fig. 6.3 that at temperatures below 0 °C, the saturation vapor pressure over ice is always less than that over liquid water, i.e., $e_{si} < e_s$. Consequently, certain conditions should favor the diffusional growth of ice more than that of liquid water droplets. We now examine these situations. The ambient vapor pressure e within the cloud can take on five distinct relationships to e_{si} and e_s, as summarized in Table 9.2.

Of the scenarios listed, the first ($e < e_{si} < e_s$) is the least interesting, because any hydrometeors present, whether frozen or liquid, will lose mass. This occurs because the vapor pressure is less than the saturation vapor pressures over both water and ice (i.e., both $S < 1$ and $S_i < 1$). If the vapor pressure is low enough, any cloudiness will completely dissipate. However, for all remaining cases, any ice present is either at equilibrium ($S_i = 1$) or growing by deposition ($S_i > 1$).

9.4.1 *Maximum diffusional growth rates of ice*

First, let us focus on the fourth scenario: that is, the environment is in exact equilibrium with respect to liquid water ($S = 1$) and thus the air is supersaturated with respect to ice ($S_i > 1$). Figure 9.16 (blue curve) shows that S_i becomes quite large at very low temperatures in this case, approaching 1.5 at −40 °C, which corresponds to a supersaturation with respect to ice of 50%!

But when the cloudy air is just saturated with respect to liquid water, the maximum growth rate of ice does not occur at the lowest temperatures. To see why, let us briefly compare the growth of spherical ice crystals (for simplicity) with supercooled water droplets by rewriting (9.28) as:

$$\frac{dM}{dt} = \frac{4\pi R(e - e_s)}{\dfrac{L_v^2 D_v e_s}{R_v T^2 K} + R_v T} = 4\pi R G'(e - e_s) \qquad (9.30)$$

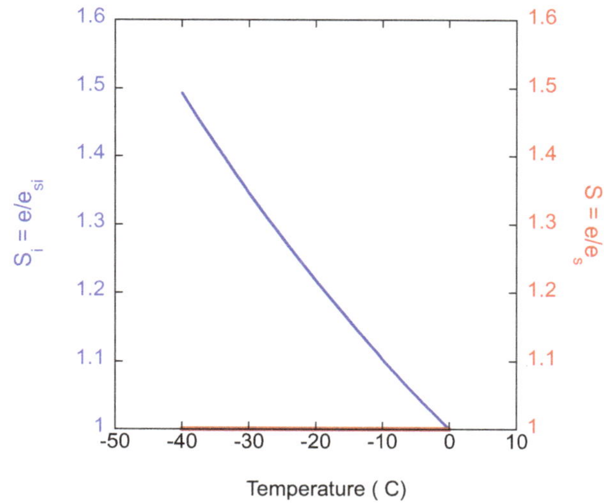

Fig. 9.16: The saturation ratio with respect to ice (S_i, blue curve) when the saturation ratio with respect to liquid water (S, red line) equals exactly one, i.e., the air is just saturated with respect to liquid water. The curves do not extend below -40 °C, because below this temperature only ice would tend to exist in the cloud.

and then writing a similar equation appropriate for spherical ice crystals:

$$\frac{\mathrm{d}M}{\mathrm{d}t} = \frac{4\pi R(e - e_{si})}{\frac{L_s^2 D_v e_{si}}{R_v T^2 K} + R_v T} = 4\pi R G_i{}'(e - e_{si}). \tag{9.31}$$

In comparing these two equations, we see that they each depend on the difference between the environmental vapor pressure, e, and their respective saturation vapor pressures, e_s or e_{si}. If ice crystals and supercooled droplets exist side by side within the same region of a cloud, the ice will always experience a larger vapor pressure difference, implying more rapid growth than that of the droplets.

So where will the vapor pressure difference for ice be maximized over the vapor pressure difference for droplets? Figure 9.17 depicts the *relative* difference in their saturation vapor pressures. At 0 °C, $e_{si} = e_s$, so the difference is zero. At very low temperatures, on the other hand, both e_s and e_{si} are extremely low, so the difference again approaches zero.

Between these two extremes, the vapor pressure difference is always nonzero but reaches a pronounced peak near -12 °C. To first order, we would expect the maximum ice deposition rate to occur at this temperature when the air is just saturated with respect to liquid water. However, we have observed that diffusional growth also increases the temper-

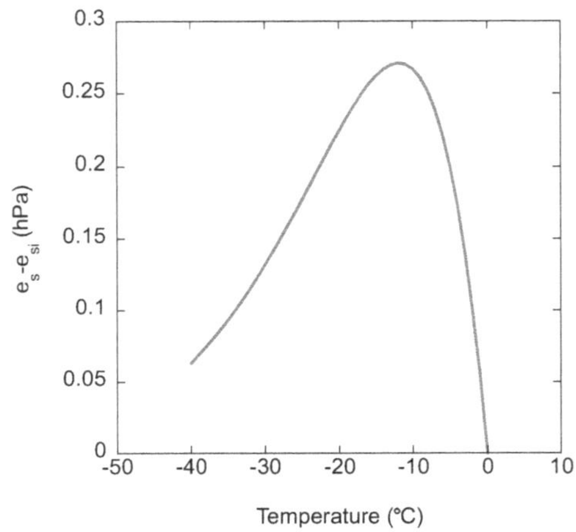

Fig. 9.17: The difference in saturation vapor pressures with respect to liquid water and ice.

ature around the crystal due to latent heating. Consequently, the actual maximum growth rate of ice occurs at the slightly lower temperature of -14.5 °C.

If we go on to consider the range of temperatures for which the saturation vapor pressure difference exceeds 0.2 hPa, it follows that *rapid diffusional growth of an ice crystal is expected in any water-saturated environment having a temperature between approximately -5 °C and -23 °C.* When growing at the higher end of that temperature range, large needles are often observed. Within the range of -12 °C to -15 °C, where the difference in saturation vapor pressures of liquid and ice is maximized, dendrites are the favored habit. This coincides with the fact that dendrites tend to have the largest maximum dimension of any individual ice crystals observed at ground level.

The assumption of saturation with respect to liquid water up to this point in our discussion is not arbitrary. On the contrary, any volume of air that already contains significant cloud liquid water—typically distributed over hundreds of cloud droplets (or even over a thousand) per cubic centimeter—will quickly reach equilibrium with those droplets either through evaporation or condensation. However, a very strong updraft might instead sustain significant supersaturation with respect to the liquid droplets, despite the constant and rapid removal of vapor via condensation. Conversely, a dying updraft or entrainment into an active updraft might decrease the air below saturation with respect to liquid water. In the next section, we discuss how the introduction of a sufficient amount of ice might also force the supersaturation to decrease below that required for liquid water.

9.4.2 The Wegener-Bergeron-Findeisen process

Now consider the particularly interesting third scenario in Table 9.2, in which $e_{si} < e < e_s$, that is, $S_i > 1$ but $S < 1$. In this case, *cloud droplets evaporate while ice particles grow by deposition.* How does such a scenario arise?

Suppose a small number of ice particles are nucleated in an initially supercooled cloud with temperatures ranging from -5 °C to -23 °C. Critical to this scenario is the observation, discussed in Chapter 7, that ice nucleating particles (INP) are far fewer in number than cloud condensation nuclei (CCN). At temperatures near, say, -20 °C, one might expect only $\sim$1 ice particle *per liter* compared to 10^2–10^3 cloud droplets *per cubic centimeter*, a ratio of 10^5–10^6. Consequently, the cloud droplets will continue to control the ambient vapor pressure and keep it very close to saturation even after the first ice appears.

However, the new ice crystals grow rapidly by deposition, as discussed in the previous section. If sufficient numbers of ice crystals are present, they can significantly deplete the water vapor, causing the vapor pressure e to fall below e_s, i.e., $S < 1$. As this occurs, cloud droplets evaporate (since e is now less than e_s) to replenish some of the vapor removed by the growing ice crystals. Meanwhile, S_i remains greater than one, so the ice crystals continue to grow by deposition. Thus, there is a continuous net transfer of water mass *from* the supercooled liquid droplets *to* the ice crystals, which is referred to as the **Wegener-Bergeron-Findeisen** (WBF) **process**. It is named after the three scientists who either initially proposed it theoretically or later verified it through observations. This process is illustrated schematically in Fig. 9.18, where the arrows indicate the direction of water vapor transfer from the droplets to the ice crystals. Consequently, the ice crystals grow larger over time, while the cloud droplets shrink in size (Fig. 9.19).

Recall that there may be $\sim 10^5$ or more cloud droplets for every ice particle. If the WBF process is allowed to continue to completion (in the absence of strong updrafts, which would otherwise keep the air supersaturated with respect to the liquid droplets), the ice crystals can acquire enough mass to begin falling out of the cloud as precipitation.

9.4.3 Importance of the WBF process

The WBF process allows the possibility of converting all supercooled cloud water into a smaller number of much larger ice crystals *solely through diffusional growth.* This is one means by which a cloud can become *glaciated*, i.e., consisting only of ice.

The WBF process can also serve as the starting point for the formation of frozen precipitation in a supercooled cloud. In particular, when pristine

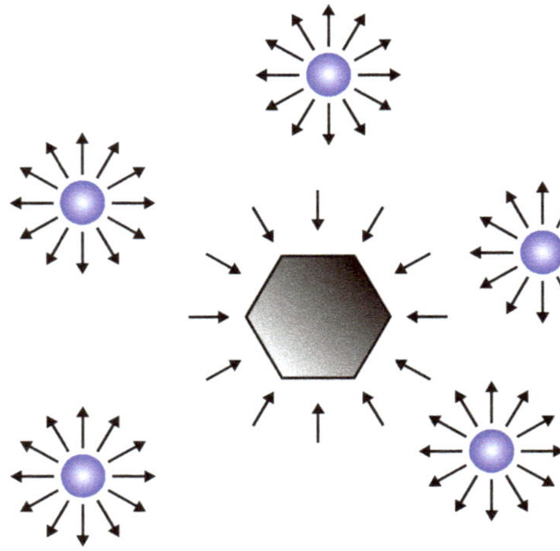

Fig. 9.18: Schematic depiction of the Wegener-Bergeron-Findeisen process, in which the environment is subsaturated with respect to liquid water but supersaturated with respect to ice; that is, $e_{si} < e < e_s$. In this case, cloud droplets (blue circles) evaporate, giving up mass to the environment in the form of water vapor, while the ice particle (gray hexagon) grows by deposition, removing vapor from the environment.

Fig. 9.19: A cloud that is simultaneously supersaturated with respect to ice and subsaturated with respect to liquid water.

(vapor-grown) ice crystals reach the surface, they have usually resulted from the WBF process. For years following its discovery, the WBF process was thought to be the dominant mechanism by which precipitation is initiated in cold clouds.

We have since learned from observations and numerical modeling that the WBF process is *not important in updrafts that can maintain supersaturation with respect to the liquid droplets.* In such cases, the ice crystals and cloud droplets both grow by diffusion from the available supersaturated air—the fifth scenario in Table 9.2. However, in clouds with weaker updrafts that produce significantly lower supersaturations and cannot keep the cloud even just saturated with respect to liquid droplets, the WBF process can be active and very important. Most stratiform clouds would fall into this category once some ice has been nucleated. Within the downdrafts of any clouds, the WBF process would also be expected to play a role in the diffusional growth of ice as long as the supersaturation with respect to ice is maintained.

The importance of distinguishing between the WBF process, in which cloud droplets evaporate at the expense of collocated ice crystals (scenario three in Table 9.2), and the case where both cloud droplets and ice crystals grow by diffusion together (scenario five, occurring in strong updrafts), lies in what happens next: *the collection processes involving ice.* Aggregation, the self-collection of ice crystals into clumps, is the dominant mechanism for producing precipitation once the cloud is glaciated by the WBF process, although it brings less precipitation mass to the ground per unit time. Riming, the collection of supercooled droplets by ice crystals, becomes the dominant mechanism when supersaturation with respect to both ice and liquid water is maintained in the cloud, often resulting in more precipitation mass at the ground per unit time. We will discuss these collection processes individually in Chapter 11 and the reasons for their differing precipitation rates in Chapter 13.

9.4.4 Hole-punch clouds

Impressive visual evidence of the WBF process can sometimes be observed when an otherwise supercooled stratiform cloud has ice particles introduced at a specific location. Typically, the cloud type is altocumulus, and a common source of these introduced ice particles is the passage of an aircraft. In such cases, the ice particles are referred to as aircraft-produced ice particles, or "*APIPs.*"

APIPs (briefly mentioned on p. 100 as a nuisance for *in situ* microphysical observations) can be generated by various types of aircraft, but the phenomenon is most commonly associated with propeller-driven aircraft and jet aircraft with turbofan engines. The key factor is the ability of the aircraft to induce sharp adiabatic drops in pressure and, consequently,

Fig. 9.20: A hole-punch cloud (or fallstreak hole), observed on 17 August 2008 about 20 km south of Linz, Austria. The undisturbed part of the cloud is supercooled altocumulus. The fallstreak emerging below the empty hole consists of snow particles that grew via the WBF process, fully consuming the original supercooled water in the layer. *Photo by H. Raab. (CCBY-SA)*

temperature, leading to the nucleation of ice crystals in a cloud layer containing supercooled water droplets.

Once ice particles are introduced, by whatever means, the WBF process is initiated and quickly glaciates that part of the cloud. The resulting ice particles fall out and leave a visible hole in the cloud, as shown, for example, in Fig. 9.20. Early observations of this phenomenon promoted interest in cloud seeding, which is briefly discussed in the next section.

9.4.5 The WBF process and the origins of cloud seeding

The WBF process has played a significant role in the historical development and understanding of cloud seeding techniques. In the case of supercooled clouds lacking an ice phase, cloud seeding aims to artificially induce or enhance precipitation by introducing ice nucleating particles (INP) with sufficiently high threshold temperatures to nucleate ice at an appropriate ratio to the available supercooled water. These newly formed ice crystals can then grow rapidly at the expense of the surrounding water droplets, which begin to evaporate. As the ice crystals grow and fall, they can either directly contribute to precipitation or collect other hydrometeors, further enhancing precipitation.

However, if too many INP are introduced into a supercooled cloud, the ratio of supercooled droplets to ice particles will become too small, preventing the ice from growing sufficiently large to fall out. Instead, the cloud will simply become glaciated and produce no precipitation. This phenomenon is known as *overseeding*.

Historically, substances like silver iodide, potassium iodide, and dry ice have been used as ice nucleating agents in cloud seeding experiments. As discussed in Sec. 8.4.1, silver iodide has one of the highest threshold temperatures of any common substance and can nucleate ice in comparatively warm supercooled clouds where most other INP are inactive.

More recently, and as mentioned in Sec. 8.4.1, *P. syringae* has been tested in cloud seeding experiments, particularly in the form of a commercial product known as Snomax®, which is primarily used for snowmaking at ski resorts. Snomax is a freeze-dried preparation of non-viable, genetically modified *P. syringae* cells that serve as effective INP.

However, the real-world effectiveness of cloud seeding remains a subject of ongoing research and debate. While the WBF process offers a theoretical framework, the outcomes are influenced by multiple factors, including cloud type, ambient conditions, and the presence of other aerosol particles. Even after decades of study, the actual impact of cloud seeding has been difficult to isolate from these other influences.

Questions for Review

1. What controls how quickly droplets grow by condensation or shrink by evaporation?
2. How does a population of cloud droplets change over time due to continued condensation?
3. How does the number of CCN in the air affect the diffusional growth of water droplets in a cloud?
4. Can raindrops form by condensation alone?
5. What controls how quickly ice crystals grow by deposition or shrink by sublimation?
6. Why do ice crystals have different shapes (habits), and under what thermodynamic conditions are different habits favored?
7. What is the electrostatic analogy, and how is it applied to vapor diffusion toward or away from ice crystals?
8. How does the positive feedback of vapor density gradients near sharp edges influence crystal growth under high supersaturation conditions, and what crystal habits result from this phenomenon?
9. What different scenarios are possible for the diffusion of water vapor when droplets and ice crystals are collocated within a cloud?

10. What is the Wegener-Bergeron-Findeisen process, under what conditions is it important, and what implications does it have for precipitation?
11. What observable evidence can reveal the presence of the WBF process in a stratiform cloud, and how has it sparked interest in cloud seeding?

Homework problems and other resources for students and instructors can be found at bit.ly/4dYBwB5, or scan this code:

Further Reading

For more information on the details of ice crystal growth [advanced]:

Lamb, D., and J. Verlinde, 2011: *Physics and Chemistry of Clouds*. Cambridge University Press

CHAPTER 10

Drop Growth by Collection

In the last several chapters, it has become clear that all hydrometeors begin their growth through the diffusion of water vapor. In the case of water droplets, this growth can be quite appreciable at first; however, it is too slow to create the drizzle and raindrops that fall to the ground as precipitation. How, then, are drizzle drops and raindrops formed within the typical lifetime of a cloud?

We saw in Chapter 4 that larger drops fall faster than smaller droplets. Larger drops can therefore overtake smaller droplets and possibly *collide* with them. If they stay together—i.e., merge into a single drop—after they have collided, we say they have *coalesced*. The result is larger drops that fall even faster and can thus potentially collide with even more small droplets.

This process is called **collision-coalescence**. It can create raindrops quite rapidly, depending on the initial cloud droplet sizes and the number of droplets available to be collected. This is, in fact, the only way that raindrops can form in *warm* clouds that lack ice. Thus, the combination of droplet nucleation, followed by growth by condensation, and finally rapid growth by collision-coalescence, is referred to as the *warm rain process*. This chain of events is shown as the leftmost path in Fig. 9.1.

10.1 Collision-Coalescence

To determine the growth rate of liquid drops from collision-coalescence, two questions need to be answered:

1. Under what conditions will two drops collide?
2. If they collide, will they coalesce (stay together)?

In this section, we first develop the concept of *collision efficiency*, which addresses the first question, followed by *coalescence efficiency*, which addresses the second. Together, these two parameters determine the overall *collection efficiency*, which describes how effectively a larger falling drop can sweep up smaller droplets in its path and add them to its own mass.

10.1.1 Collision efficiency

We will discuss drop collisions as a function of the size of a larger drop, often referred to as the *collector drop*, and a smaller *collected droplet*. To determine whether these two drops will collide, we first need to consider the airflow around drops of various sizes. For now, we will consider how the drops fall in *non-turbulent air*, such that the air motions carrying the drops within the cloud are uniform. The collector drop, being larger, always falls faster than the collected droplet, so in a frame of reference that follows the collector drop, the collected droplet appears to be moving upward toward it, at a speed equal to the difference in their terminal fall speeds. Streamlines (lines containing arrows showing the direction of the airflow relative to the collector drop) are useful to visualize the relative movement of a pair of drops.

Figure 10.1 shows streamlines representing the airflow around collector drops of various sizes. These streamlines are influenced by the collector drop size, denoted by its radius R, its fall speed, which is a function of its size, and the viscosity of the air. The streamlines are relatively simple for the smallest collector drops shown at the upper left of the figure, but they become more complex as the collector drop size increases.

Starting with the smallest cloud droplets of less than 10 μm radius, the flow around them is *laminar*, meaning it is not turbulent, and streamlines are deflected only by gentle "bending" around the droplet as it falls. As the drop size increases to larger cloud droplets and drizzle drop sizes up to 500 μm radius, the streamlines pack more closely around the drop and even develop small, counter-rotating vortices above the falling drop. For large drizzle drops (and including smaller raindrops) exceeding 500 μm radius, these counter-rotating vortices become stretched over a larger distance above the falling drop. For large raindrops greater than 1 mm (1000 μm) radius, the counter-rotating vortices break up into a chaotic, turbulent wake above them.[1]

[1] We are still considering the drop falling in non-turbulent air here; the *local* turbulence shown in some of the streamlines in Fig. 10.1 develops from the disturbance of the surrounding air by the falling drop. Later, we will address the effects of drops falling in a cloud with its own turbulent air motions.

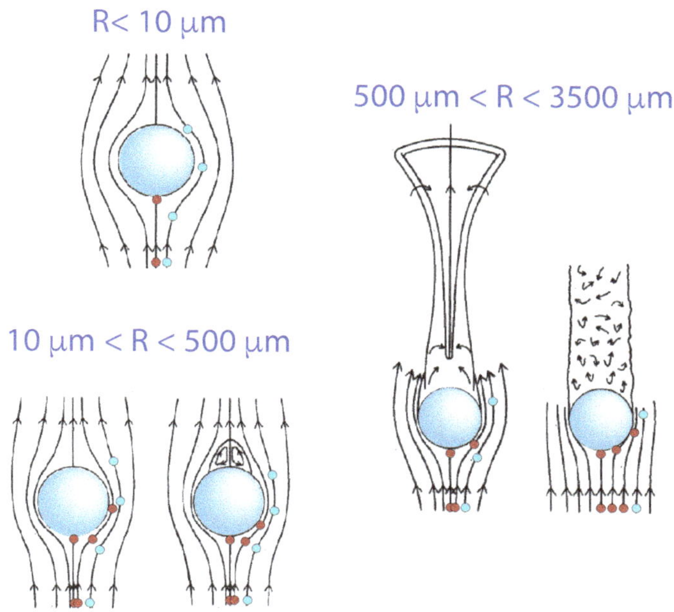

Fig. 10.1: Streamlines of the flow of air around falling drops of progressively larger radii, R, starting at upper left, then to lower left, then to right. Smaller droplets in the path of each falling drop may collide with it (shaded red) or go around it (shaded blue), depending on the airflow at their locations. *Adapted from Beard (1976).*

Now consider what happens if smaller droplets are placed at various locations within the path of the falling collector drop. Droplets shaded red in Fig. 10.1 will collide with the collector drop because their paths intersect, while others (shaded blue) follow the airflow around the collector drop and never collide with it, even if they were originally immediately below the collector drop. Droplets below the collector drop do not necessarily follow the streamlines exactly. Small droplets, due to their low inertia relative to viscous forces, are more likely to closely follow the streamlines around the collector drop. For larger droplets with greater inertia, this is not necessarily the case, particularly where the streamlines around the collector drop are sharply curved.

In general, a combination of factors such as relative velocities, air resistance, and turbulence can influence the actual path taken by the droplets as they approach the collector drop. These factors can significantly alter the likelihood of drop collisions.

We quantify the frequency of droplet collisions by comparing the geometries of the two droplets and the amount of airspace within which collisions will occur. If the drop streamlines beneath the collector droplets in

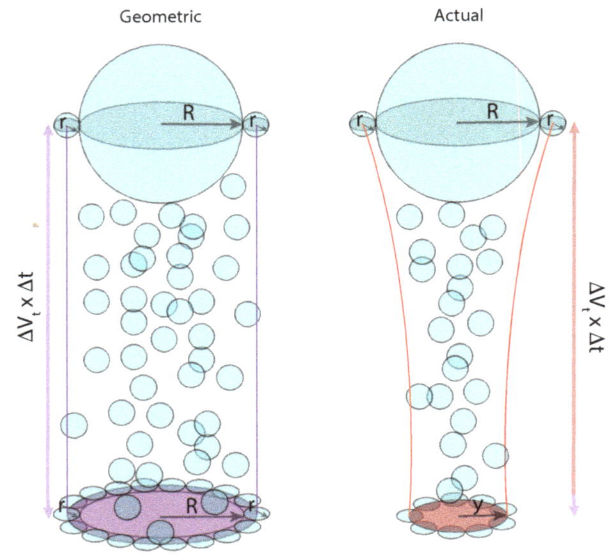

Fig. 10.2: *Left:* Illustration of the idealized geometric sweepout volume (within purple vertical lines) and geometric cross-section (area shaded purple) of a larger collector drop of radius R and smaller droplets of radius r that would be collected by the larger drop. *Right:* Actual sweepout volume (within red vertical curved lines) and actual collision cross-section (shaded red) within which droplets of radius r will collide with the collector drop of radius R, due to the bending of the streamlines around the collector drop. The depth of both sweepout volumes is given by the difference in fall speeds of the collector and smaller drops (ΔV_t) multiplied by some time interval Δt.

Fig. 10.1 *did not bend*, then any smaller droplets in the "shadow" of the collector droplet, i.e., immediately beneath it, as well as those that just graze the collector droplet on the sides, would collide with it (Fig. 10.2, left).

The **geometric cross-section**, represented by the shaded purple circle, is the idealized area beneath the collector drop that allows a collision to occur. It is determined by the geometry of the two drops and is given by $\pi(R+r)^2$, where R is the radius of the collector (larger) drop, and r is the radius of the smaller droplet.

The **sweepout volume** is the maximum volume of the cylinder swept out by a pair of drops of radii R and r as they fall relative to each other. It can be expressed as the geometric cross section multiplied by the difference in the fall speeds of the two drops ($\Delta V_t = V_{tR} - V_{tr}$):

$$V_{\text{swept}} = \pi\,(R+r)^2\,\Delta V_t\,, \tag{10.1}$$

where V_{tR} is the fall speed of the collector drop and V_{tr} is the fall speed of the smaller droplet. The sweepout volume V_{swept} has units of volume per unit time. The depth of this volume is simply the difference in the drop fall speeds multiplied by some time interval Δt. Note that if $R = r$, i.e., if the two drops are identical in size, their fall speeds will be identical ($\Delta V_t = 0$), and the drops *will never collide* because one can never "catch up" with the other.

It is also important to note that the sweepout volume depends only on the *difference* in the fall speeds of a pair of drops. Thus, it is *not necessary* for either of the drops to be descending through the cloud. Drop collisions can occur even if both drops are ascending within an updraft! The larger collector drop ascends more slowly than the smaller droplet because each drop ascends at a rate equivalent to the updraft speed minus its fall speed. Assuming the updraft speed carrying both drops is the same, the smaller droplet can catch up to the larger drop and potentially collide with it.

However, as suggested by Fig. 10.1, some of the smaller droplets in this sweepout volume will actually *be carried around the larger drop* rather than collide with it, even though they were originally in its path, due to the curvature of some of the streamlines around the collector drop.

So, we define a new quantity called the **collision efficiency**, equal to the ratio of the *actual area* of the cross-section beneath the collector drop within which droplets will collide with it (red circle of radius y beneath the collector drop shown on the right in Fig. 10.2) to the idealized geometrical cross-section (purple circle beneath the collector drop shown on the left), i.e.,

$$E_c = \frac{\pi\, y^2}{\pi\, (R + r)^2} = \frac{y^2}{(R + r)^2} \tag{10.2}$$

where R is the radius of the collector drop, r is the radius of the smaller droplet, and y is the radius of the circle within which all drops of size r will collide with the collector drop of size R. Note that y itself is a function of both R and r: the streamlines around the collector drop depend on its size, and the degree to which the smaller droplets follow those streamlines depends on their size. Thus, for any given pair of drop sizes R and r, a unique value of the collision efficiency E_c exists.

The collision efficiency E_c represents the fraction of smaller droplets in the immediate path of a falling collector drop that actually collide with it. If the collision efficiency is much less than one, few droplets in the path of the collector drop will collide with it, resulting in less frequent collisions and slower growth of the collector drop. Conversely, if the collision efficiency is close to one, most droplets in the path of the collector drop will collide with it, leading to more frequent collisions and potentially faster growth of the collector drop.

Based on a combination of laboratory measurements and theoretical calculations, the collision efficiency E_c has been determined for most combinations of drop sizes. These values are plotted in the left panel of

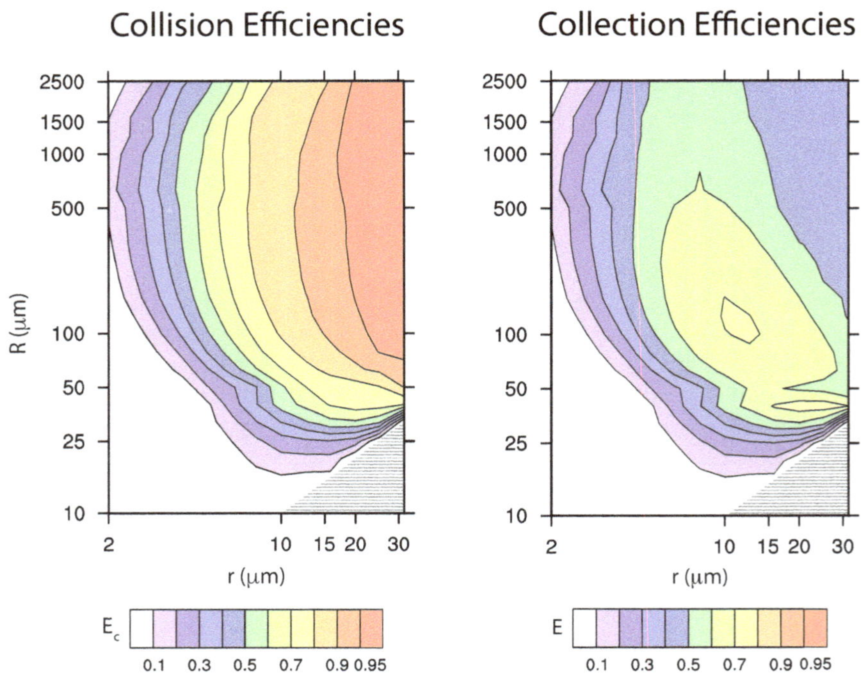

Fig. 10.3: *Left:* Collision efficiencies E_c as a function of collected droplet radius r and collector droplet radius R based on laboratory studies of Klett and Davis (1973) and Beard and Ochs (1984). *Right:* Collection efficiencies E that include the laboratory results of coalescence efficiency by Beard and Ochs (1984). Hatched gray triangles indicate regions where $R < r$.

Fig. 10.3, contoured with values between 0.1 and 0.95. They reveal the following general behavior:

1. For a given size of the smaller droplet, as the radius of the collector drop increases up to a few hundred micrometers (i.e., moving upward along the lower half of the plot), E_c also generally increases.
2. For a given size of the collector drop, as the smaller droplet size increases (i.e., moving toward the right side of the plot), E_c also generally increases.
3. Similar-sized drops do not collide, i.e., $E_c = 0$ when $R = r$ (hatched gray triangle in the lower-right corner of the plot).
4. E_c values are very low (less than 0.1) for collector drops less than $\sim$20 μm in radius.

From this analysis, we see that larger sizes are required for both the collector and the smaller drops to have frequent drop collisions. Thus, we expect the frequency of collisions to increase greatly over time in the cloud, as the collector drops grow by collecting smaller droplets, or as the smaller droplets grow by condensation to sizes where they collide effectively with larger drops.

10.1.2 Coalescence efficiency

Having addressed the frequency of drop collisions, we must now consider the second question posed at the beginning of this section: the likelihood that the drops will stay together, or **coalesce**, once they have collided. Three outcomes are possible: (i) the drops will collide and coalesce; (ii) the drops will collide but then rebound; or (iii) the drops will collide, momentarily coalesce, but then shear apart.

The first outcome (coalescence) is easy to understand: the two drops make full contact and merge, and surface tension then ensures that they remain together despite any residual kinetic energy from the collision. The second outcome (rebound) is less obvious and is believed to occur when a thin barrier of air prevents the two drops from making full contact. Finally, the third possible outcome (momentary joining followed by breakup) results when surface tension is insufficient to absorb the excess kinetic energy remaining from the collision.

To consider these possible outcomes in our calculations, we define another ratio, the **coalescence efficiency**, as the fraction of collisions that result in coalescence. In this textbook, we will use the symbol E_c' to denote the coalescence efficiency.

Laboratory studies show that the coalescence efficiency *is always less than 1.0* but peaks near 0.9 for collector drops of 100–300 μm radius and smaller collected droplets of 10–15 μm radius. Collisions involving smaller or larger collector drops and/or smaller or larger collected droplets are often followed by breakup, sometimes more than 50% of the time ($E_c' < 0.5$). For collector drops less than 50 μm radius, measurements are limited, so in practice, these collisions are typically assumed to result in perfect coalescence ($E_c' = 1$)."

10.1.3 Collection efficiency

To simplify calculations, the collision efficiency E_c and the coalescence efficiency E_c' are combined into a single **collection efficiency**:

$$E \equiv E_c \cdot E'_c. \tag{10.3}$$

Both E_c and E_c' are assumed to be functions only of the sizes of the collector and collected drops, so the same is true for the collection efficiency E.

The right panel of Fig. 10.3 summarizes collection efficiencies compiled from laboratory studies. Maximum values of E sometimes exceed 0.7 but are always less than 0.8; the greatest values occur near $R = 100\ \mu$m and r=10 μm. As stated earlier, these values are for drops colliding in still air or in uniform flow fields. They do not account for the effects of small-scale turbulent motions inside a cloud, which we will briefly consider later.

10.2 Mathematical Models of Drop Collection

How quickly can cloud drops grow to produce raindrops? Different mathematical models have been developed to address this question, as outlined in the following sections.

10.2.1 Collection kernel

The **collection kernel** K is a quantity used in various drop collection models. Its purpose is to transform the geometric volume of air swept out per unit time (V_{swept}) into a reduced volume per unit time that accounts for collection efficiency $E < 1$. In other words, we can use K instead of V_{swept} when calculating the rate of collection by a collector drop without separately accounting for variations in E, because both E and V_{swept} are functions of the radii of the collector and collected drops. The collection kernel is thus given by

$$K = V_{\text{swept}}E = \pi(R+r)^2 \Delta V_t E\ , \tag{10.4}$$

where R is the collector drop radius, r is the smaller droplet radius, ΔV_t is the difference in their fall speeds, and E is their collection efficiency, which is also a function of R and r. The units of K are m^3 s^{-1}, as it represents the volume of air (per unit time) containing droplets collected by the larger drop.

A unique K exists for each drop pair with radii R and r, due to its dependence on the geometric cross section $\pi(R + r)^2$ as well as on E and ΔV_t, which are also functions of R and r. Thus, precalculated values (often referred to as "lookup tables") of K can be generated and utilized by numerical models to expedite collision-coalescence calculations.

10.2.2 Continuous collection model

The simplest mathematical model of drop collections is called the **continuous collection model**. It approximates the collected cloud droplets as bulk "cloud water." In other words, the collected cloud water is represented as a smooth distribution in space rather than as discrete, individual droplets.

The concept behind this model is shown in Fig. 10.4: it assumes that the collected droplets are uniformly distributed in space like a continuous volume of water, simply represented by a cloud water content L (in g m^{-3})

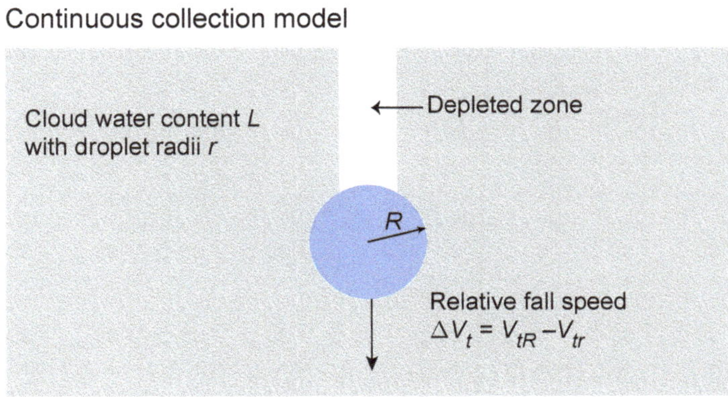

Fig. 10.4: Conceptual drawing of the continuous collection model. The much larger collector drop of radius R falls at a faster fall speed V_{tR}, and the much more numerous and much smaller cloud droplets of radius r fall at a much slower fall speed V_{tr}. The radius of the depleted cloud water zone in this idealized model is determined not by the collector drop radius R, but rather by the collection kernel K defined by (10.4).

composed of droplets with some assumed average radius r. The equation governing this model is then

$$\frac{\mathrm{d}M}{\mathrm{d}t} = KL, \tag{10.5}$$

where M is the mass of a collector drop and K is given by (10.4). Note that K depends upon the difference in the fall speeds of the collector and collected drops, the cross-sectional area of the collector drop, and the collection efficiency. Thus, as the collector drop grows larger, its rate of growth increases sharply with time.

By making some simplifying approximations, we can find an expression for the functional dependence of the collector drop radius on time, $R(t)$. Starting with the left-hand side of (10.5), we eliminate M using (9.14) as we did in converting (9.13) to (9.17):

$$\frac{\mathrm{d}M}{\mathrm{d}t} = \frac{\mathrm{d}}{\mathrm{d}t}\left[\rho_l \frac{4}{3}\pi R^3\right] = 4\pi\rho_l R^2 \frac{\mathrm{d}R}{\mathrm{d}t}, \tag{10.6}$$

to yield

$$4\pi\rho_l R^2 \frac{\mathrm{d}R}{\mathrm{d}t} = KL\,. \tag{10.7}$$

Expanding K using (10.4), we have

$$4\pi\rho_l R^2 \frac{\mathrm{d}R}{\mathrm{d}t} = \pi(R+r)^2 \Delta V_t E L. \tag{10.8}$$

Canceling like terms on both sides of this equation, assuming $R \gg r$ (so r and V_{tr} can both be neglected) and setting $V_{tR} = kR$ where k is a constant,[2] we find:

$$4\rho_l \frac{\mathrm{d}R}{\mathrm{d}t} = kREL. \tag{10.9}$$

Holding E and L constant, and placing all terms related to R on the left-hand side, we can integrate from some initial drop radius R_0 to $R(t)$ over a time t:

$$\int_{R_0}^{R(t)} \frac{\mathrm{d}R}{R} = \frac{k}{4\rho_l} EL \int_0^t \mathrm{d}t'. \tag{10.10}$$

Performing the integration and simplifying, we find the result

$$R(t) = R_0 \exp\left[\frac{kEL}{4\rho_l} t\right]. \tag{10.11}$$

which shows that the rate of drop growth by continuous collection of water drops *exponentially increases in time*, even with the unrealistic assumption that the collection efficiency is constant as the collector drop grows. We previously saw (Eq. 9.20) that the rate of increase in the droplet radius by diffusion only had a square-root dependence on time, a much slower rate of increase.

This behavior is even more evident from the results of calculations using the continuous collection model with more accurate representations of drop fall speeds and collection efficiencies, as shown in Fig. 10.5. In the top panel of that figure, for collector drops with initial radii of 30 μm or 50 μm falling through a cloud with a constant liquid water content L of 1.5 g m^{-3}, consisting of droplets with 9 μm radii, the growth rate of each collector drop increases exponentially over time. Furthermore, an initially larger collector drop will grow faster than a smaller one due to all the factors within the collection kernel K being larger.

In this example, the speed at which collision-coalescence can produce raindrops is impressive, *much faster* than condensation alone. A drop of 50 μm radius can grow into a raindrop of 1 mm radius in 20 minutes, even when collecting droplets in a modest liquid water content of 1.5 g m^{-3}. It is not uncommon for continental cumulus congestus clouds, for example, to contain nearly three times that value of L before collision-coalescence begins, in which case the collector drop would grow three times as fast. Under the same conditions, a 30 μm radius drop requires slightly longer,

[2]Note that the assumption that $V_{tR} = kR$ only holds within the size range of larger cloud drops and drizzle drops, per (4.14).

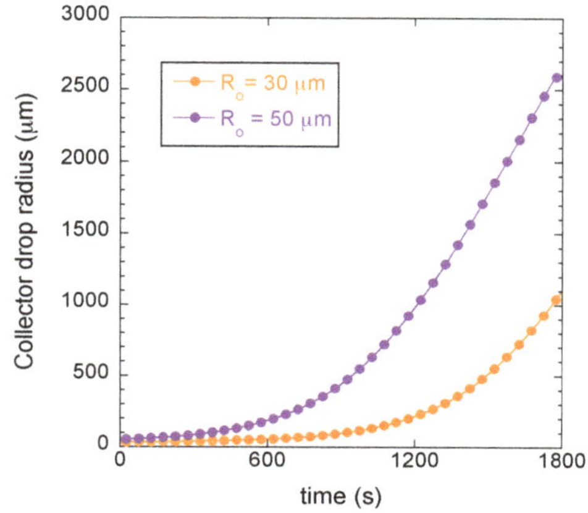

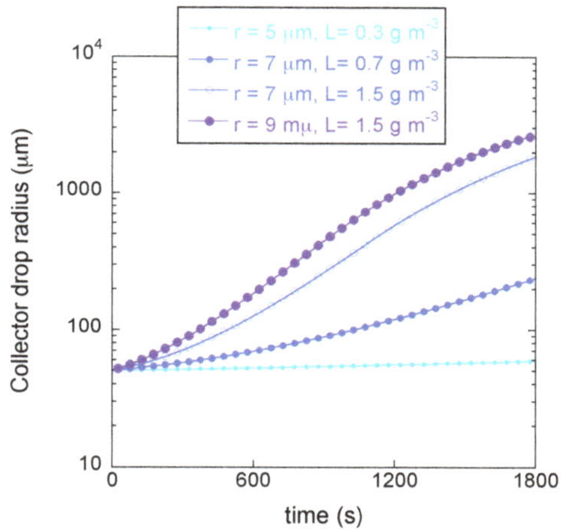

Fig. 10.5: Sample calculations performed with the continuous collection model as given by (10.5). *Top:* Growth of two different collector drops with an initial size of 30 μm (orange) or 50 μm (purple), falling through a cloud water content of $L = 1.5\ g\ m^{-3}$ that consists of droplets 9 μm in radius. *Bottom:* Growth of an initial 50 μm collector drop falling through a liquid water content L consisting of droplets of radius r, as labeled. Note the log scale on the y-axis.

about 30 minutes, to grow to 1 mm radius due to the dependence of K on the collector drop size.

In the bottom panel of Fig. 10.5, we can see the significant influence of the liquid water content L on the continuous collection growth rate for a 50 μm collector drop. In these calculations, L is varied from 1.5 to 0.3 g m^{-3}. The collector drop grows much faster in a region of higher L, and much slower in a region of lower L.

The rate of growth is also dependent on the size of the smaller collected droplets, as shown when r is increased from 7 to 9 μm in radius while holding L constant. When r is smaller, the geometric cross-section is also smaller, resulting in fewer collisions. The collection efficiency is also reduced when the collected droplets are smaller (Fig. 10.3). Both factors decrease the rate of growth of the collector drop with decreasing collected droplet size, although this effect is less significant than that of changing the overall liquid water content L.

The applicability of this simple model is limited, however. Due to the assumption that the collected cloud water is distributed uniformly in space, *any collector drop of size R (or mass M) will grow at exactly the same rate.* Among other issues, this eliminates the possibility that drops of initially similar size could give rise to a more realistic, broader distribution of sizes after a period of growth.

In reality, collisions between drops are random, discrete occurrences. Through sheer chance, some drops will experience more collisions than average, and therefore more growth, while others experience fewer collisions than average, and therefore less growth. The continuous collection model cannot reproduce these real-world variations.

In fact, tests with this model show that it is only a good approximation of drop growth by collision-coalescence when $R \gg r$ and when the number concentration of collector drops N is a small fraction of the number concentration of collected cloud droplets n (i.e., $N \ll n$). These conditions are valid in some situations, but not always. When they do not hold, the continuous collection model underestimates the rate of drop growth, yielding an "average" value that is insufficient in many cases to make accurate predictions of the onset of rain, in part due to its inability to predict a realistic increase in the variety of collector drop sizes over time.

10.2.3 Quasi-stochastic collection model

Due to the limited applicability of the continuous collection model, more sophisticated mathematical models of collision-coalescence, called **quasi-stochastic collection models**, are often required. Quasi-stochastic models improve upon the continuous collection model in several ways:

- They consider droplets as discrete entities, so collection of a fraction of a droplet is not possible.
- They allow for non-uniform spacing of the droplets in the cloud.
- They permit collection of all possible sizes of drops.

We consider these three different aspects in the following sections.

i) Discrete droplet collection

Let us first consider a simple, albeit unrealistic, cloud where we have a set of collector drops with initial mass M_0 that collect smaller droplets, each having mass m. The smaller droplets are uniformly distributed within the cloud.

Let's rewrite the right-hand side of the continuous collection equation to give

$$\frac{\mathrm{d}M}{\mathrm{d}t} = KL = Kmn, \tag{10.12}$$

where K is still the same collection kernel given by (10.5) and n is now the number concentration of the smaller droplets of mass m in the cloud.

Suppose that, given the sizes of the collector drops and the smaller droplets in our hypothetical cloud, the quantity $Kn\Delta t = 0.5$. For the continuous collection model, this quantity is interpreted as the number of smaller droplets collected by the larger drop over each time interval Δt. Each collector drop would thus acquire the mass of 0.5 collected droplets over that interval.

The continuous collection model ignores the fact that, in reality, either an entire droplet is collected or it is not. For this model, the quantity $Kn\Delta t$ represents an average rate of growth over all the collector drops. Consequently, all collector drops starting with the same mass M_0 will grow at exactly the same rate, as shown by the brown spheres in Fig. 10.6.

A different result is obtained when we allow only *whole* droplets to be collected by the larger collector drop. For now, we'll refer to this as the "discrete model," represented by the purple spheres. In this model, $Kn\Delta t$ is interpreted as the *fraction of the number of collector drops* that will collect one of the smaller droplets. It then follows that for a value of 0.5, half of the collector drops must collect an entire droplet, and the other half must collect none, for each time interval Δt.

As shown in the second row of Fig. 10.6, after one Δt, the discrete model produces some variability in the sizes of the collector drops: half of them will have a new mass of $M_0 + m$, while the other half will not have grown at all and will still have mass M_0. After another Δt (third row), the process repeats, and the variability in collector drop sizes increases.

ii) Non-uniform drop spacing

We now add another specification to our simple, hypothetical cloud by assuming that the small droplets are randomly spaced rather than uniformly spaced. Measurements in clouds have verified that drops can have great variability in their spacing, being clustered together in some locations or spaced much farther apart in others. This variability has been documented

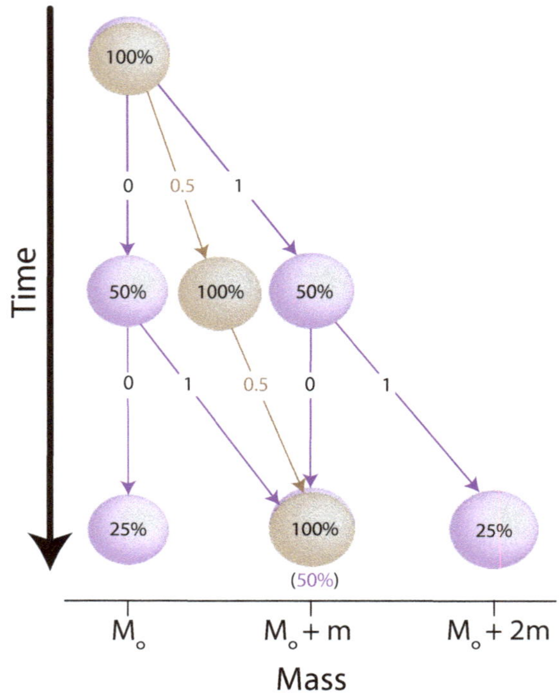

Fig. 10.6: A comparison of the continuous collection model (brown spheres) and a discrete model (purple spheres) after two time steps. Both models start at the upper left with 100% of the collector drops having a mass of M_0. The smaller collected droplets each have a mass of m. The average number of drop collections per time step is 0.5. Each sphere represents the number of the original collector drops that have a given size after each time step as a result of collecting the number of droplets labeled along each downward arrow. *After Young (1975).*

down to length scales of a few millimeters within some clouds. The "random nature" of the collection process is now partly represented here: some collector drops can be favorably located so that they will have more of the smaller droplets in their paths to collect compared to other collector drops, even those of the same size.

Figure 10.7 shows that the quasi-stochastic model accelerates the growth of some collector drops even more than the discrete model examined earlier in Fig. 10.6, due to its consideration of *both* the discreteness of drop collection and the random spacing of the smaller droplets in the cloud. The results of the continuous collection model have not changed, because it cannot consider variability in the spacing of the collected droplets.

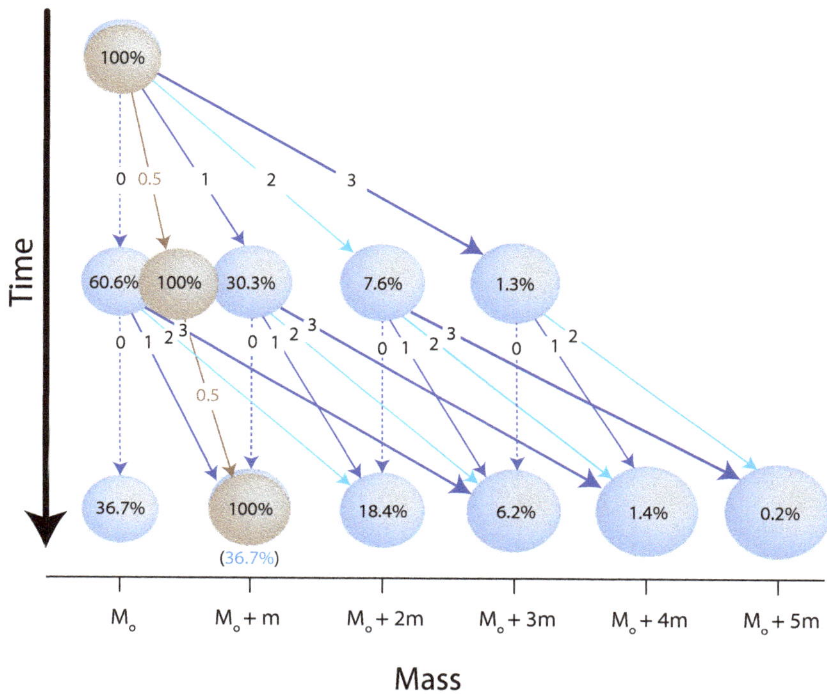

Fig. 10.7: A comparison of the continuous collection model (brown spheres) and a quasi-stochastic model (blue spheres) over time $2\Delta t$. Both models start at the upper left with 100% of the collector drops having an initial mass of M_0 and the collected droplets (not shown) having a mass of m. In this example, the average number of droplet collections per Δt is 0.5. Arrows are labeled by the number of droplets collected over each Δt. The quasi-stochastic model assumes the smaller droplets are spaced randomly according to a Poisson distribution. Collection events involving fewer than 0.1% of the collector drops are not shown. *After Young (1975).*

In this case, the quasi-stochastic model predicts that some of the collector drops will collect *no* droplets over one Δt (we might call these drops "unlucky"), while some others will collect one whole droplet, two whole droplets, or even be so "lucky" as to collect three whole droplets.[3] Thus,

[3]Here the spatial distribution of the smaller droplets is assumed to follow a *Poisson probability distribution.* As a result of our earlier assumption that the average number of smaller droplets collected by each collector drop is equivalent to 0.5 per Δt, a Poisson distribution then dictates that 60.6% of the collector drops will collect no droplets, 30.3% will collect one droplet, 7.6% will collect two droplets, and 1.3% will collect three droplets per Δt.

the variability in collector drop sizes increases in time even faster than for the discrete model, producing a few collector drops that have increased their mass by $3m$ after a single Δt, and some by $5m$ after two Δt.

It is important to understand that the total water mass is still conserved for both sets of calculations shown in Fig. 10.7. This is true because the quasi-stochastic model predicts two contrasting outcomes: while over 26% of the collector drops have collected more droplets than the continuous collection model, more than 36% of the collector drops remain "very unlucky" and collect no droplets at all. However, *to correctly predict the onset of rainfall, we must take into account the fastest rate at which growth by collision-coalescence might occur.*

iii) The stochastic collection equation

The final advantage of quasi-stochastic models over the continuous collection model is that they permit collections among all different sizes of drops, not just those designated as "collector drops." We have seen that for a quasi-stochastic model, drops originally of the same size can grow by collision-coalescence at different rates, so tracking the number of drops of a given size is essential.

Consider an idealized cloud that contains drops with masses that are multiples of some unit mass m. The drop population has been placed into bins of increasing mass numbered from 1 to J, and n_j denotes the number concentration of drops having a mass of j times m. The **stochastic collection equation** (SCE) that considers all possible drop collections for this idealized cloud can then be expressed as:

$$\frac{\mathrm{d}n_j}{\mathrm{d}t} = \frac{1}{2}\sum_{i=1}^{j-1} n_i n_{j-i} K_{i,j-i} - n_j \sum_{i=1}^{J} n_i K_{i,j} \tag{10.13}$$

where K is the collection kernel for the two drop bins given in the subscript, equivalent to (10.4).

The first term on the right-hand side of this equation represents an *increase* in the number of drops in bin j due to the collision-coalescence of drop pairs smaller than those in bin j, easily identifiable in our idealized cloud by bin subscripts that add up to equal j. The second term on the right-hand side represents the *decrease* in the number of drops in bin j due to their collection with any other drop, which would then move them to a bin of greater mass.

iv) Example calculation using the SCE

As an example, let us examine the growth of drops into and out of one size bin within the quasi-stochastic model, shown for illustration as the blue bin in Fig. 10.8. Suppose $j = 7$ in (10.13). According to this equation, we first add up the collections between all possible pairings of bins 1 through 6. However, we need to multiply that number of drops by 1/2 to avoid double-counting; i.e., collections between bins 1 and 6, 2 and 5, and 3 and 4 are counted twice due to the indexing. Finally, we subtract any drops from bin 7 that have collected drops from other bins, as they are now too large to remain in bin 7.

Although only one step is shown in Fig. 10.8 for simplicity, the algorithm actually starts with $j = 1$ and proceeds until $j = J$ to calculate the new number concentration of drops n_j for all bins. In reality, the bins are more finely spaced than shown (often distributed logarithmically in mass) and extend to much larger values as the drops grow over time.[4]

It is not immediately obvious from (10.13) how the assumption regarding the spatial distribution of the collected drops is incorporated into the SCE. As an illustration, let us examine the second term on the right-hand side. The quantity

$$n_j \sum_{i=1}^{J} n_i K_{i,j} \tag{10.14}$$

represents (per unit time) drops in bin j collecting drops from any of the other bins. Suppose that for a given bin i, the quantity $n_i K_{i,j} \Delta t = 0.5$. If, for example, the drops in bin i are assumed to be randomly distributed in space according to a Poisson distribution, then a value of 0.5 sets the different probabilities within the Poisson distribution for the collection of a certain number of drops. The fraction $F(x)$ of the collector drops that would collect x drops from bin i is then given by

$$F(x) = \frac{0.5^x}{x!} e^{-0.5}. \tag{10.15}$$

Inserting values of x as 0, 1, or 2, for example, yields values of $F(x)$ as 0.606, 0.303, and 0.076, respectively. Multiplying these fractions by n_j in (10.14) means that 60.65% of the drops in bin j would collect no drops from bin i, 30.3% would collect one drop from bin i, 7.6% would collect two drops from bin i, and so on. The assumption that the droplets are spaced according to a Poisson distribution is not built into the SCE, allowing for other assumptions about the spatial distribution of the drops to be made and implemented.

From the SCE, we find that a population of cloud drops can develop into raindrops quite rapidly once collision-coalescence begins in earnest.

---

[4]The algorithms for determining the placement of drops into new bins as they grow to larger masses, ensuring the conservation of water mass, preventing numerical errors, and maintaining computational efficiency are too complex to be detailed here.

$$\frac{dn_j}{dt} = \frac{1}{2}\sum_{i=1}^{j-1} n_i n_{j-i} K_{i,j-i} - n_j \sum_{i=1}^{J} n_i K_{i,j}$$

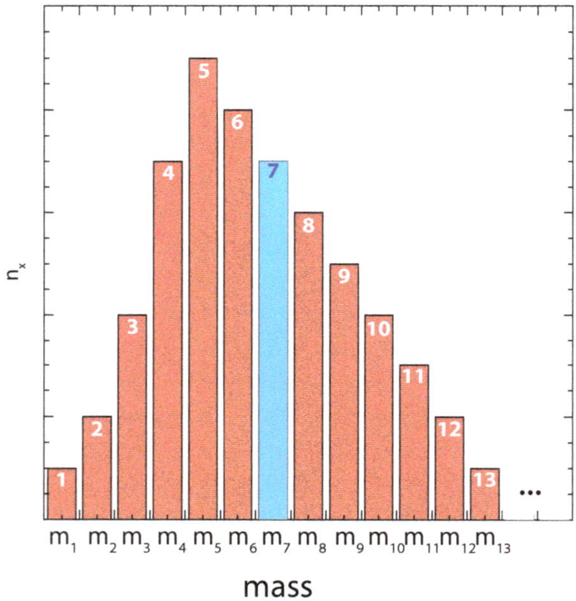

dn_j/dt= 1/2 x collections between bins i, j-i
- collections between bin j and all other bins

j=7: dn_7/dt= 1/2 x collections between [(bins 1,6)+ (bins 2,5)+ (bins 3,4)+
(bins 4,3)+ (bins 5,2)+ (bins 6,1)]

- collections between [(bins 7,1) + (bins 7,2) + (bins 7,3) +
(bins 7,4) + (bins 7,5) + (bins 7,6) +
(bins 7,7) + (bins 7,8) + (bins 7,9) +
(bins 7,10)+ (bins 7,11)+ (bins 7,12)+
(bins 7,13)+ ...]

Fig. 10.8: An illustration of how (10.13) is used with a distribution of drops binned according to their mass. The bin numbers (white labels) correspond to the indices of i, j, or $j - i$. The ordinate n_x is the number of drops in each bin. This example shows the algorithm when the blue bin shown in the distribution ($i = 7$) is considered. Collections among drops smaller than bin 7 may increase the number in bin 7, while collections among drops in bin 7 with any other drops would produce larger drops than those in bin 7.

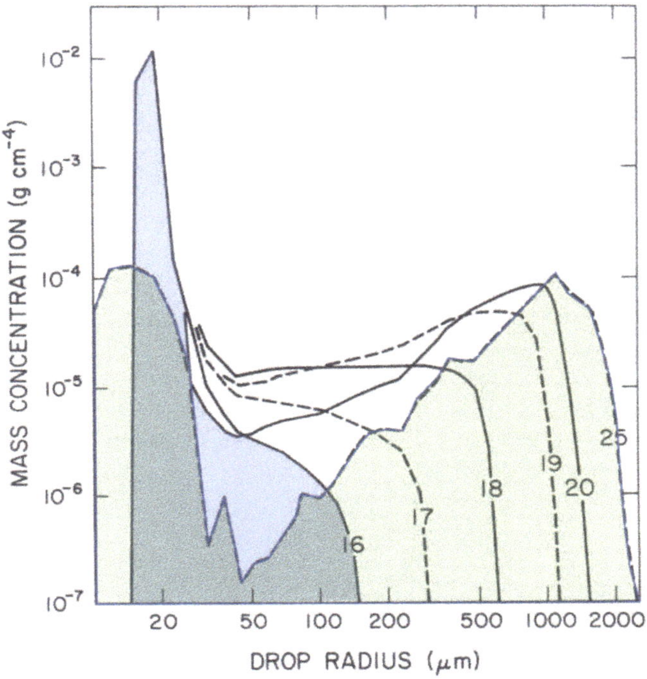

Fig. 10.9: The evolution of a drop population (shown by mass concentration rather than by number concentration on vertical axis) predicted by quasi-stochastic collision-coalescence in a maritime cloud parcel experiencing a 3 m s^{-1} updraft. Numbered labels denote time in minutes. By 25 minutes, some of the largest raindrops break up into smaller drops, expanding the drop size distribution toward smaller sizes, as discussed in Sec. 10.4. *Adapted from Young (1993). © Oxford Publishing Limited. Reproduced with permission of the Licensor through PLSclear.*

In Fig. 10.9, two different times are shown from the results of a simplified cloud model that tracks the growth of a population of drops. Initially, the model grows the droplets by condensation, followed by quasi-stochastic collision-coalescence. At 16 minutes into the computations, the liquid water mass distribution is still dominated by cloud droplets (shaded in blue). However, after less than 10 minutes of additional collision-coalescence, the mass distribution is dominated by raindrops (shaded in green).

This, then, is how raindrops are produced within the typically short lifetime of a warm cloud: cloud droplets first grow by condensation, and once they are large enough, collision-coalescence rapidly transforms them into raindrops.

Just one more complication in this series of events remains to be addressed. After all, many warm clouds never produce raindrops that precipitate to the ground because the conditions for collision-coalescence to be initiated are never attained.

10.3 The Onset of Collision-Coalescence

For collision-coalescence to begin, a variety of drop sizes is necessary so that faster-falling larger drops can capture slower-falling smaller droplets. In other words, a *broad* drop size distribution is required to initiate collision-coalescence. Additionally, collection efficiency values (Fig. 10.3) suggest that some drops must first surpass a threshold of about 25 μm radius in order to start collecting other drops with sufficient frequency.

However, in Sec. 9.1.2, we saw that the size distribution of a droplet population growing by condensation alone *narrows* with time. Specifically, as the smaller droplets increase their radii more quickly by condensation than the larger ones, the spread in droplet sizes—and thus fall speeds—quickly decreases. If diffusional growth leads to similar sizes for all droplets and if it often takes too long for any of them to grow to 25 μm by condensation initially, how can collision-coalescence even get started in a cloud? We examine multiple possibilities in this section.

10.3.1 The advantage of maritime clouds

As we have discussed multiple times in this book, air masses over oceans typically contain far fewer aerosol particles, and thus fewer CCN, than over land. This results in maritime clouds containing fewer cloud droplets and thus higher supersaturations than continental clouds, if considering similar updraft speeds. Therefore, cloud droplets growing by condensation tend to be larger in maritime clouds than in continental clouds. In fact, many maritime clouds have no difficulty in producing the drop sizes needed to initiate collision-coalescence (exceeding 25 μm radius) in reasonable amounts of time. For this reason, we can often observe shallow, warm maritime clouds producing some rainfall, albeit in the form of showers (Fig. 10.10).

Continental clouds, particularly those in heavily polluted urban areas, contain far more cloud droplets. As a result, the supersaturation in these clouds is lower (when comparing clouds with similar updraft speeds), and therefore the rate of droplet growth via condensation is slower. These high droplet number concentrations can limit droplet sizes to less than 20 μm in radius, preventing the initiation of collision-coalescence. And yet, the larger drops required to undergo collision-coalescence are sometimes observed in warm continental clouds, or within the warm part of deeper mixed-phase continental clouds. Some possible ways that these large drops might be created are discussed next.

Fig. 10.10: A small, warm cumulus cloud (less than a couple of kilometers deep) over the Caribbean Sea producing precipitation by the warm rain process. *From Rauber et al. (2007).*

10.3.2 Possible broadening mechanisms

Clouds must exhibit broad droplet size distributions for collision-coalescence to begin in earnest. The left panel of Fig. 10.11 illustrates that when using the diffusional growth equation in an adiabatic parcel model, the *predicted droplet size distribution is very narrow*. In a real cloud, however, as depicted in the right panel, the *measured droplet size distribution is always much broader*, even in continental clouds where it might not yet be wide enough to initiate collision-coalescence.

What is occurring in real clouds that our calculations of diffusional growth are missing? Several candidate mechanisms have been identified. One or more is probably active in any given cloud but may not be entirely successful in creating the drop sizes and variability required for the onset of collision-coalescence. The relative importance of these mechanisms remains uncertain but quite likely varies widely from cloud to cloud. In the following sections, we briefly highlight these possible mechanisms for broadening cloud droplet size distributions, which are summarized schematically in Fig. 10.12.

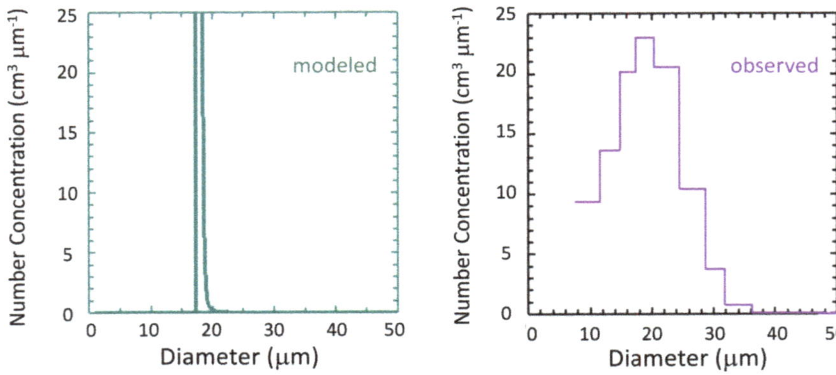

Fig. 10.11: *Left*: A droplet size distribution calculated using an adiabatic parcel model that simulates diffusional growth as the parcel rises to just over 1 km above the cloud base. *Right*: An observed droplet size distribution in the updraft of a small cumulus cloud at a similar distance above the cloud base. Both the model and the observations had the same cloud base conditions.

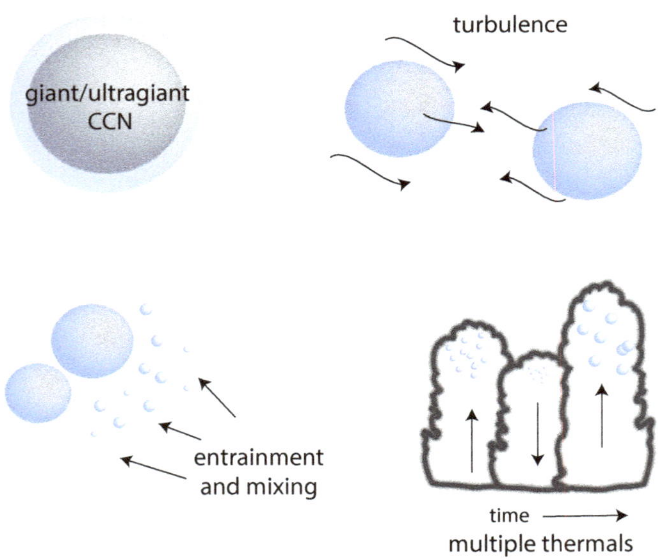

Fig. 10.12: Summary of common mechanisms for broadening the drop size distribution in a cloud to initiate collision-coalescence: giant/ultragiant CCN (top left); enhanced collisions due to turbulence (top right); inhomogeneous mixing after entrainment (bottom left); and successive thermals (bottom right).

i) Giant and ultragiant CCN

In Sec. 8.1, we saw that some aerosol particles can be quite large, greater than 1 μm in radius (*giant*), or sometimes even greater than 10 μm in radius (*ultragiant*). CCN of these sizes nucleate larger droplets, having a "head start" compared to droplets formed on smaller CCN. These CCN could explain the larger droplets sometimes measured in clouds, including those that could attain or exceed a 25 μm radius to initiate collision-coalescence. Indeed, some ultragiant aerosol particles are already larger than this size when first introduced into the cloud!

However, recall that the largest CCN are also the fewest in number. In certain cases, such as in polluted clouds over some cities, or clouds over the ocean where sea spray on a windy day may loft many large salt particles into the air, there might be enough giant/ultragiant CCN to account for the larger drops observed, including those necessary to initiate collision-coalescence in the cloud. However, in most other scenarios, studies have suggested that these giant/ultragiant CCN are simply too rare to have a significant impact.

ii) Entrainment and mixing

The predicted drop size distribution shown in Fig. 10.11 was produced using the diffusional growth equation within an *adiabatic* parcel. However, as discussed in Chapter 3, many clouds do not contain purely adiabatic parcels. Instead, they contain parcels that have been partially diluted by environmental air entrained during their ascent.

Specifically, entrainment introduces drier air into the cloud, and once mixed with the cloudy air, the thermodynamic properties of a cloudy parcel are altered by reducing both its temperature and its liquid water content. Importantly, not all ascending parcels in a cloud are equally affected by entrainment.

Entrainment and subsequent mixing of the entrained air are thus important sources of *variability* within the cloud. They can reasonably cause variability in droplet sizes by producing different "pockets" within the cloud with different temperatures and supersaturations. After all, when air from outside the cloud is mixed into its interior, it can

- dilute the overall droplet concentration,
- shrink droplets by evaporation,
- completely evaporate some droplets, and/or
- nucleate new (and thus smaller) droplets by introducing new CCN into the cloud.

The degree to which any of the effects listed above can be realized depends on how the entrained air mixes with the cloudy air. Initially, the entrained air undergoes *inhomogeneous mixing*, meaning that the overturning of large eddies causes gradients in relative humidity (or supersaturation) within the cloud in clearly separated zones. However, as these large eddies break down into smaller and smaller eddies, *homogeneous mixing* dissipates the gradients, and all the cloud droplets in that mixed region of the cloud experience the same decreased supersaturation or even subsaturation.

The ramifications of this mixing process for cloud droplet sizes are illustrated in Fig. 10.13. When the entrained air is first mixed inhomogeneously with the cloudy air, the cloud droplets encountering the dry air first will completely evaporate (the drops evaporate faster than the time required to fully mix in the entrained air), while other cloud droplets farther from the dry air will remain unaffected.

Thus, inhomogeneous mixing causes "pockets" within the cloud containing very few droplets. Model calculations suggest that these droplets can then grow faster by condensation due to the now higher supersaturation, especially if they re-enter the updraft, experiencing a much higher supersaturation than before the entrainment occurred. In this way, *inhomogeneous mixing can create larger cloud droplet sizes than would otherwise be possible in an adiabatic cloudy parcel.* Although not shown in Fig. 10.13, some new CCN are also introduced into the cloud by entrainment, which may nucleate additional cloud droplets, thereby extending the droplet size distributions toward smaller sizes as well. This broadening of the droplet size distribution by entrainment and inhomogeneous mixing is thought to be common in warm cumuli across all environments.

In time, homogeneous mixing tends to shrink the entire droplet population toward smaller sizes (Fig. 10.13). By this stage, the gradients due to the entrained air have mostly dissipated, so all cloud droplets in the mixture experience the same reduced supersaturation. The homogeneous stage of mixing does broaden the droplet size distribution, but only toward smaller sizes, making it of limited use in initiating collision-coalescence.

iii) Cloud turbulence

Another mechanism for broadening the droplet size distribution and potentially initiating collision-coalescence is small-scale turbulence, which we have ignored up to this point. Recall that the streamlines shown in Fig. 10.1, as well as the collection efficiencies depicted in Fig. 10.3, were considered to be valid for still air or within a uniform updraft or downdraft lacking turbulence. However, one could easily imagine that small-

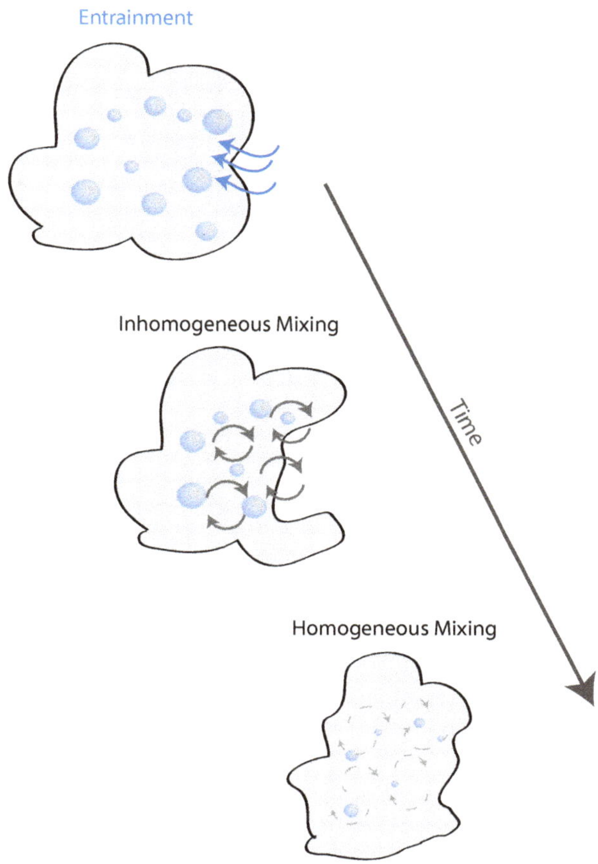

Fig. 10.13: Schematic of a cloudy parcel entraining dry air and mixing it internally as a function of time. After the air has been entrained into the cloud (top, blue arrows), inhomogeneous mixing (center, gray arrows) evaporates the droplets that first encounter the dry air, while the remaining cloud droplets farther away are unchanged. In time, the mixing becomes more homogeneous (bottom, dashed gray arrows), creating more uniform conditions across the cloudy parcel, so all the remaining droplets are partially evaporated (and if the air is subsaturated, they may completely evaporate).

scale turbulence, giving rise to varying cloud velocities at the droplet scale, might alter numerous aspects of the collision-coalescence process.

Specifically, turbulence can enhance the collision-coalescence process in three distinct ways:

1. Turbulence changes the relative velocities between pairs of droplets, enabling collisions that would otherwise be unlikely.

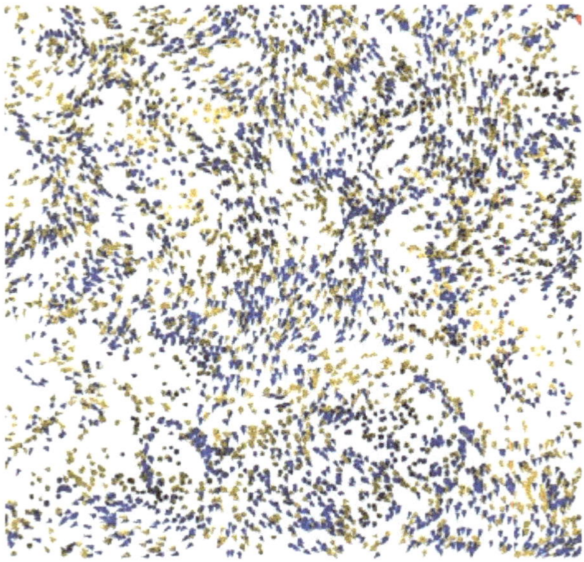

Fig. 10.14: Example of the spacing of droplets 20 μm in radius (gold) and droplets 10 μm in radius (blue) in a turbulent wind field using DNS. Symbols showing positions of droplets are not scaled to droplet size. Turbulence within the model (designed to mimic real clouds) produces tiny eddies that spin and centrifuge the droplets outward, causing a higher concentration of droplets in some regions. The model domain shown here is 16.77 cm on each side. *From Ayala et al. (2008). © Deutsche Physikalische Gemeinde. Used with permission.*

2. Turbulence causes droplets to cluster in specific regions of the cloud through a phenomenon known as *preferential concentration*. This creates small areas with extremely high droplet number concentrations (as shown in Fig. 10.14), significantly increasing the likelihood of collisions.

3. Turbulence alters the collision efficiencies between droplet pairs through hydrodynamic interactions.

All three of these changes can result in more frequent droplet collisions than would otherwise occur in still air. Small-scale turbulence can even cause droplets of identical size to collide if it "pushes" one droplet toward another.

Via any of these effects, turbulence can effectively lower the barrier to initiating collision-coalescence in clouds. Specifically, it can eliminate the necessity for some droplets to have a radius larger than 25 μm, or for the droplet size distribution to be particularly broad. However, the turbulence must be at sufficiently high levels to produce these effects, and many types of clouds contain little or no turbulence.

Research on the effects of turbulence is ongoing. Both laboratory experiments and specialized microscale models that incorporate turbulence, known as *direct numerical simulation* (DNS), have produced noteworthy findings. For example, in moderately turbulent clouds, the *turbulent collection kernel* can be up to three times greater than the conventional collection kernel (10.4) that does not account for turbulence. This enhancement is especially pronounced for certain droplet pairs: either when the droplets are nearly the same size, or when the collected droplets are less than 20% of the size of the collector drop.

Thus, cloud turbulence appears to be a likely candidate for promoting the initiation of collision-coalescence, particularly in clouds with stronger turbulence, such as deeper, more vigorous cumulus clouds.

iv) Successive thermals

Some clouds, especially cumulus, tend to "pulse" as new cloud turrets ascend near or emerge from beneath older, dying turrets, as shown schematically in Fig. 1.3. These older turrets may not completely evaporate before the new turrets emerge through them. Thus, old turrets can leave behind some remnant droplets, which, when engulfed by a new, stronger emerging cloud turret, enjoy a significant "head start" on their diffusional growth. This broadens droplet size distributions in the new turret and possibly produces droplets with sizes capable of initiating collision-coalescence. Although it is difficult to demonstrate this effect with aircraft observations, numerical modeling has suggested that this mechanism can lead to the broadening of the droplet size distribution toward larger sizes.

v) Surfactant effects

Certain CCN contain chemicals that, when liquefied by droplet nucleation, produce a surface coating over the droplet called a *surfactant*. These coatings can significantly slow diffusional growth temporarily. However, eventually, the surfactant breaks apart, potentially leading to sudden, explosive growth of the cloud droplet by condensation. These surfactant effects can alter the normal progression of the narrowing of the cloud droplet size distribution over time. This results in a greater variety in droplet sizes, which is necessary for the onset of collision-coalescence. However, it is unknown how commonly this phenomenon occurs.

vi) Radiation broadening

Clouds with longer lifetimes, particularly stratiform clouds, can experience significant cooling at their tops via the emission of longwave radiation to space. Any enhanced cooling at the cloud top will result in higher supersaturations there, which can accelerate the diffusional growth of some cloud droplets to the sizes needed to initiate collision-coalescence. This effect is not as important in shorter-lived clouds like cumulus, as radiative cooling is a comparatively slow process and does not have as much time to act, but it is likely important in stratocumulus clouds that can produce drizzle.

10.4 Raindrop breakup

Once some raindrops are produced within a cloud, the number of collector drops may also be increased by the process of **raindrop breakup**. Raindrops can break up by two different means: *spontaneous breakup* or *collisional breakup*. In either case, drops smaller than the original raindrop are produced, as shown, for example, in the final mass distribution curve in Fig. 10.9 (shaded green).

Spontaneous breakup occurs when a large raindrop exceeds 4.5 to 5 mm in diameter and becomes hydrodynamically unstable. This instability arises because, for larger raindrop sizes, surface tension is no longer strong enough to oppose the aerodynamic forces that act to distort its shape as it falls. High-speed cameras show evidence of waves building up on the surfaces of these very large raindrops, and these waves can break them up into multiple fragments, limiting the maximum sizes of raindrops we tend to observe in nature.

Collisional breakup, on the other hand, involves more than a single raindrop. It occurs during the collision-coalescence process, typically when two drizzle-sized drops collide. Due to the kinetic energy imparted by the collision, the drizzle drops can each break up, resulting in 1 to 25 additional "satellite drops," as has been observed in laboratory experiments studying these collisions.

In general, raindrop breakup not only limits the largest sizes that raindrops can attain, but also assists further raindrop production within the cloud. The result of raindrop breakup is smaller drops with slower fall speeds than the drizzle or raindrops from which they originated, which may then allow them to ascend within the cloud updraft once again. If they are sufficiently large, they might grow further by collision-coalescence. As a result of raindrop breakup, there can be many more collector drops than the original one or two raindrops that broke apart. This enhanced number of collector drops can then "scavenge" more liquid water from the cloud. Ultimately, raindrop breakup helps bring more water mass from the cloud to the ground as precipitation.

10.5 The Warm Rain Process

Despite the various mechanisms that can broaden the droplet size distribution and generate larger cloud droplets to initiate collision-coalescence, most warm clouds lacking ice still fail to produce rain within their lifetimes. You have undoubtedly walked outside beneath a field of small cumulus humilis overhead without being concerned about needing an umbrella!

Some warm clouds fail to produce rain due to insufficient liquid water content, which can result from their shallow depths, the erosive effects of entrainment, or their short lifetimes. Other clouds may lack active or sufficiently strong mechanisms for producing droplets capable of initiating collision-coalescence.

The series of microphysical processes involving only water drops (no ice) that lead to the production of raindrops is referred to as the **warm rain process**. These processes include (i) droplet nucleation and activation with the assistance of CCN, (ii) their growth by diffusion of water vapor (condensation), (iii) their growth by collision-coalescence to form raindrops, and possibly (iv) raindrop breakup. Figure 10.15 illustrates these key steps. We finish this chapter by first discussing factors that encourage the success of the warm rain process in producing rainfall at ground level, followed by its possible influence on ice processes in mixed-phase clouds.

10.5.1 Importance in warm clouds

It is widely accepted that *rainfall created by the warm rain process is common in warm maritime clouds due to their low droplet number concentrations.* Fewer droplets are nucleated in maritime clouds compared to continental clouds because there are fewer CCN in the air over the oceans than in the air over land. The water vapor available for condensation is thus distributed over fewer droplets in the maritime clouds, and they can grow to larger sizes.

In turn, these larger cloud droplets can then initiate collision-coalescence more quickly, producing raindrops within the short lifetime of a shallow cumulus cloud or within the transient, weak updrafts of a stratocumulus cloud. Observations have shown warm rain falling from warm cumuli over the Caribbean Ocean in as little as 15 minutes from the first appearance of the cloud.

On the other hand, *warm continental clouds rarely produce rainfall by the warm rain process.* The greater number of CCN in the air over land leads to higher droplet number concentrations in these clouds, reducing supersaturations and thus growth by condensation. As a consequence, the large drops necessary to initiate collision-coalescence often cannot form within the lifetime of a shallow warm cumulus cloud. In a longer-lived warm

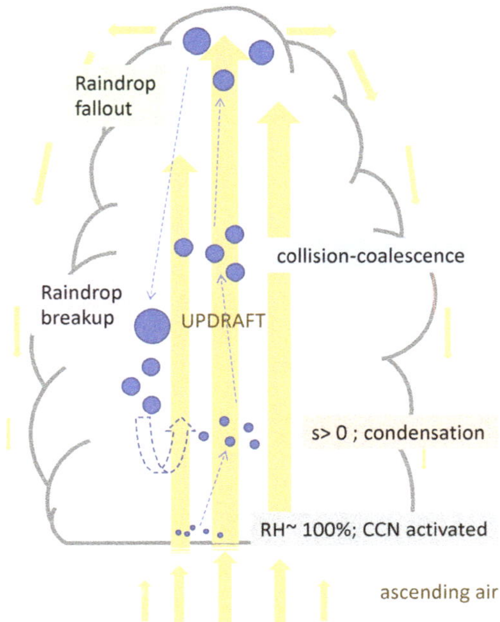

Fig. 10.15: Overview of the key steps in the warm rain process. Dashed arrows show the progression of the liquid hydrometeors in the updraft of the cloud. The possible additional step of raindrop breakup can create smaller drops that are again able to be carried upward in the updraft to grow by collision-coalescence. Yellow arrows are simplified depictions of the air motions within and outside the cloud (ignoring entraining eddies at the cloud edge).

stratocumulus cloud, a small amount of drizzle might be produced, but rarely in quantities sufficient for appreciable rainfall.

10.5.2 Importance in mixed-phase clouds

Deeper clouds are characterized by colder cloud tops that generally contain some frozen hydrometeors. The greater amounts of rainfall that these clouds can create are then attributable to processes involving ice. This finding applies to both maritime and continental clouds.

This does not imply that the warm rain process is irrelevant in deeper mixed-phase clouds. On the contrary, the warm rain process can be active in any deep cloud that has a *warm cloud base*, a *strong updraft*, a *high liquid water content*, and a *long lifetime*, if they are sufficient to overcome any disadvantage due to the number of CCN ingested.

In these cases, raindrops formed by the warm rain process will not immediately fall out of the cloud. Instead, the strong updrafts of deep mixed-phase clouds can transport them upward, far above the 0 °C isotherm

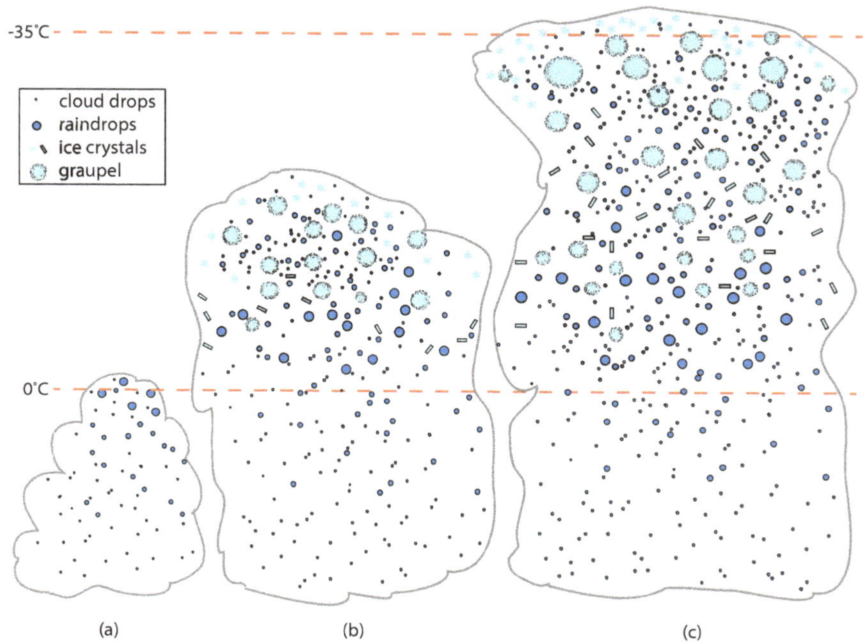

Fig. 10.16: Schematic depiction of the warm rain process and its possible enhancement of ice processes as a deep cumulus cloud grows. In the initial stage (a), a warm cumulus cloud with a strong updraft contains only cloud drops and raindrops formed by the warm rain process. As the cloud deepens and extends to lower temperatures (b), warm rain having formed in the warmer part of the cloud below is carried within the strong updraft to regions less than 0 °C. It can continue to grow by collision-coalescence with other supercooled drops, it can freeze to become graupel, or it can be collected by other frozen hydrometeors. At an even later and deeper stage when the cloud top contains mostly ice (c), most of the raindrops created by the warm rain process in the lower parts of the cloud have frozen to contribute to the production of large graupel particles or even hailstones.

in Fig. 10.16. These supercooled raindrops can still undergo growth by collision-coalescence until they freeze. Once frozen, they can begin riming smaller supercooled droplets, producing large graupel particles in less time. They can also be collected by large graupel or hailstones. Thus, the formation of warm raindrops at lower altitudes in a deep mixed-phase cloud can accelerate the growth of frozen hydrometeors at higher altitudes as the cloud deepens. This interaction can ultimately increase the amount of rainfall reaching the ground in these cases, as discussed in the following chapters.

Questions for Review

1. How are droplet collisions quantified?
2. When cloud drops do collide, how do we quantify the frequency with which they coalesce?
3. What different models are used for the collision-coalescence process, and what assumptions are made for each type?
4. How do larger drops contribute to the formation of precipitation compared to smaller droplets?
5. What are the impediments to the initiation of collision-coalescence in a cloud, and what are some ways they can be overcome?
6. Is the collision-coalescence process sufficient to explain raindrop formation within the typical lifetime of clouds?
7. How does collision-coalescence differ in clouds over the ocean versus those over land?
8. What steps make up the warm rain process?
9. Is the warm rain process only important in clouds with tops warmer than 0 °C?

Homework problems and other resources for students and instructors can be found at bit.ly/4dYBwB5, or scan this code:

Further Reading

For more information on the interpretation of different collision-coalescence models [advanced]:

GIllespie, D. T., 1975: Three models for the coalescence growth of drops. *Journal of the Atmospheric Sciences*, **32**, 600–607

CHAPTER 11

Ice Growth by Collection

The depositional growth of ice crystals can be faster than the condensational growth of water droplets, because the supersaturation with respect to ice achieved in clouds can be an order of magnitude greater than that with respect to liquid water. However, the growth of ice by deposition alone can only yield solitary ice crystals.

Deposition can produce neither the aggregates or rimed ice crystals observed falling from thick stratiform clouds in cold seasons nor the graupel particles or hailstones falling from deep mixed-phase clouds in warm seasons. All these hydrometeors contain significantly more mass (often orders of magnitude more) than single ice crystals. Clearly some sort of collection process involving ice is also necessary to produce these frozen hydrometeors.

Aggregation, shown as the rightmost path in Fig. 9.1, includes the self-collection of ice crystals, snowflakes, or the collection of individual crystals by snowflakes. Larger ice crystals fall more quickly than smaller ice crystals, and aggregates fall faster than individual ice crystals. Thus, aggregates and larger ice crystals can overtake smaller ice crystals and possibly *collide* with them. If they stay together, they have *aggregated*. Aggregate fall speeds tend not to increase much with increasing size (unless they are also rimed), but their geometric cross-section does increase with size, allowing them to collect ice crystals at a higher rate as they grow larger. Aggregation is prevalent in fully glaciated clouds or the glaciated regions of clouds with weak vertical air motions, and it is responsible for surface snowfall in colder areas of the globe.

Riming is the process by which supercooled drops are collected by ice crystals, graupel particles, or hailstones. The more riming that occurs, the more massive the frozen hydrometeor becomes and the faster it falls, accelerating the collection process. Riming occurs in the mixed-phase regions of clouds with stronger updrafts. It is the primary mechanism by which showers of rimed snow, graupel, or hail are observed at ground level. The raindrops formed from the melting of rimed hydrometeors contribute significantly to heavier surface rainfall over the Earth. This microphysical chain of events is depicted in the center of Fig. 9.1.

Both riming and aggregation processes almost always occur to some extent during the life cycles of warm-season single-cell and multi-cell storms. These processes are also common in many midlatitude cool-season storms, including snowstorms.

We now examine each of these collection processes in detail.

11.1 Aggregation

As we did for collision-coalescence in the previous chapter, two questions are again important, this time for *aggregation*:

1. Under what conditions will two ice crystals (or an aggregate and an ice crystal, or two aggregates) collide?
2. If they collide, will they "stick together," i.e., *adhere* to one another?

For several reasons, obtaining answers to these questions is more difficult for ice crystals and aggregates than it is for the collision-coalescence of water drops. First, ice crystal habits are generally not spherical. Due to their complex and highly variable geometry, their fall speeds are less predictable, and their collision cross-sections are more challenging to define.

Moreover, once aggregation begins, the aggregates quickly become asymmetrical, often to a significant degree, as demonstrated by the images shown in Fig. 11.1 of four different aggregates taken from various viewing angles.[1] It is impossible to predict how the shape of a given aggregate will evolve in time, unlike a water droplet growing by collision-coalescence, because the aggregate shape depends upon *where* ice crystals collide with the aggregate surface and stick. The great variety of possible aggregate shapes and bulk densities also makes prediction of their fall speeds extremely difficult.

In short, due to all these difficulties, quantification of the aggregation process requires that substantial simplifications be made.

[1] These observations lend support to the old adage that "no two snowflakes are alike".

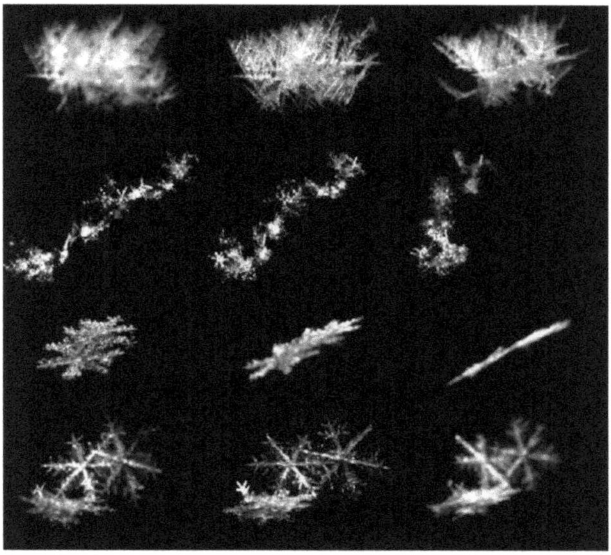

Fig. 11.1: Examples of four different aggregates (rows) from three different viewing angles (columns) of natural aggregates observed in Alaska. *Adapted from Jiang et al. (2019). © American Meteorological Society. Used with permission.*

11.1.1 Sweepout volumes

The sweepout volume, in turn, depends upon the geometric cross-section of the aggregate and its fall speed.

Because ice crystals are not spherical and aggregates can develop into widely varying and unpredictable shapes, some approximations must be made in order to specify a geometric cross-section for their respective sweepout volumes. For example, detailed calculations may assume a circular disk for a single ice crystal, or an oblate spheroid shape (Fig. 9.13) with a constant aspect ratio for an aggregate, to approximate its sweepout volume as shown by the sketched cylinders in Fig. 11.2. When the aggregate shape is known (for example, in a laboratory or if an assumed shape is used in a numerical model), the square root of the aggregate horizontal area may also be used as a representative value of a "radius" for a circular cross-sectional area.

The fall paths for the ice crystal and aggregate shown here must also be simplified; they would in reality have a variable horizontal component (for example, refer back to Fig. 4.16 for planar ice crystals), yielding highly asymmetrical sweepout volumes. This horizontal component is instead just factored into the magnitude of an average ice crystal or aggregate fall speed. For aggregates consisting of just a few individual crystals, an av-

Fig. 11.2: Simplified definition of the sweepout volumes (cylinders) for a single ice crystal (left) or an aggregate (right).

erage fall speed of 1 m s^{-1} might be used, but observed fall speeds for a given aggregate mass can vary by 50% or more from the mean value. Thus, calculations using the average fall speeds will underestimate the growth rate of some aggregates, and overestimate it for others.

11.1.2 Aggregation efficiencies

Another difficulty is that the efficiencies with which aggregates collect ice crystals (or other aggregates) are poorly known. In the case of water drops, collision efficiencies can be theoretically predicted, observed in the laboratory, or numerically simulated simply as a function of their spherical radius. Using similar methods for individual crystal habits or aggregates, even if assuming a given maximum dimension, remains impractical.

Large deviations can occur in the streamlines around falling, fluttering, non-spherical ice crystals or irregularly shaped aggregates. As a result, extensive theoretical calculations, numerical simulations, or cloud chamber observations would be required to produce some "average" collision efficiencies. Such values may not be very representative due to the simplifying assumptions that would have to be made with any of these methods.

It is similarly difficult to quantify the likelihood that an aggregate and an individual crystal will stay together after colliding, often referred to as the **sticking efficiency**. This efficiency represents the fraction of collisions that result in their adhesion (staying together). Like drop coalescence efficiencies, some inferences about sticking efficiencies can be made based on observations of the total **aggregation efficiency**, E_a, which is simply the collision efficiency multiplied by the sticking efficiency.

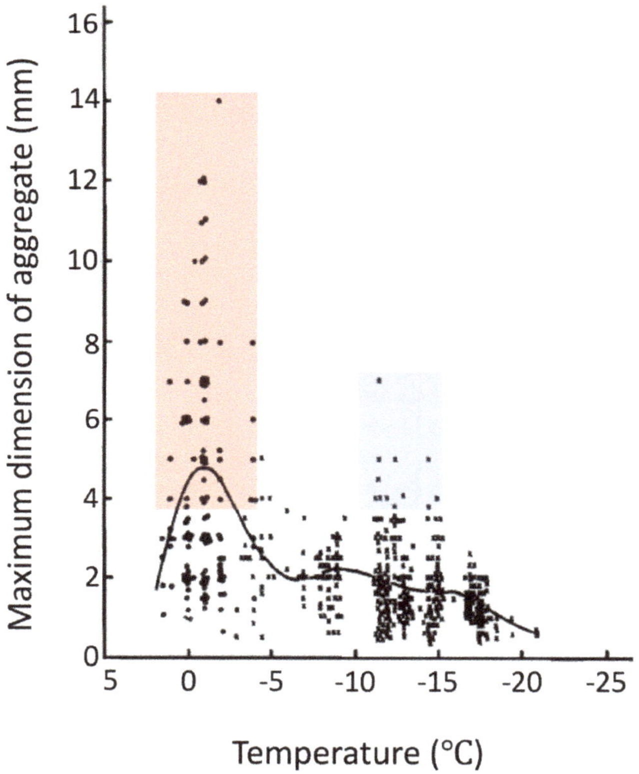

Fig. 11.3: Observed aggregate sizes as a function of air temperature, including both ground-based and aircraft measurements. Best-fit line to all data is shown by the black curve. Pink rectangle highlights the much larger aggregates occurring near 0 °C; blue rectangle highlights some larger aggregates occurring between −12 and −15 °C, as discussed in the main text. *Adapted from Hobbs et al. (1974). ©1974 John Wiley & Sons, used with permission.*

Shown in Fig. 11.3 are observations of maximum aggregate sizes versus the air temperature at which they were observed. The general trend is for maximum aggregate sizes to increase with increasing temperature until 0 °C. Once the aggregates begin to melt at temperatures greater than 0 °C, the larger ones can fragment into smaller pieces. Hence, the largest aggregates are often observed at air temperatures near 0 °C, as highlighted by the pink rectangle in Fig. 11.3. This author has observed aggregates that are double the maximum size shown there. Your personal experience as an observer on the ground may have already made you aware of this tendency; the largest aggregates are often seen when temperatures near the ground are just below or above the freezing point.

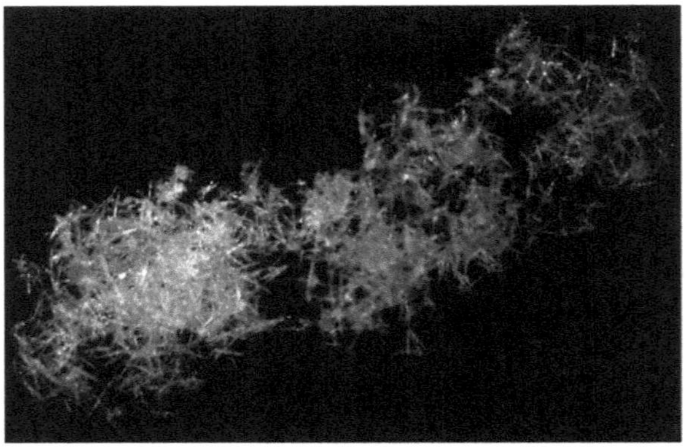

Fig. 11.4: An aggregate consisting mostly of needles, but also containing a few dendrites. *Image from the Multi-Angle Snowflake Camera, courtesy of S. Yuter.*

This tendency results from the crystal surfaces transforming to a semi-liquid state as the air temperature approaches 0 °C, thereby becoming more "sticky." When crystals collide under these conditions, a process referred to as **sintering** can bind them together by refreezing the semi-liquid layer between them. Thus, due to an enhanced sticking efficiency, the overall aggregation efficiency is expected to be higher at temperatures near 0 °C.[2]

A secondary peak in maximum aggregate sizes is also apparent in the observations highlighted by the blue rectangle encompassing temperatures between −12 and −15 °C. It results from the predominance of dendrites in that temperature range, as covered in Sec. 9.2.3. Colliding dendrites (and other branched crystals, to a lesser extent) tend to stick together more easily after collisions than other habits because their branches become entangled, like the opposite ends of a plastic Velcro® strip. This process is called **mechanical interlocking**.

Needles, another habit type observed in many aggregates, may also owe some of their enhanced sticking efficiencies to mechanical interlocking. They can become entwined with other needles, dendrites, or aggregates, as shown in Fig. 11.4. Thus, the overall aggregation efficiency for needles might be higher due to their inherently greater sticking efficiency.

---

[2]This is why it is much easier to pack snow into a firm snowball when its temperature is at most a few degrees below freezing.

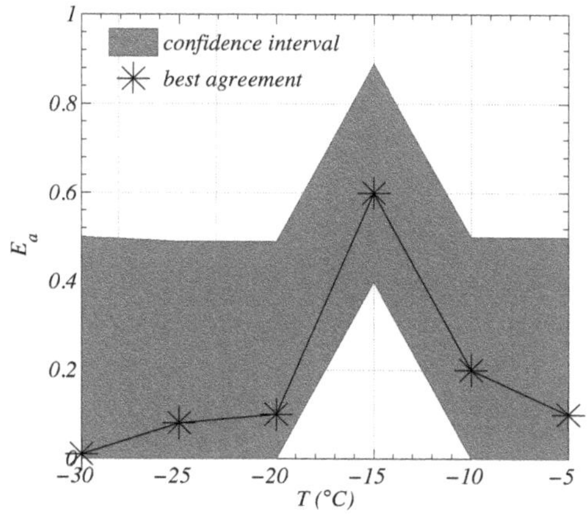

Fig. 11.5: Aggregation efficiencies as a function of air temperature, based on aggregate sizes observed in the cloud chamber shown in Fig. 5.1, combined with a numerical aggregation model. The line with asterisks denotes where the observations and model estimates align most closely. The gray-shaded confidence interval shows 25th to 75th percentiles of the data. *Adapted from Connolly et al. (2012). (CC-BY)*

Aggregate sizes tend to be much smaller as air temperatures fall below −17 °C. The reduced prevalence of branched crystal habits at these temperatures is the likely cause of this sharp decline, which consequently reduces the frequency of mechanical interlocking, and thus also reduces aggregation efficiencies at lower temperatures.

Aggregation efficiencies have been estimated using simplified one-dimensional aggregation models constrained by observations made within a cloud chamber or within a cloud sampled by a research aircraft. With either method, one can measure aggregate sizes and fall speeds as they descend through different vertical layers and measure the ice water content through which the aggregates are falling. By plugging these values into an aggregation model, the model results that best match the observations can then be used to estimate the only remaining variable: the aggregation efficiency.

Figure 11.5 shows the results of a comparison between cloud chamber observations and numerical calculations. In this study, only the initial stages of aggregation (i.e., before the shapes become very complex) among 1-mm-diameter crystals were considered. The temperature was held constant during the experiment, and the aggregation efficiencies among the crystal habits at that temperature were estimated. Multiple experiments were conducted at five-degree temperature intervals, with the temperature held constant for each experiment.

Aggregation efficiencies were small (0.1 to 0.2) among columns growing by diffusion at temperatures of −5 and −10 °C. Dendritic crystals growing by diffusion at −15 °C exhibited a peak in aggregation efficiency, with a value of 0.6. Experiments at a temperature of −20 °C showed a significant decrease in aggregation efficiency to values near 0.1 (where ordinary plates were growing). The aggregation efficiency decreased to nearly negligible values by −30 °C.

Although these trends appear robust with respect to past studies, including the trends shown in Fig. 11.3, the dark shaded region in Fig. 11.5 indicates the considerable uncertainty in these quantitative values of aggregation efficiencies, due to the reasons discussed earlier. Clearly, more research is needed.

It should also be kept in mind that once ice crystals and aggregates are falling from the cloud and encountering higher temperatures than those where they originated, the aggregation efficiencies shown in Fig. 11.5 will not necessarily hold. For example, individual dendrites, as well as aggregates of dendrites, will continue to have appreciable aggregation efficiencies due to mechanical interlocking as they fall into regions with temperatures greater than −15 °C. Additionally, all aggregates, consisting of any ice crystal habit, will experience an increase in aggregation efficiency due to the process of sintering near 0 °C.

11.1.3 Mathematical models of aggregation

As discussed in the preceding sections, quantifying aggregation rates is significantly more complex than collision-coalescence rates for water droplets. For purely pedagogical purposes, we can start from an equation like the continuous collection equation (10.13) used for water droplets and address some of the necessary complexity later. Thus, we begin with:

$$\frac{\mathrm{d}M}{\mathrm{d}t} = \pi(R+r)^2 \Delta V_t E_a I = V_{\mathrm{swept}} E_a I = KI \,, \tag{11.1}$$

where M is the aggregate mass, R and r are the radii of the aggregate and collected ice crystals, respectively (assuming both are spherical), ΔV_t is the difference in fall speeds between the aggregate and the collected ice crystals, E_a is the aggregation efficiency, and I is the *ice water content* of the cloud, expressed in units of mass per unit volume.[3] The sweepout volume per unit time by the aggregate is indicated by V_{swept}, and the collection kernel is given by K.

---

[3]The ice water content I is the frozen equivalent of the liquid water content, L, used in Eq. (10.5).

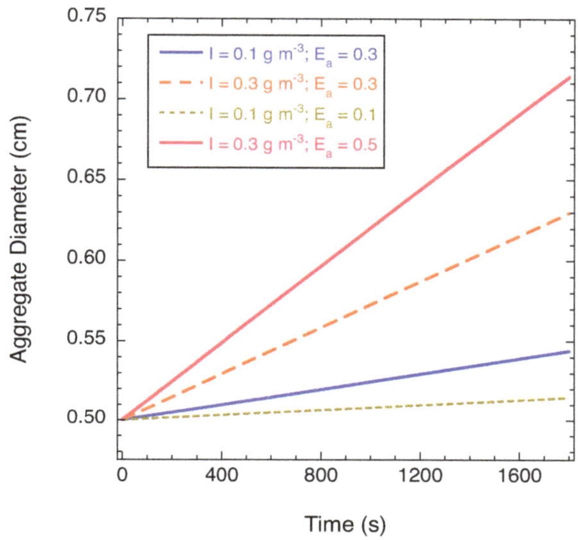

Fig. 11.6: Aggregate growth by collection of single ice crystals using (11.1) for various combinations of the ice water content I and the aggregation efficiency E_a. For all calculations, aggregates were assumed to be oblate spheroids with the maximum dimension labeled here as diameter and the minimum dimension assumed to be two-thirds of the diameter. For simplicity, ΔV_t was held constant at 90 cm s^{-1}, and the aggregate bulk density was held constant at 0.1 g cm^{-3}.

Calculations using (11.1) are shown in Fig. 11.6, where an aggregate is modeled as an oblate spheroid with constant aspect ratio and constant bulk density. The difference in fall speeds between the aggregate and smaller ice crystals, as well as the aggregation efficiency, is held constant within each set of calculations. As a result of these simplifications, the growth of the aggregate diameter appears linear over time.

The ice water content through which the aggregate falls is of paramount importance to its growth rate: aggregation will proceed more quickly in areas of greater ice water content compared to areas in the cloud having less ice to collect, all other factors being equal (comparing the solid blue and dashed orange curves in Fig. 11.6). However, the efficiency with which it collects this ice is also important. When the aggregation efficiency is reduced to 0.1 for the same value of I (now comparing the solid blue line to the dotted gold line), the aggregate grows much more slowly, and when the aggregation efficiency is increased to 0.5 (comparing the dashed orange line to the solid red line), it grows more quickly.

Relative to collision-coalescence, where the growth rate increases exponentially in time (Fig. 10.5), the mass growth rate by aggregation predicted by the continuous collection equation is much smaller. Here, after

30 minutes of aggregation, and even considering the advantageous higher values of both ice water content and aggregation efficiency (red curve), the aggregate has increased in diameter by about a couple of millimeters. This might seem like substantial growth, but if we calculate the equivalent initial and final drop sizes (i.e., by melting the aggregate into a water drop), the difference in the drop diameters during these 30 minutes would be about 0.9 mm, a little less than a 3-fold increase in mass for this aggregate. Referring back to the collision-coalescence calculations of Fig. 10.5, a 3 mm raindrop was produced during 30 minutes of growth from an initial collector drop having a diameter of 100 μm, a difference of about 3 mm growth over 30 minutes, equivalent to an increase in its mass by a factor of 27,000! The mass growth rate by aggregation is much slower as a result of the small ice water content that is common in clouds undergoing aggregation, the small values of aggregation efficiency, and especially the much smaller increase in fall speed as the aggregate grows larger.

Many models, including some numerical weather prediction models, utilize the continuous collection model for aggregation. In this approach, the collection kernel depends on representative "average" aggregate and ice crystal radii (often assuming they are spherical). If other shapes are assumed, the collection kernel can be adapted to use averages of the horizontal cross-sectional areas of the aggregates (A_a) and ice crystals (A_i):

$$K = \left({A_a}^{1/2} + {A_i}^{1/2} \right)^2 \Delta V_t E_a \,. \tag{11.2}$$

However, improved agreement between calculations and observations of aggregation requires the use of a quasi-stochastic model that considers the random nature of the collection process. Compared with its use for collision-coalescence of water drops, this model is even more complicated here. To account for the multiple values of aggregate or crystal fall speeds and sizes (due to variable densities given their respective masses), a *probability-based* collection kernel K^* can be used in a quasi-stochastic collection equation.

For the collection of an aggregate or ice crystal of mass m_2 by an aggregate or ice crystal of mass m_1, we simplify the matter by assuming spherical crystals and aggregates. Thus, we can write the new collection kernel as:

$$\begin{aligned} K^*(m_1, m_2) = \pi \int_0^\infty \int_0^\infty \int_0^\infty \int_0^\infty & E_a (R_1 + R_2)^2 (V_{t1} - V_{t2}) \\ & \times P(R_1, V_{t1} \mid m_1)\, P(R_2, V_{t2} \mid m_2)\; dR_1\, dV_{t1}\, dR_2\, dV_{t2} \,, \end{aligned} \tag{11.3}$$

where $P(R, V_t \mid m)$ is the (assumed) underlying probability distribution of all possible radii and fall speeds in the distribution of individual crystals or aggregates, given an ice crystal or aggregate of mass m. This equation would be integrated over all possible radii ($\mathrm{d}R$) and fall speeds ($\mathrm{d}V_t$) for a given mass m in a bin.

As one might guess, however, there is significant uncertainty in specifying the appropriate underlying distribution of sizes and fall speeds for a given crystal or aggregate mass. Given both this uncertainty and the additional uncertainty in aggregation efficiencies and fall velocity variability for a given mass value, complex representations of aggregation like (11.3) appear to be of little additional value, at least in most cloud, numerical weather prediction, and climate models.

For this reason, some operational numerical models use an algorithm called *autoconversion*,[4] a technique by which single ice crystals are simply "converted" to aggregates in the model once they reach a certain mass or size threshold, or when the total ice crystal population in a region of the modeled cloud reaches some total mass threshold. Currently, more detailed calculations using quasi-stochastic models and statistical representations of aggregate and ice crystal fall speeds are only performed in some numerical models used for research.

11.1.4 Breakup

The largest aggregate sizes, often near 2 cm (though they can sometimes be twice that size), are self-limited due to **fragmentation**: aggregates can break up spontaneously or as a result of collisions with other aggregates. Unfortunately, there is currently very little quantitative information regarding aggregate breakup.

Fragmentation can occur at weaker points in an individual aggregate as its size increases, especially when it falls into unsaturated air. This allows thinner parts of the joined crystals within the aggregate to sublimate, similar to removing the hinges from a door frame.

Fragmentation due to collisions can often be observed by a ground observer during heavy snowfall. When large aggregates collide, parts of one or both aggregates can break off into smaller pieces, and the likelihood of this breakup increases as the total surface area of the aggregate increases.

---

[4]Using a formula for the autoconversion of cloud droplets to raindrops to represent the process of collision-coalescence was once quite common in numerical models. However, improved quantification of its dependencies, gained through years of research, has allowed its replacement with more sophisticated representations. It is anticipated that with ongoing research, the use of autoconversion to represent the complex process of aggregation may also eventually be replaced by more complex, knowledge-based numerical algorithms.

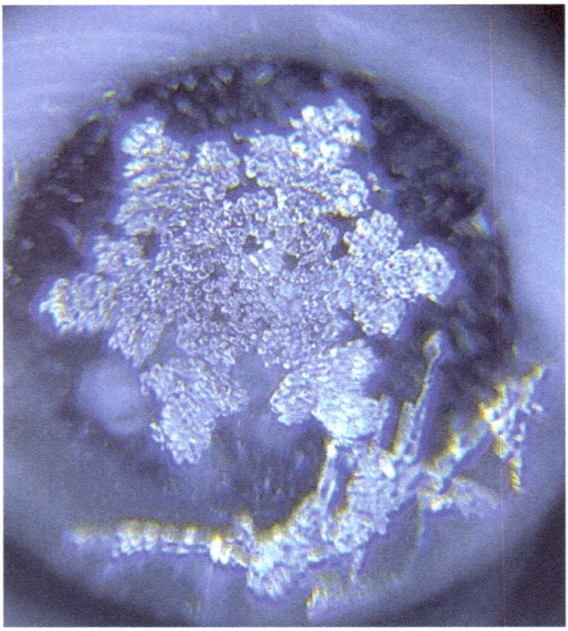

Fig. 11.7: Magnified photo of a lightly rimed sector plate crystal (center) and a lightly rimed aggregate of needles (bottom) during a natural snowfall.

11.2 Riming

Riming requires a frozen hydrometeor to collect supercooled water droplets, and thus this process can only occur in regions of the cloud with temperatures below 0 °C. The bumpy layer of frozen droplets on the ice particle surface, referred to as *rime*, is visible in the magnified image shown in Fig. 11.7. The frozen hydrometeor collecting the supercooled water droplets is sometimes referred to as "the rimer." The rimer can be an individual ice crystal, an aggregate, or even frozen drizzle drops or frozen raindrops. If any of these hydrometeors acquires so much rime that its original shape becomes indistinguishable, the hydrometeor is then referred to as a *graupel* particle. If a graupel particle rimes further such that it exceeds 5 mm in diameter, it is then called a *hailstone*.

This possible progression of the riming process is indicated by the steps leading into the central microphysical pathway of Fig. 9.1. For the remainder of this chapter, we examine this process, from the initial stages of riming on ice crystals to the production and growth of graupel particles, and in the most extreme cases, the formation of hailstones.

11.2.1 Mathematical model of riming

We can first start with an equation similar to the continuous collection equation, combining (10.4) and (10.5) that were used for water droplets. For a frozen hydrometeor of mass M and radius R, falling relative to smaller supercooled liquid drops of radius r that make up a supercooled liquid water content L:

$$\frac{dM}{dt} = \pi\,(R+r)^2 \Delta V_t E_r L = V_{\text{swept}} E_r L, \tag{11.4}$$

where ΔV_t is the difference in fall speeds between the frozen hydrometeor and the supercooled liquid drops, and E_r is their *riming efficiency*, discussed later in Sec. 11.2.2.

Calculations of riming are often performed by approximating the frozen hydrometeor as spherical, which is reasonable for frozen raindrops as well as many graupel particles and hailstones. Thus, the geometric cross-section $\pi(R+r)^2$ used in (11.4) remains appropriate. However, for the onset of riming of an ice crystal, simpler approximations for the cross-sectional area of the downward-pointing face of the falling crystal can be employed.[5] For aggregates, the cross-sectional area of an oblate spheroid is more appropriate.

Like other collection processes, it is not necessary for the frozen hydrometeor to actually be descending through the cloud. It is the *relative difference in fall speeds* between the frozen hydrometeor and the supercooled drops that matters in (11.4). Thus, riming can occur within the strong updraft of a cloud where both types of hydrometeors are ascending. In fact, we will discuss later that this is a preferred location for riming due to the high supercooled liquid water content present there.

It should also be recognized that for the riming process, the frozen hydrometeors are usually much larger and fewer in number than the collected supercooled drops. Thus, the continuous collection equation provides a reasonably accurate basis for calculating riming growth, as discussed for collision-coalescence.

In the next section, we examine how efficiently various types of frozen hydrometeors can collect the supercooled water drops within their respective sweepout volumes, i.e., their riming efficiencies.

---

[5]For plates and dendrites, the basal face points downward (on average) as they fall. Thus the $\pi(R+r)^2$ term in (11.4) might still be a sufficient representation of the geometric cross-section. For columns, the prism face points downward (on average), and its geometric cross-section could be represented as the column length l times the column diameter d, with the radius of the collected droplets added to each dimension.

11.2.2 Riming efficiencies

To determine appropriate values for the **riming efficiency** E_r in (11.4), we can start by asking essentially the same two questions as for the other collection processes:

(i) Will supercooled drops collide with the frozen hydrometeor?

(ii) Will they stay together after the collision?

The first question pertains to the collision efficiency of the frozen hydrometeor and the supercooled drops within its path. For riming, however, the second question is not really an issue. Except for the riming growth of hailstones in very wet conditions examined later in this chapter, we can often safely assume that any supercooled drops colliding with the frozen hydrometeor will freeze upon impact. Hence, in most cases, *the riming efficiency is assumed to be equal to the collision efficiency.*

i) Ice crystals

Estimates of ice crystal-drop collision efficiencies have been obtained from detailed, but idealized, numerical simulations of the trajectories of water drops within the geometric sweepout volumes of individual falling ice crystals. An example of the results of such calculations is shown in Fig. 11.8 for plates, broad-branched crystals, and columns.

For any crystal habit, as its size increases, a broader range of drop sizes can collide with it, as shown by the widening of the shaded region of a given color along the x-axis. Additionally, the efficiency with which any particular drop size is collected also tends to increase with crystal size, as indicated by the increasing height of the shaded region of a given line color along the y-axis.

The results also show that to collect smaller cloud droplets, individual ice crystals must achieve sufficiently large sizes, similar to how the onset of collision-coalescence requires the presence of larger cloud drops to collect smaller droplets. The sizes required for riming efficiencies to become appreciable differ markedly; these differences arise from variations in the air streamlines around different crystal habits. Small columns of 67 μm length can begin colliding with supercooled droplets just a few micrometers in diameter, while plates require a width of 160 μm or larger, and branched plates require widths exceeding 200 μm. The collision efficiencies of plates or broad-branched plates with supercooled droplets do not exceed 0.5 until the maximum dimensions of the crystals reach 200 to 250 μm, whereas columns reach comparable collision efficiencies with a length of half that size ($\approx$ 100 μm).

The larger sizes of plate-like crystals required for riming to begin are due to the dynamic pressure that forms at the center of the bottom of the crystal (e.g., the center of the hexagonal basal face) as it falls. Small

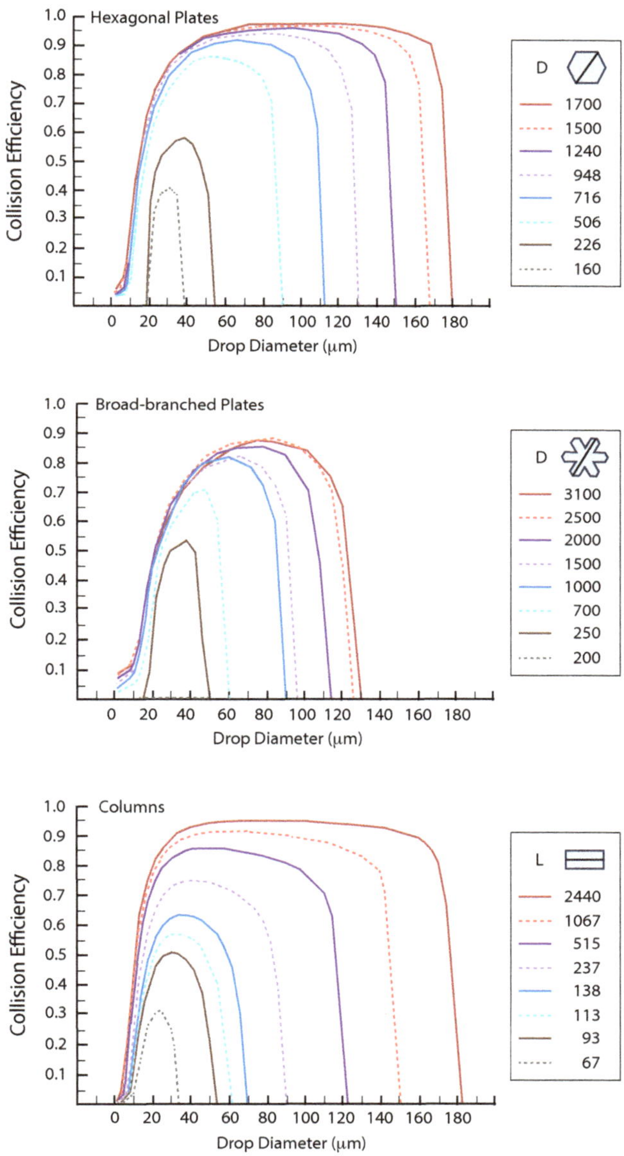

Fig. 11.8: Numerically simulated collision efficiencies for (top) hexagonal plates, (center) broad-branched plates, and (bottom) columns, with supercooled drops of various sizes. Each curve represents the largest crystal dimension (diameter D or length L as shown on crystal shape) in micrometers as labeled in the legend. The thickness of each crystal (not shown) also increases with D or L. *Data from Wang and Ji (2000).*

droplets that would have collided with the crystal at this location are instead deflected outward toward the crystal edges and possibly around the crystal altogether. Additionally, the gaps of air between the broad-branched plates allow some droplets to pass between the branches without being collected, lowering the collision efficiencies even more compared to hexagonal plates.

Figure 11.8 shows that collision (riming) efficiencies can be quite high, approaching 1.0 for 1 mm plates and 0.9 for broad-branched plates and columns. Cloud drops must attain diameters exceeding approximately 50 μm for these highest efficiencies; their larger inertia prevents them from being swept around the crystal by the airflow around it. When the drops become too large, however, their fall speeds become comparable to those of the ice crystal, and no collisions occur.

Unlike the onset of collision-coalescence, where multiple hypotheses have been proposed to explain how the large drop sizes needed to initiate collision-coalescence might be attained, ice crystals can grow by deposition in a reasonable amount of time to achieve the sizes needed to begin riming. The supersaturation with respect to ice in clouds is often sufficiently high, and the non-spherical shapes of the crystals require less mass than water drops to achieve the necessary maximum dimensions.

ii) Aggregates

For aggregates, the riming efficiency is largely unknown due to the diverse shapes they can assume. Experiments in a cloud chamber, involving the collection of small droplets by disks with holes of increasing size and number, suggest that riming begins earlier than for solid disks of equivalent size. This implies that aggregates might have higher riming efficiencies than individual plate crystals of similar dimensions. However, due to the lack of quantitative information on the riming efficiency of aggregates, it is often assumed to be equal to 1.0 in model calculations.

iii) Graupel

Theoretical calculations have been made for the collision (riming) efficiencies of spherical graupel particles (Fig. 11.9) in combination with different sizes of supercooled drops. In general, the growth of graupel particles requires supercooled droplets to be at least a few micrometers in radius to attain collision efficiencies of 10% ($E_r = 0.1$) or more. Once the supercooled droplets exceed a radius of 10 μm, collision efficiencies range from 40% to 90%, with larger graupel having higher values. For the largest drop sizes shown here, many graupel particles can rime at nearly 100% efficiency as long as the drops have fall speeds less than that of the graupel.

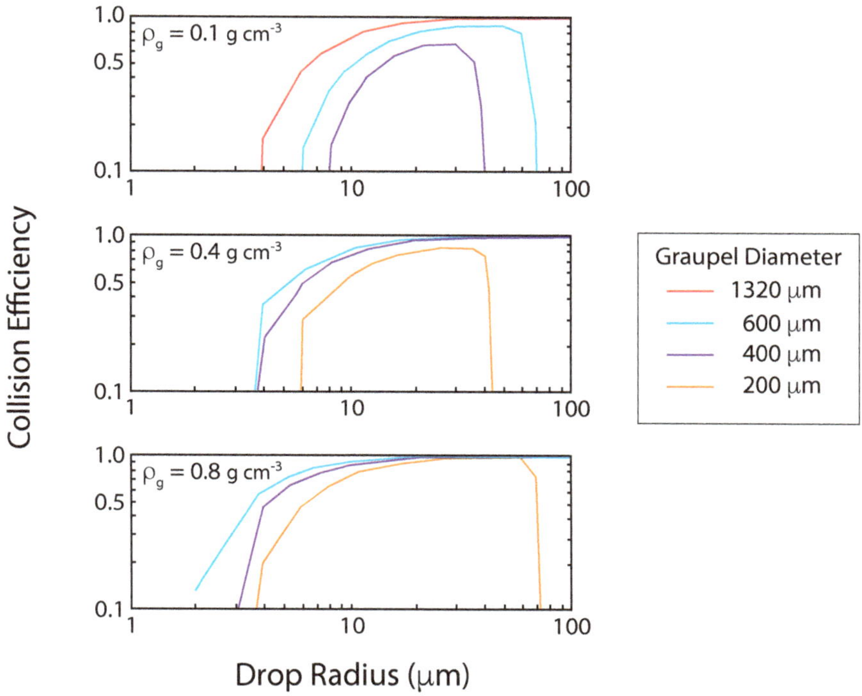

Fig. 11.9: Theoretical calculations of collision (riming) efficiencies of graupel particles (diameters shown in legend) with supercooled drops of various sizes. The calculations were performed assuming different bulk densities of graupel particles ρ_g as labeled in each panel. The cutoff at larger drop sizes occurs when the drop fall speed exceeds the graupel particle fall speed. *Data from Khain et al. (2001).*

The densities of the graupel particles affect collision efficiencies due to their influence on fall speeds and thus on the streamlines around the falling particles. As shown in the three different panels of Fig. 11.9, as the density of a graupel particle increases (moving from the top panel toward the bottom panel), most droplets smaller than 10 μm in radius exhibit increased collision efficiencies. In all cases, drops with a radius of 10 μm or larger are expected to collide—and thus rime—with nearly perfect efficiency, except for the smallest and least dense graupel particles, at least until the drop fall speed matches that of the graupel particle.

It is noteworthy that all three panels trend toward having collision efficiencies of 100% for the larger sizes of graupel particles and supercooled drops, again with the exception of the smallest and lowest-density graupel. Thus, while Fig. 11.9 only considers graupel particles less than 1.32 mm in diameter, graupel exceeding this size would be expected to capture supercooled drops greater than 10 μm in radius with nearly perfect efficiency ($E_r \approx 1$).

iv) Frozen drops

Frozen drops can also grow by riming. In a convective cloud with a sufficiently strong updraft, drizzle or raindrops can be carried from lower, warmer levels to higher, colder levels in the cloud where they can freeze. Once they freeze, they may then begin riming by collecting smaller supercooled drops. In this case, riming efficiencies would be considered equivalent to the collision efficiencies for similar sizes of a pair of liquid drops (Fig. 10.3 left panel), until they become more graupel-like. At that point, riming efficiencies would be slightly smaller, as indicated by the values shown in Fig. 11.9.

v) Hailstones

The riming efficiency for hailstones is typically assumed to be 100%. This means it is often assumed that any supercooled drop in the path of a falling hailstone will be captured. This assumption is based on both theoretical considerations, due to the extreme fall speeds of hailstones compared to those of the drops, and on some very limited experimental work in wind tunnels. However, as we will see later in this chapter, when hailstones are very large or are growing in regions of the cloud containing large amounts of supercooled liquid water, the riming efficiency can be modified to account for the fact that not all of the collected water will be retained and frozen on the hailstone surface.

11.2.3 Riming growth of graupel

Graupel can be quasi-spherical or conical in shape and are white (not transparent) in appearance (Fig. 11.10). They typically have densities much lower than solid ice,[6] ranging from 0.1 to 0.5 g cm^{-3}. These characteristics result from aspects of the growth by riming.

i) Graupel shapes

The possible conical and quasi-spherical shapes of graupel particles result from variations in the preferred locations where supercooled drops attach during the riming process. It is generally accepted that spherical graupel forms due to random tumbling of the particles as they fall. Consequently,

---

[6] When graupel particles originate as frozen raindrops, their bulk densities can approach 0.85 g cm^{-3}, nearly the same as solid ice.

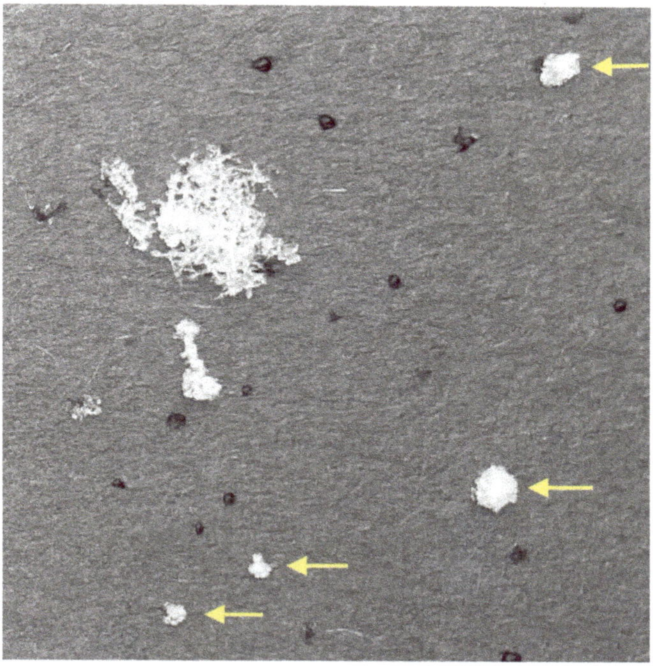

Fig. 11.10: Photograph of natural graupel particles (indicated with yellow arrows) showing their opaque appearance. The larger hydrometeors in the upper left of the image are lightly rimed aggregates. The large graupel particle shown in the lower right corner has an approximate diameter of 3 mm. *Photo courtesy of A. Pappas.*

rime builds up more or less equally over their entire surfaces. For simplicity, most numerical models assume graupel particles are spherical in shape.

The origin of the conical graupel shape remains uncertain, as it is unclear where the rime is accumulating. Small conical graupel particles fall with their flat bases pointed downward (Fig. 4.20). Some studies suggest that the original riming crystal resides near the top of the conical graupel particle, leading to the conical shape when rime accumulates only at its base, fanning outward over time. Other studies propose that counter-rotating vortices in the wake of a falling crystal or graupel particle preferentially deposit some rime on its top side, while more rime still accumulates at its base, resulting in the conical shape. This latter hypothesis appears to be supported by streamlines from idealized numerical simulations of falling graupel.

Regardless of their origin, as conical graupel particles grow larger, they begin to undergo horizontal oscillations in their direction of fall, as observed in wind tunnels and suggested by numerical simulations. As the particles grow beyond several millimeters in diameter, the increasing oscillations eventually cause them to tumble as they fall, thereby altering their shapes to be more spherical.

ii) Dry growth

As supercooled water droplets freeze onto the surface of a frozen hydrometeor, latent heat is released, warming its surface. When this latent heat can quickly diffuse away from the frozen hydrometeor surface to the surrounding air, it does not substantially change the surface temperature of the frozen hydrometeor. This scenario is referred to as **dry growth**. Dry growth is favored for graupel and other frozen hydrometeors undergoing slower rates of riming.

The opaque appearance and low density of graupel are attributable to dry growth. Because the surface temperatures of the graupel particles always remain below 0 °C, the collected supercooled drops freeze immediately upon impact, allowing the resulting layer of rime to contain air trapped between the individual frozen drops. These air spaces both decrease the density of the rime and cause it to appear white. Both the low density of the rime and the roughness of the particle surfaces reduce the fall speeds of graupel particles relative to raindrops of similar sizes.

iii) Calculations of graupel growth rates

Using (11.4), we can examine the riming growth rates of spherical graupel under various conditions, the results of which are shown in Fig. 11.11. The growth rates here do not appear linear over time as they did for the aggregation process, because the fall speeds of the graupel particles are not constant. Their increase with size is too significant to ignore. The larger the graupel particle, the faster it falls and therefore the more rapidly it can collect supercooled water, accelerating its growth as it increases in size.

As we have seen before with the continuous collection equation and other collection processes, the amount of water mass available to be collected—in this case, supercooled liquid water content L—is very important. Growth rates are faster for greater L, as illustrated by comparing the solid dark blue and dashed green curves.

The effect of the density of the graupel particle is also partly represented by comparing the solid dark blue and solid light blue curves in Fig. 11.11. The density is assumed to be smaller in the calculations represented by the light blue curve, so the diameter of that particle increases faster over time under the same conditions as the dark blue curve. This happens because the same amount of rime occupies more "space"; that is, more air gaps between the rimed drops exist when the bulk density of

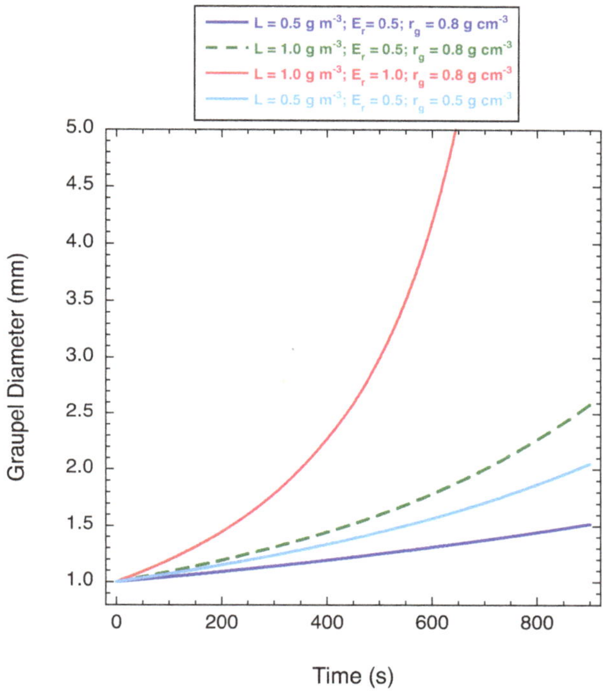

Fig. 11.11: Graupel growth by riming using (11.4), for various combinations of the liquid water content L, the riming efficiency E_r, and the graupel density ρ_g as labeled. For all calculations, graupel were assumed to be spheres with density and riming efficiency held constant for each set of calculations. The fall speeds of the graupel were assumed to follow a line fit to the data shown in Fig. 4.20.

the rime is lower, causing the rime to spread into a larger spherical graupel particle. However, these calculations do not account for the additional effect of the lower density on the fall speed of the graupel—less dense hydrometeors of the same size fall more slowly than those of higher density. Therefore, the growth shown for the light blue curve should actually be a little less.

As we have seen with other collection processes, higher efficiencies also accelerate the rate of riming growth. Comparing the dashed green and solid red curves, the very high riming efficiencies possible for graupel particles (approaching 1.0) can greatly increase their growth rate, particularly due to the large increases in fall speeds when riming efficiencies are high. In fact, for the solid red curve with a riming efficiency of 1.0, the calculations were halted because the maximum size of graupel (5 mm diameter, by definition) was attained before 15 minutes. Once their diameters exceed 5 mm, graupel particles are then considered hailstones, and other complications of riming growth must be considered, as explained in the following sections.

The striking difference in these riming calculations, compared to those for aggregation (Fig. 11.6), lies in the significantly higher mass growth rate of riming. Here, for the graupel particle with the most favorable conditions (higher L and higher E_r, shown by the solid red curve), the difference in the melted equivalent drop diameter from the start to the end of the calculation is roughly 1.9 mm *over a period of just 11 minutes*. In contrast, aggregation required *30 minutes* to achieve a comparable increase in the melted equivalent drop size (and thus a comparable increase in mass)!

If the graupel particle remains in the cloud for extended periods, it grows even more rapidly once it attains the size of a hailstone. We now address hailstone growth by riming and the reasons for the remarkable mass growth rates that can be achieved in such cases.

11.2.4 Riming growth of hailstones

The processes responsible for the development of hail are, in many ways, just extreme forms of those that yield graupel. For at least a portion of their time in the storm, developing hailstones will experience higher cloud liquid water content and higher updraft speeds than are typical for graupel. But when very high mass growth rates occur due to riming, the nature of the riming can change from dry growth (described earlier) to *wet growth*, addressed next.

i) Wet growth

The large sizes of hailstones compared to ice crystals or graupel particles allow them to collect supercooled water at a much higher rate. They are also produced within regions of the cloud that have more supercooled water mass available for collection than either of these smaller rimers.

When the hailstone riming rate is sufficiently high, the latent heat released during the freezing of the drops onto the hailstone surface cannot be rapidly diffused away. As a result, *the surface temperature of the hailstone can temporarily rise to 0 °C or slightly above*. In this case, the collected drops will then no longer immediately freeze upon impact with the hailstone surface. Such a scenario is called **wet growth**.

When a hailstone undergoes wet growth, some of the unfrozen water may initially be pushed across its surface by the relative airflow around the hailstone. This liquid water can then fill any open spaces of air in the uneven rime before the latent heat dissipates and the collected water finally freezes. Once the water freezes, it results in a layer of clear, high-density ice. Owing to the lack of air bubbles, the density of these layers is essentially near that of pure ice, approximately 0.9 g cm^{-3}. Additionally, hailstones undergoing wet growth may have *lobes* on their surfaces. These lobes are likely formed by collected liquid water traveling across the hailstone surface to preferred areas where it finally freezes after covering longer paths.

Other collected drops may actually be **shed** from the hailstone surface, meaning they roll off entirely before having a chance to freeze, particularly under extreme wet growth conditions. This shedding process consequently somewhat slows its growth rate.

For calculations of wet growth, the riming efficiency must be adjusted to be less than the collision efficiency of the hailstone and supercooled drops, to account for the water that is possibly shed. However, this value is not well known. A few wind tunnel experiments have estimated that 70 percent of the collected but unfrozen water can still be retained as a "spongy" layer when it is soaked into the rime at the hailstone surface, equivalent to a riming efficiency of ~0.7. Similar studies of larger hailstones suggest that this value could be smaller, but limited experiments have not yet converged upon a riming efficiency value under these conditions.

Experiments in cloud chambers have shown that the drops shed from hailstones can actually be quite large, even if the collected water originally consisted of much smaller drops. In this way, hailstone shedding can be an additional source of raindrops in the storm.

ii) Dry versus wet growth of hailstones

The alternating layers of clear and opaque ice within many hailstones (Fig. 11.12) suggest that they can undergo dry growth at times and wet growth at other times. The critical factor is the hailstone surface temperature. If the hailstone surface temperature remains below 0 °C while riming, the hailstone undergoes dry growth, but if it increases to 0 °C or slightly above, the hailstone will begin wet growth. The factors critical to determining the hailstone surface temperature during riming include the hailstone size, the magnitude of the supercooled liquid water content surrounding the hailstone, and the local air temperature. The transition from one type of riming growth to the other is represented by the so-called **Schumann-Ludlam limit**, which estimates the boundary between the dry and wet growth regimes based on these conditions.

The Schumann-Ludlam limit and conditions for wet and dry growth of hailstones with diameters ranging from 6 mm to 6 cm are depicted in Fig. 11.13. Under conditions to the left of the Schumann-Ludlam limit, dry growth occurs because the rate of collection of supercooled liquid water by the hailstone is insufficient to raise its temperature above 0 °C. Along and to the right of the Schumann-Ludlam limit, however, the hailstone

Fig. 11.12: Image of the famous Coffeyville, Kansas hailstone (here cut in half to expose its interior) that fell in 1970; it was the largest documented hailstone in the United States up until 2010. The Coffeyville hailstone weighed 757 g ($\sim$1.7 lbs) and had a diameter over 14 cm (5.5 inches). *Photo from the University Corporation for Atmospheric Research (UCAR) via* opensky.ucar.edu.

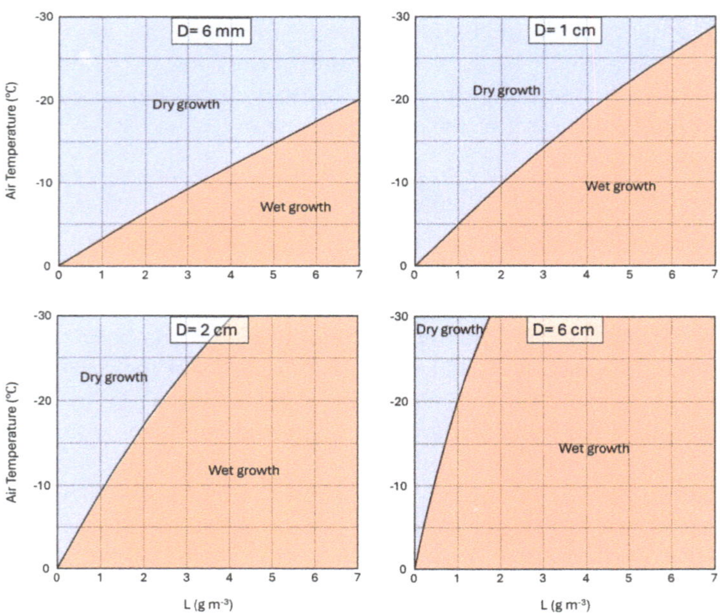

Fig. 11.13: The Schumann-Ludlam limit (solid black line) for four different diameters of hailstones (labeled at the top of each panel), as a function of their surrounding air temperature and (supercooled) liquid water content L. Wet growth will occur (with possible shedding) within the conditions over the entire red shaded region; dry growth will occur within the conditions over the entire blue shaded region. *After Young (1993).*

experiences wet growth, and some of the collected water might even be shed.

We see that the larger the hailstone, the less restrictive the conditions required for it to undergo wet growth. The collection rate of the supercooled water is a function of the hailstone fall speed, which generally increases with hailstone size. Therefore, larger hailstones will experience the conditions for wet growth over a range extending to lower air temperatures and smaller supercooled liquid water content. Accordingly, the 6 cm hailstones in the bottom right panel can undergo wet growth over a much broader range of conditions than the 6 mm hailstones in the top left panel.

For any hailstone size, wet growth can be achieved if the ambient cloud temperature is sufficiently warm and the liquid water content is sufficiently high. The closer the surrounding cloud temperature is to 0 °C, the less latent heating is required to bring the hailstone surface to 0 °C. Thus, the likelihood of wet growth for all hailstone sizes increases as the air temperature approaches 0 °C and decreases as the air temperature approaches −30 °C. Additionally, the greater the supercooled liquid water content surrounding the hailstone, the faster its rate of collection and hence the faster the rate of latent heating of the hailstone surface, potentially bringing its temperature to 0 °C. Therefore, dry growth is often restricted to regions with low supercooled liquid water content.

It is therefore common for larger hailstones (e.g., those 2 cm or greater in diameter) to undergo *some* wet growth, but not throughout their entire duration within the storm. Initially, the hailstone is often too small to experience wet growth unless the air temperature is very near 0 °C. In later stages, the hailstone's trajectory may pass through regions of the storm with varying amounts of supercooled liquid water, resulting in alternating periods of wet and dry growth. The shedding of collected water on the hailstone surface may allow it to cool down sufficiently to experience dry growth temporarily, until enough liquid water is again collected to raise the hailstone's surface temperature to 0 °C.

iii) Calculations of hailstone growth rates

For a hailstone with multiple layers of varying density, like that shown in Fig. 11.12, a representative average density must be used when calculating its mass. Additionally, large hailstones undergoing wet growth tend to have *lobes* on their surfaces as shown in the same figure. These lobes are likely formed by collected liquid water traveling across the hailstone surface to preferred areas where it finally freezes after covering longer paths. These lobes can increase the drag force on the falling hailstone, decreasing its fall speed and potentially reducing its tendency to undergo further wet

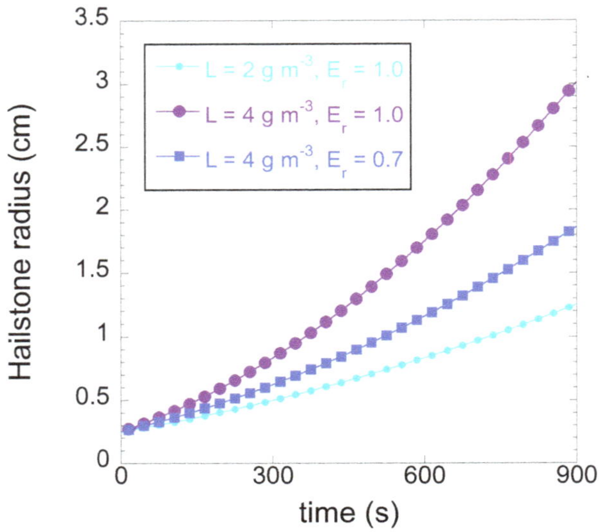

Fig. 11.14: Hailstone growth as a function of time using (11.5) with parameters as noted in the legend. All calculations consider spherical hailstones with initial radii of 0.25 cm. The hailstone densities are held constant at 0.9 g cm^{-3}, and their fall speeds follow those shown in Fig. 4.21.

growth. We can now use the limits of wet and dry growth to calculate possible extremes of hailstone growth by riming.

For hailstones, the continuous collection model represented by (11.4) can be simplified further. Hailstones are typically significantly larger in size than the supercooled drops they collect, and they fall much faster than the supercooled drops. Thus, we can neglect the size r of the collected drops and the fall speeds of the collected drops, yielding:

$$\frac{dM}{dt} = \pi R^2 V_t E_r L = V_{\text{swept}} E_r L \,. \tag{11.5}$$

So for hailstones—and even the largest and most dense graupel particles—the sweepout volume V_{swept} no longer depends on either the collected drop sizes or their fall speeds.

As we have seen from other calculations using the continuous collection equation, Fig. 11.14 shows that the rate of growth by riming depends greatly on the surrounding supercooled liquid water content, L. Hailstones passing through regions of small L will not grow as quickly as those of similar sizes passing through regions of large L (comparing the light blue curve with small circles to the purple curve with larger circles). Because different hailstones can grow in different parts of the storm with varying supercooled liquid water contents, one often sees a variety of hailstone sizes within the hailfall from a single storm (Fig. 11.15).

Fig. 11.15: Photo showing visibly different hailstone sizes and types of riming growth within a single hailfall event. Clear hailstones, or clear layers within hailstones, indicate mostly wet growth. Cloudy hailstones, or cloudy layers within hailstones, indicate dry growth. *Photo by Petr Brož. (CC-BY).*

As a hailstone grows, it sweeps out an ever larger volume per unit time due to its greater fall speed, collecting increasingly more supercooled liquid water. Thus, the growth rates shown in Fig. 11.14 are significantly larger than those for graupel (Fig. 11.11) when converted to similar units. Note that these calculations are carried out for 900 seconds (15 minutes). Even considering the moderate growth curve using a lower riming efficiency, the increase in mass over these 15 minutes is more than a factor of 400! For larger hailstones, this factor would be even greater, far surpassing any other collection growth rates we have examined.

The effects of dry and wet growth can also be seen in the same figure. A hailstone in a region of lower L (light blue curve) could easily grow during these 15 minutes through dry growth the entire time (at an air temperature less than $-15\,°C$), so the use of a perfect riming efficiency, representative of dry growth, seems reasonable. However, the hailstone growing in large L (purple curve with large circles) would likely switch to wet growth conditions as its size increased, making that calculation less realistic as time increases. Repeating the same calculation but using a riming efficiency of 0.7 to allow for shedding during wet growth (dark blue curve with squares) tempers the rate of growth and is likely more realistic. Thus, hailstones

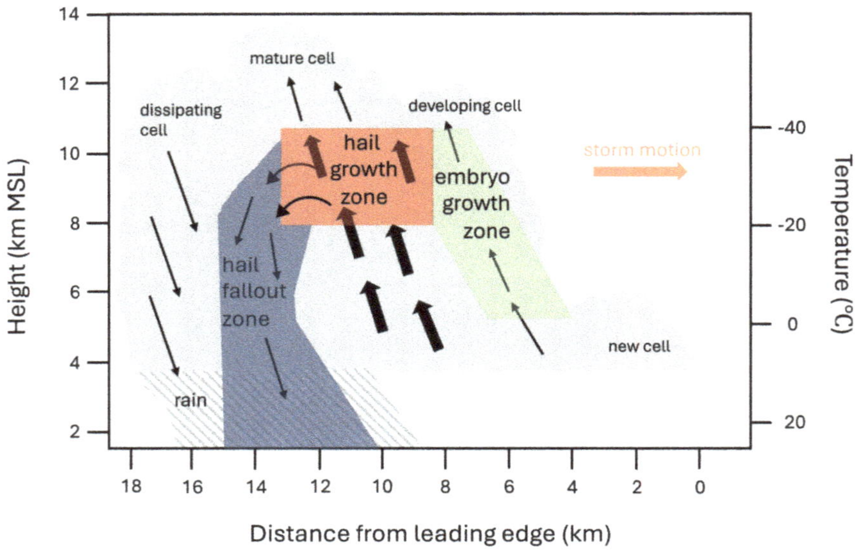

Fig. 11.16: Schematic of a vertical cross-section through a multi-cell thunderstorm, with zones thought to be important for hailstone growth highlighted: embryo growth zone (green), hail growth zone (red), and hail fallout zone (blue). Updrafts and downdrafts are annotated with arrows; wider arrows denote stronger vertical velocities. *After Browning and Foote (1976) and Young (1977).*

occupying very favorable growth regions of the cloud would not grow as quickly, although their growth rates would still be much faster than those of hailstones undergoing dry growth.

11.2.5 Production of large hailstones

Thunderstorms (cumulonimbus clouds) capable of producing hailstones that reach the ground tend to have lifetimes of multiple hours. *Multicellular thunderstorms* consist of clusters of individual storms at various stages in their evolution. While an individual storm within the cluster may have a lifetime of less than an hour, the mature and dissipating storms help to generate new, developing storms, promoting a longer lifetime for the cluster overall. In contrast, although *supercell thunderstorms* consist of a single storm, their unique dynamical motions enable them to have lifetimes of several hours. Simpler *pulse thunderstorms,* with lifetimes of an hour or less, produce hailfall less commonly. A sample schematic, loosely based on an observed multicellular hailstorm, is shown in Fig. 11.16. It helps illustrate different "zones" relevant to the production of large hailstones.

The so-called **hail growth zone** refers to the region in a thunderstorm where the riming growth of a hailstone can potentially be extreme. This region contains a strong updraft on the order of tens of meters per second and supercooled liquid water contents (with temperatures below 0 °C) exceeding several grams per cubic meter. A region of high L enhances the hailstone growth rate, as shown by (11.5). A strong updraft helps to counteract the increasing fall speed of the hailstone as it grows, allowing it to remain in regions of high L for a longer time and grow to larger sizes.[7] The top of the hail growth zone would be near the -40 °C isotherm, as all supercooled liquid water would have been eliminated by homogeneous freezing above this level.

Hailstones can continue to grow outside of the hail growth zone, but the growth rates will not be as extreme as within the hail growth zone. For example, once the fall speed of a hailstone within the hail growth zone exceeds the local updraft speed, it will fall out of the hail growth zone into weaker parts of the updraft with smaller supercooled liquid water content. As the hailstone descends into the lower part of the storm with temperatures exceeding 0 °C, it will begin to lose mass by melting (addressed in Chapter 12), which is critical in determining the hailstone sizes that reach the ground.

Although many thunderstorms contain a favorable hail growth zone, very few produce large hailstones within it. Two prerequisites are thought to be necessary: (i) the number of hail embryos within the hail growth zone must not be too large, and (ii) the residence time of the hailstones in the hail growth zone must be sufficiently long. We now examine these two factors in greater detail.

i) Hailstone embryos

A hailstone must begin as a frozen hydrometeor less than 5 mm in diameter that first starts to grow by riming. This initial frozen hydrometeor is often referred to as a potential **hailstone embryo**. In principle, any frozen cloud drop, ice crystal, or aggregate could rime sufficiently to become a hailstone embryo, but because the rime obscures the original habit, it can only be identified as a graupel particle. Frozen raindrops, on the other hand, are often similar in size to graupel particles when they begin riming in earnest and can be more easily discerned.

Hailstones collected on the ground, when dissected, tend to contain one of two general embryo categories: graupel particles or frozen raindrops, as shown in Fig. 11.17. Analyses of hailstones collected worldwide suggest that any given hailstorm may contain both types of hailstone embryos, sometimes in nearly even proportions. However, hail embryos

[7] Recall that once the hydrometeor fall speed exceeds the local updraft speed, it begins to fall out of the storm.

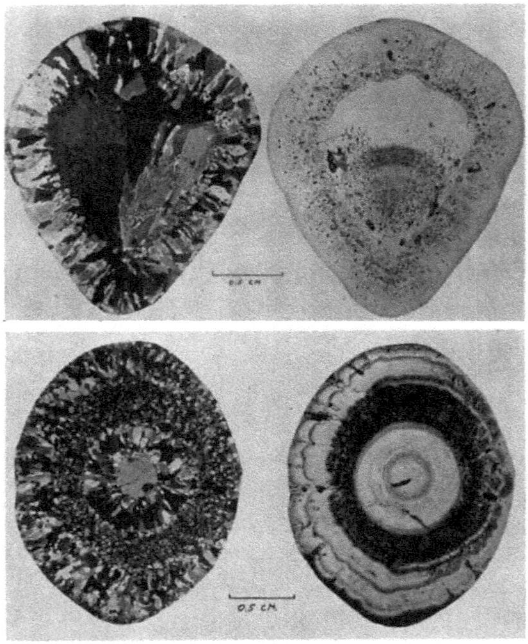

Fig. 11.17: Photos of the cross-sections of two different hailstones under a microscope. Top set of images shows a hailstone containing a conical graupel embryo; bottom set of images shows a hailstone containing a frozen raindrop embryo. For each hailstone, photo on the left is taken under polarized light to reveal crystal sizes characteristic of dry or wet growth. Photos on the right are taken under regular light and show the embryo and layers as seen by the human eye. Scale shown on each set of photos is for 0.5 cm length. *Adapted from Knight and Knight (1970).*

consisting of graupel appear to be more prevalent in thunderstorms with cooler bases, while hail embryos consisting of frozen raindrops tend to dominate in thunderstorms with warmer bases. The latter may result from a stronger warm rain process in those storms, where raindrops are produced in the lower part of the storm and then carried within the updraft above the 0 °C isotherm, where they freeze.

Hail embryos are then produced in a region of the thunderstorm we will refer to as the **embryo growth zone** (Fig. 11.16), where slower updraft speeds promote riming in the form of dry growth. The embryo growth zone may be located within the newly developing cells of a multicellular thunderstorm, as shown in this figure, or along the flanks of the main updraft in a single-cell storm. The internal storm dynamics then transport some embryos from the embryo growth zone into the hail growth zone, where they can undergo much faster riming rates due to the greater supercooled liquid water content there.

ii) Hail embryos within the hail growth zone

The number of hail embryos introduced into the hail growth zone likely has a significant effect on the maximum sizes of hailstones that can grow there. If *very few* hail embryos are introduced into the hail growth zone, much more supercooled liquid water is available for each hailstone to potentially grow quite large. On the other hand, having *many* hail embryos within the hail growth zone limits the amount of supercooled water available for each of them to collect, reducing the maximum size that any of the hailstones can attain. We can envision this scenario, for example, as a reduction in the supercooled liquid water content L in (11.5) that can be used to evaluate the growth rate for each individual hailstone. Having a plethora of hail embryos in the hail growth zone will likely produce hailstones too small to survive melting before falling to the ground, contributing to rainfall rather than hailfall.

Past hail seeding experiments have attempted to use this principle, called "beneficial competition," to introduce more hail embryos into the hail growth zone of a thunderstorm. The intent was to sufficiently reduce hailstone sizes so that they would completely melt before reaching the ground. However, the results of these efforts have been largely unsuccessful or, at best, inconclusive, possibly due to the difficulty in getting the seeding material into the right place in the storm updraft, at the right time, and for extended periods.

Many or even most graupel particles or frozen raindrops serving as hail embryos within a given storm may fail to enter the most advantageous growth conditions within the hail growth zone. Such potential hail embryos may simply grow too large in the embryo growth zone and fall out of the weaker updraft there, into warmer levels below, where they melt into raindrops before reaching the ground.

You may wonder why an embryo growth zone within hailstorms might be needed at all. After all, hailstone growth will be maximized within the hail growth zone, so why couldn't embryos be created there as well? The answer is that *time* is first needed for these embryos to grow to larger sizes.

Recall (4.16), which describes the trajectory of any hydrometeor. In an updraft, the hydrometeor will ascend at a rate determined by the difference between the updraft speed and the hydrometeor's fall speed. At the time that hail embryos are produced—as frozen droplets, ice crystals, or aggregates collect supercooled droplets—their fall speeds will initially be less than a meter per second. Therefore, if they enter the hail growth zone immediately, they will quickly be carried above and out of the hail growth zone by an updraft exceeding 25 m s^{-1}. They will be too small to remain in the hail growth zone for very long, preventing them from growing to large sizes. Thus, hail embryos must first grow in a much weaker part of the updraft before being introduced into the stronger updraft within the hail growth zone.

It is also currently thought that the speed of the updraft within the hail growth zone needs to increase with height in order to grow large hailstones. Initially, a hail embryo introduced into the hail growth zone would be small, perhaps a few millimeters in diameter. If it encounters an updraft that is very strong, it will be carried through and above the hail growth zone too quickly, limiting its ability to grow large by riming. Thus, a slower updraft speed near the bottom of the hail growth zone seems favorable. As the hailstone grows and its fall speed increases, it must encounter greater updraft speeds to remain suspended in the hail growth zone and not fall out. Therefore, an updraft speed that increases with altitude within the hail growth zone seems necessary for the growth of very large hailstones. The largest hailstones may have growth trajectories where their increasing fall speeds are nearly equal to the local updraft speed, effectively suspending them within the hail growth zone for an extended period.

Once the hailstone's fall speed exceeds the local updraft speed at its location, it will fall out of the hail growth zone and into the "hail fallout zone" (Fig. 11.16). It will continue to grow by riming, however, as it descends through any regions containing supercooled drops. The rate of growth can increase substantially for a larger hailstone within the hail fallout zone. Once it falls to an altitude within the storm where the temperature exceeds 0 °C, it will begin to melt. The amount of melting experienced by the hailstone will also determine its maximum size at the ground, but melting affects large hailstones less because they fall much faster through the warm air than smaller hailstones do. Melting will be addressed in more detail in Chapter 12.

11.2.6 Breakup

Rimed hydrometeors colliding with one another tend to bounce apart,[8] sometimes breaking apart the hydrometeor with lower mass. Quantifying the frequency of such collisions and breakup is an area of current research. Under some conditions, collisions among graupel particles may also produce tiny ice crystals shed from their rough surfaces, which is another proposed mechanism of secondary ice production, discussed in Chapter 7.4.

---

[8]There are some exceptions. Lightly rimed branched crystals or aggregates can still collide and stick together by mechanical interlocking. Rimed ice crystals or aggregates may also stick to a wet hailstone surface and become integrated into the outermost layer of ice.

Unlike other hydrometeors, hailstones do not undergo any significant breakup process within the storm because their number concentrations are, on average, too low to experience collisions with other hailstones. However, large hailstones, particularly those with lobes, have been observed to break apart when they make contact with the ground or other objects near the ground. This tendency can result in the underestimation of the largest hailstones produced by a given storm.

Questions for Review

1. Why is quantification of aggregation for snow more difficult than it is for collision-coalescence for rain?
2. At what temperatures are the largest aggregates observed, and why?
3. Why is the simple continuous collection equation not generally valid to represent the aggregation process?
4. Can aggregation also be important for rainfall reaching the ground?
5. What conditions are necessary for the onset of riming of a frozen hydrometeor?
6. What is dry growth and how does it produce graupel?
7. What are the key differences in the conditions that result in dry growth and wet growth of hailstones?
8. Why is the continuous collection equation a reasonable representation of the growth of graupel particles and hailstones by riming?
9. What factors are currently thought to influence the production of large hail within a thunderstorm?

Homework problems and other resources for students and instructors can be found at bit.ly/4dYBwB5, or scan this code:

Further Reading

For newer developments in modeling aggregation [advanced]:

Sulia, K. J., Z. J. Lebo, V. M. Przybylo, and C. G. Schmitt, 2021: A new method for ice–ice aggregation in the adaptive habit model. *Journal of the Atmospheric Sciences*, **78 (1)**, 133–154

For more on hail and hailstorms:

Knight, C. A., and N. C. Knight, 2001: Hailstorms. *Severe Convective Storms, Meteorological Monographs*, **28 (50)**, 223–248

For information on hail and climate change:

Raupach, T. H., and Coauthors, 2021: The effects of climate change on hailstorms. *Nature Reviews Earth & Environment*, **2 (3)**, 213–226

CHAPTER 12

Melting, Evaporation, and Sublimation

In this chapter, we conclude our examination of the microphysical chain of events that culminates in precipitation reaching the ground (Fig. 9.1). Hydrometeors precipitating through the cloud, and especially once they exit the cloud, often undergo further phase changes and may lose some of their mass.

We begin by addressing the melting of frozen hydrometeors, a common occurrence in warmer seasons or climates. In the mid-latitudes and even in tropical regions, much of the rainfall reaching the ground results from the melting of graupel particles or aggregates as they fall into warmer air.

We then address other phase changes, such as the evaporation of raindrops and the sublimation of frozen hydrometeors, that occur when these hydrometeors are subjected to subsaturated air outside or beneath the cloud. In drier environments, these processes can significantly reduce surface precipitation, induce strong downdrafts that reach the ground, and substantially cool the air brought to the ground.

12.1 Melting of Frozen Hydrometeors

The melting of frozen hydrometeors primarily depends upon the air temperature to which they are exposed. Therefore, we begin by examining how the vertical temperature profile beneath a precipitating cloud can influence the phase of precipitation encountered at the ground.

12.1.1 Influence of the vertical temperature profile

The precipitation that initially falls from the cloud results solely from the various microphysical processes operating within it. However, as precipitating hydrometeors fall, the vertical temperature structure within and beneath the cloud can completely alter their phase by the time they reach the ground. Let us consider some examples where frozen precipitation encounters different vertical temperature profiles as it falls, as shown in Fig. 12.1.

We first consider the scenario depicted in panel (a), in which the temperature within the cloud and all the way to the ground is entirely below 0 °C. In this case, the snow created within the cloud will produce snowfall at the ground. This is common during winter months at many mid- and high-latitude locations, as well as high-altitude locations. In warmer seasons or tropical locations, cold near-surface temperatures rarely, if ever, exist, and snowfall at the surface is correspondingly rare.

Now suppose that the vertical temperature structure through and beneath a deep mixed-phase convective cloud is as shown in Fig. 12.1b. Graupel produced in the upper part of the cloud, where the temperature is below 0 °C, falls into a deep and very warm layer of air that extends from the lower part of the cloud to the ground, and completely melts. An observer at the ground thus experiences rainfall. *Rainfall at the surface that originated as graupel aloft is typical of the warmer seasons of the year in most regions worldwide.*

Note that the warm layer shown in panel (b) is very warm. If this cloud were precipitating small hailstones rather than graupel, they would still likely melt before reaching the ground. Larger hailstones, however, are less likely to melt completely, considering their greater frozen mass and fall speeds. It is thereore not uncommon to observe hail on the ground from cumulonimbus clouds in warm seasons or regions, even when the subcloud air temperatures are well above freezing.

The vertical profile of temperature in the lower atmosphere is not always so straightforward. The bottom panels in Fig. 12.1 depict two examples where a warm layer of air overlies a much colder layer near the ground. The type of precipitation reaching the surface depends critically on the depth of this cold layer.

In panel (c), the snow first falls into a deep warm layer and melts. The resulting drizzle/raindrops subsequently fall into a colder layer of air with a temperature less than 0 °C all the way to the ground. This situation can arise, for example, when a warm air mass overrides an existing cold air mass in advance of a warm front. If the cold air layer above the ground

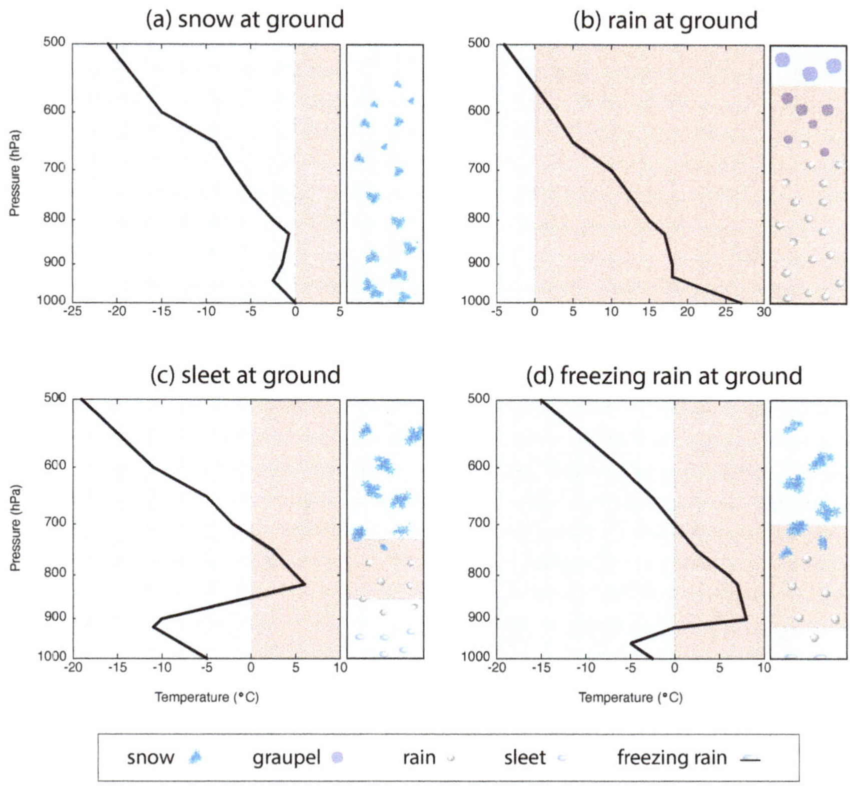

Fig. 12.1: Examples of the influence of the vertical temperature profile on the type of surface precipitation. For each case, left panel shows a vertical temperature profile; right panel depicts the locations of frozen and liquid hydrometeors falling through the corresponding temperature profile. (a) Snow falling within or from a cloud would remain snow all the way to the ground, because the air temperature is less than 0 °C throughout the layer. (b) Graupel falling within a deep mixed-phase cloud encounters a warm layer extending to the ground that completely melts it into rain. (c) Snow within a mixed-phase cloud falls into a warm layer that initially melts it into rain, but a deeper cold layer below allows the rain time to freeze and produce sleet at the ground. (d) Snow in the upper colder regions of a deep cloud melts as it falls into the warmer layer immediately below, but a very shallow layer of cold air right above the ground supercools the rain so that it freezes on contact with objects at the ground.

is deep enough, as in this case, then the drizzle/raindrops have time to refreeze, and they will arrive at the ground as sleet (also known as ice pellets).

A more hazardous situation for society, depicted in Fig. 12.1d, arises when the cold layer of air near the ground is not deep enough to freeze the drizzle or raindrops, but it is deep enough to supercool them. In this case, the drizzle drops or raindrops reach the surface as *freezing drizzle/rain*; they almost immediately freeze on contact with any exposed surface that is also colder than 0 °C. This process can lead to a growing layer of transparent ice that renders roads and sidewalks treacherously slick and, with sufficient accumulation, can topple or destroy trees, power poles, and other structures.

Figure 12.2 depicts an actual temperature profile associated with a destructive freezing rain event that affected several states in December 2002, including North Carolina, where the sounding was taken. Here, a pronounced warm layer between 0.9 and 3.5 km altitude is evident; its depth of 2.6 km ensures that all frozen hydrometeors are completely melted by the time they reach the lower level. Below 0.9 km, a much colder atmospheric layer exhibits subfreezing temperatures all the way to the surface, with temperatures as low as −5.7 °C near 0.5 km altitude. Assuming raindrop fall speeds of $\sim$5 m s^{-1}, the raindrops were exposed to these temperatures for around 3 minutes—plenty of time to supercool them, but not cold enough to actually freeze them.[1]

In summary, the phase of precipitation at the ground is ultimately dictated by the temperature profile of the air through which it falls. During our discussion, we have often referred to a warm layer in the vertical temperature profile that would melt frozen hydrometeors, but we did not specify its depth. In the next section, we address some factors that affect the rate at which hydrometeors melt, and hence govern the depth of the warm layer required to completely change their phase from ice to liquid.

12.1.2 General considerations

The melting of frozen hydrometeors is not instantaneous, nor does its onset necessarily occur as soon as the surrounding air temperature reaches 0 °C. We examine these aspects of melting in the sections that follow.

---

[1]Recall from our discussion of ice nucleating particles in Sec. 8.4 that the number of these particles capable of nucleating ice at a temperature of −5.7 °C is very small.

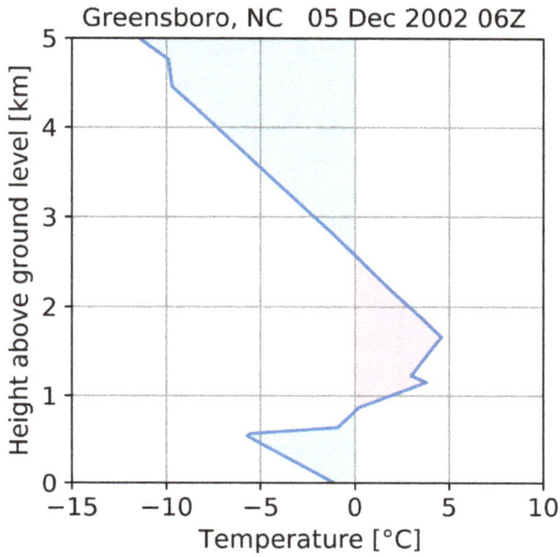

Fig. 12.2: The observed atmospheric temperature profile on 5 December 2002 at 06 UTC over Greensboro, North Carolina. A severe ice storm (freezing rain) was in progress at that time, affecting that location and several neighboring states.

i) Size effects

Compared to the melting of ice crystals and aggregates, the melting of sizable frozen hydrometeors —such as frozen raindrops, graupel particles, and hailstones—requires more time due to their greater mass. This is analogous to observing two ice cubes of different sizes at room temperature; the smaller one melts completely in a shorter amount of time. The surface-area-to-mass ratio also matters: for example, if two graupel particles are of equal diameter but one has a higher density, that particle will require more time to melt completely.

However, in the atmosphere, the fall speed of the hydrometeor also plays a role. Larger and heavier frozen hydrometeors fall faster, allowing them less time to melt before reaching the ground compared to smaller and lighter frozen hydrometeors. This factor also increases the likelihood of larger frozen hydrometeors reaching the ground as some form of ice.

Melting of a frozen hydrometeor initiates at its surface, similar to how an ice cream cone begins to melt on a warm day. The melting rate is then governed by heat transfer through its liquid layer (Fig. 12.3), where the latent heat of fusion is absorbed. This involves several processes: (i) heat conduction through the surrounding air and the liquid layer; (ii) latent heating or cooling due to condensation onto or evaporation from the liquid layer; and (iii) possible heat exchange through collision and coa-

Fig. 12.3: Schematic of a spherical ice particle melting in warm air having a wet-bulb temperature greater than 0 °C.

lescence with cloud droplets if the hydrometeor is still within the cloud. These factors can be integrated into an energy budget model to predict, for example, the melting rate of spherical raindrops, graupel particles, or hailstones.

ii) Relative humidity effects

If frozen hydrometeors begin to melt outside the cloud, i.e., in environmental air having a relative humidity less than 100%, their surface liquid layer is subject to evaporation. The effective temperature experienced by the liquid layer is then not the conventional air temperature, but rather the *wet-bulb temperature*, the temperature achieved by a wet surface when exposed to adequate ventilation.

The wet-bulb temperature results from simultaneously cooling the air due to latent heat uptake and increasing the dewpoint via evaporation until the temperature matches the new dewpoint temperature. At low relative humidity, the wet-bulb temperature can be substantially lower than the true air temperature, which can slow or even prevent the melting of frozen hydrometeors, even when the conventional air temperature exceeds 0 °C.

The delay in the melting of a frozen hydrometeor due to the evaporative cooling of its wet surface is shown by the red curve in Fig. 12.4, depicting combinations of air temperature and relative humidity for which the wet-bulb temperature is exactly 0 °C.[2] Conditions above that curve lead to melting, while conditions below it do not. For example, if the air sur-

[2]It is actually the wet-bulb temperature with respect to the ice phase T_{wi} that determines the onset of melting for an initially frozen hydrometeor. However, the difference between T_w and T_{wi} is usually less than a few tenths of a degree, with the latter being the slightly colder value. By comparison, the overall effect of T_w on the onset of melting can be as much as 10 °C, as shown in Fig. 12.4, so its use in place of T_{wi} does not introduce a significant error.

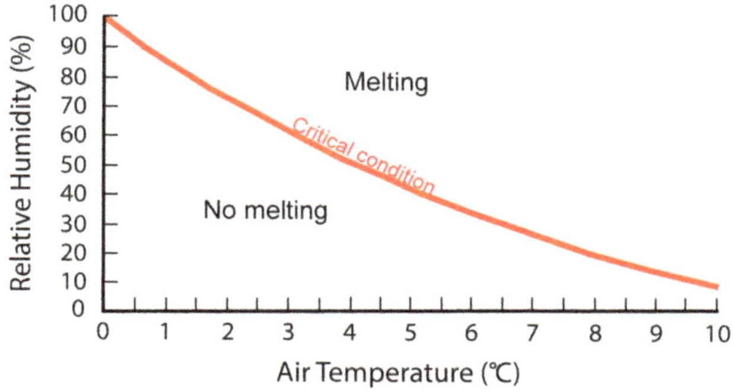

Fig. 12.4: Ambient air temperature and relative humidity required for the melting of frozen hydrometeors to begin. *Adapted from Rasmussen and Pruppacher (1982). © American Meteorological Society. Used with permission.*

rounding a falling frozen hydrometeor has a relative humidity of 40% or less, it must also have a temperature of 5 °C or more to initiate melting.[3]

In contrast, frozen hydrometeors falling within a cloud into air at or exceeding 0 °C will immediately begin to melt. The relative humidity of the cloudy air surrounding the frozen hydrometeors is 100% or slightly greater, preventing any evaporative cooling of their liquid surfaces as they melt.

12.1.3 Melting of graupel

As discussed elsewhere in this book, much of the Earth's heavier rainfall results from the melting of graupel produced within deep mixed-phase clouds. Melting graupel can produce raindrops of highly variable sizes that depend not only on the original particle sizes but also on their bulk densities. Lower-density graupel particles, when melted, will yield raindrops or even drizzle drops that are much smaller than the original sizes of the graupel, due to the higher density of liquid water. For example, the melting of a 1-mm-diameter graupel particle with a density of 0.2 g cm^{-3} would result in a drizzle drop only 600 μm in diameter. For larger graupel particles, which also tend to have larger bulk densities, the size difference after melting may not be as pronounced. A 5-mm-diameter graupel particle with a density of 0.85 g cm^{-3}, for example, would produce a raindrop 4.7 mm in diameter when melted.

---

[3]This effect can sometimes be observed during a snowfall event in which the air temperature is several degrees above 0 °C, but precipitation continues to fall as snow. If the relative humidity of the air then increases above the critical value for that temperature, the snow will begin to melt and reach the surface as wet snow or possibly rain.

12.1.4 Melting of hailstones

In contrast to graupel particles, hailstones have greater fall speeds and thus have less time to melt before reaching the ground. When the 0 °C wet-bulb isotherm is closer to ground level, either due to high surface elevation or a cooler air mass, the depth over which melting can occur is reduced, and hailstones are more likely to reach the surface without melting completely.

For example, the annual maximum of hailstorm frequency in the U.S. is not in the Central Plains—where the frequency of *large* hailstones at the ground is highest—but rather in Colorado. This is because of two key factors: (a) the general decrease in temperature with altitude means that the 0 °C isotherm in Colorado is often closer to the surface as compared to the lower elevations of the Central Plains, and (b) the air just east of the Rocky Mountains has lower relative humidity, pushing the 0 °C wet-bulb isotherm even closer toward ground level.

One focus of current research is determining how hailfall severity or frequency might change in a future warmer climate. Changes in convective storm *frequency* are projected to vary regionally, but convective storm *intensity* is expected to increase worldwide, including the possibility of these storms producing larger hailstones. A major factor to then consider is the extent to which hailstones might melt as they fall toward the ground in such a climate. We can examine, using climate model projections, how the average height of the *melting level* (i.e., the average height of the wet-bulb 0 °C isotherm) might change over the summer months. An increase in the average melting level height corresponds to a greater depth over which hailstones can melt before reaching the ground.

Fig. 12.5 shows an example of the projected *difference* in the melting level height over the U.S. between the end of the 21st century and the end of the 20th century, for the summer months of May, June, and July, when deep convection is most frequent. The projections shown here are from a single global climate model, assuming that humankind fails to curb its emissions of carbon dioxide at all, i.e., a "worst-case scenario."

In general, all the months show an increase (positive differences) in melting level height over the entire U.S. compared to typical values at the end of the 20th century. The most extreme increases in melting level height tend to occur in regions where hailstorms are less frequent (at least currently), such as the Northeastern U.S. and parts of the West Coast. In regions that do typically experience hailstorms, the projected increases in melting depth are often 1,000 m or more and may be sufficient to completely melt small hailstones as they fall to the ground. Multiple studies suggest that the frequency of "pea-sized" hailstones reaching the ground may be greatly reduced in the future due to anthropogenic climate change.

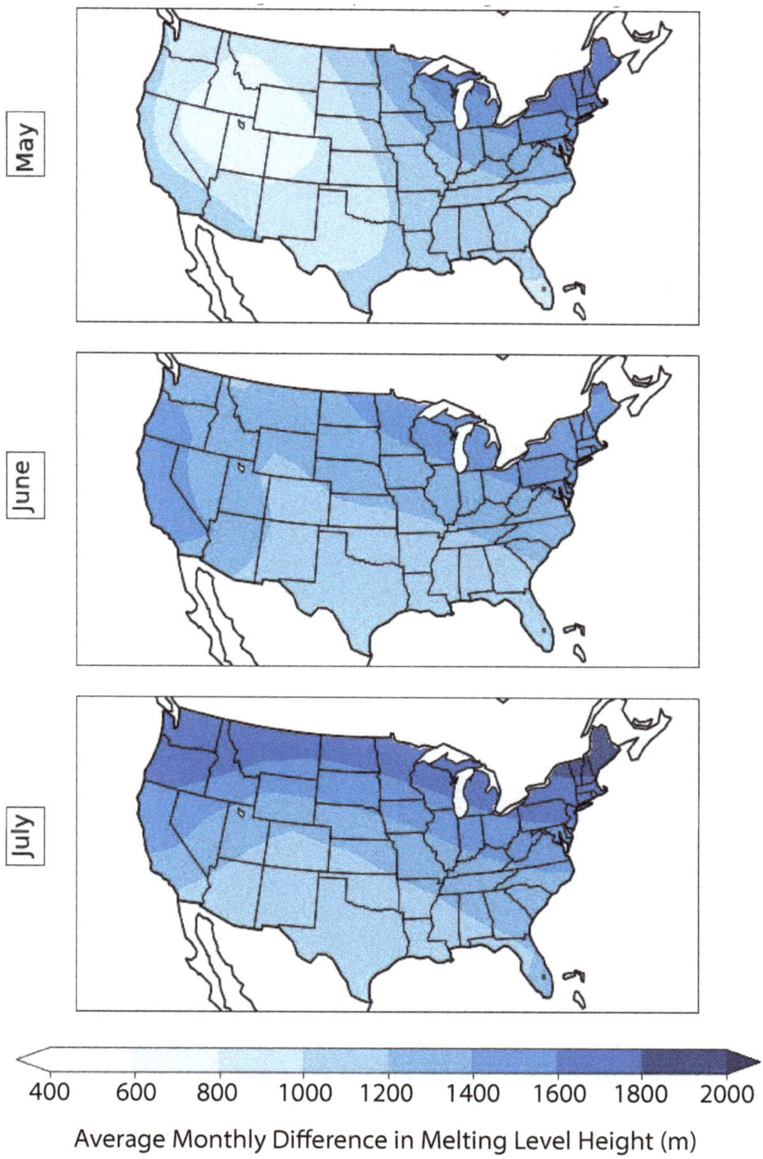

Fig. 12.5: Increases in monthly averaged melting level height (height of 0 °C wet-bulb isotherm) over the continental U.S. from the end of the 20th century to the end of the 21st century, as predicted by a global climate model that assumes the most extreme anthropogenic warming, i.e., that no decrease in the increasing trend of CO_2 will occur by that time. *Courtesy of M. Majetic and A. Elizondo.*

Unfortunately, large hailstones would be minimally impacted by these increases in melting depth because their fall speeds are significantly greater. For example, a hailstone 2.5 cm (1 inch) in diameter, with a fall speed of about 25 m s^{-1}, would only require about 60 seconds to travel through an additional 1,500 m of melting depth, which would not be sufficiently long to completely melt it before reaching the ground.

12.1.5 Melting of aggregates and the radar bright band

The melting of an aggregate is more complicated than for a quasi-spherical frozen hydrometeor like a frozen drop, a graupel particle, or a hailstone. As the aggregate melts, its geometry changes significantly, shrinking and compacting toward the more spherical shape of a drizzle drop or a raindrop. Meanwhile, its fall speed can increase by a factor of five or more, from that of an aggregate ($\sim$1 m s^{-1}) to that of a drizzle drop or small raindrop ($\sim$5 m s^{-1}). The evolving hydrometeor geometry, density, and fall speed as aggregates melt make practical mathematical modeling challenging.

Under certain circumstances, these changes give rise to a phenomenon in radar meteorology known as a **bright band**, a quasi-horizontal band of localized higher reflectivity within a field of otherwise lower reflectivity, as shown in Fig. 12.6. (Another example was shown in Fig. 5.15.) Recall from Sec. 5.3.5 that the strength of a radar echo is proportional to the reflectivity factor Z, which is primarily sensitive to the size of the hydrometeors and, to a lesser extent, their number concentration. Additionally, the radar echo is influenced by the fact that frozen hydrometeors coated with even small amounts of liquid water are significantly more reflective of radar microwave energy than fully frozen hydrometeors.

Let us now consider the evolution of falling ice crystals and aggregates within a cloud, starting well above the 0 °C isotherm. Initially, the ice crystals or aggregates may be relatively small, but (further) aggregation increases their sizes during their descent. Due to their smaller sizes and the fact that they are still completely frozen, their radar reflectivity is rather low at this stage.

In the example shown in Fig. 12.6, above 4 km height, the radar reflectivity is 20 dBZ or less, but increases to approximately 30 dBZ between 2.5 and 3 km height due to aggregation. As these aggregates approach and then pass below the 0 °C isotherm (near 2 km), they become significantly stickier, especially once a film of liquid water begins to form on their surfaces. This enhances the aggregation efficiency, as discussed in Chapter 11, forming even larger aggregates. The small but increasing liquid water surface due to the onset of melting also makes them more reflective. Consequently, the radar reflectivity increases greatly just a short distance below the 0 °C isotherm.

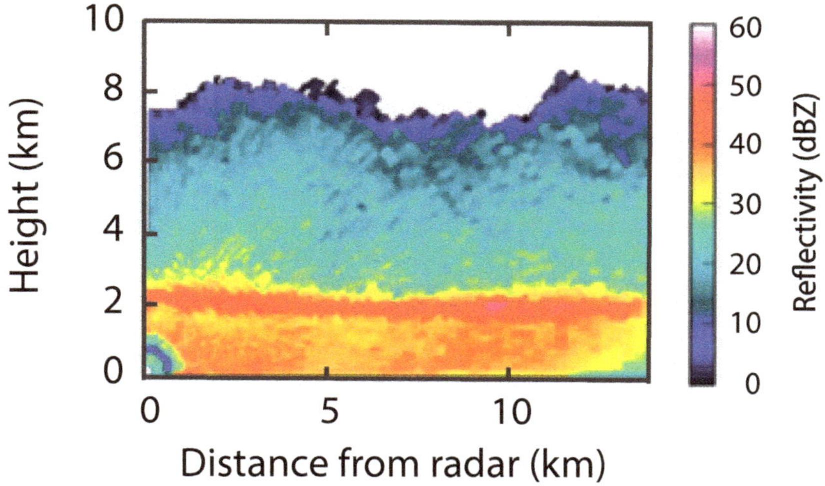

Fig. 12.6: Radar reflectivity from a ground-based, vertically scanning radar, showing a bright band at 2 km height. Height of the 0 °C isotherm is 2.25 km. *Adapted from Sandford et al. (2017). © American Meteorological Society. Used with permission.*

Thus, in Fig. 12.6, the reflectivity suddenly increases to over 40 dBZ at an altitude near 2 km, roughly 10 dBZ higher than seen just above this altitude. The combination of (a) larger aggregate sizes, (b) their small coatings of liquid water, and (c) their small fall speeds is responsible for this sudden increase in radar reflectivity at and directly below the 0 °C isotherm.

In the final stage of aggregate melting, often within a few hundred meters after passing the 0 °C isotherm, melting is completed so that only raindrops remain (below $\sim$2 km in Fig. 12.6). Their diameters are much smaller than those of the aggregates higher up in the cloud, and their fall speeds may be up to five times larger than those of the original aggregates. These faster falling raindrops spread out vertically relative to the slower falling aggregates higher in the cloud, decreasing their number concentration. Decreases in both hydrometeor size and number concentration correspondingly reduce the radar reflectivity factor relative to the zone of melting onset and enhanced aggregation above, demarcating the bottom of the bright band.

A radar bright band is observed only within precipitation falling from stratiform clouds with weak updrafts. When moderate to strong updrafts and downdrafts occur in a cumuliform cloud, this orderly downward progression of aggregation and melting is disrupted. The process is often dominated by the riming and/or melting of graupel and hail, which obscures the bright band feature on a radar image.

12.2 Evaporation or Sublimation of Falling Precipitation

Hydrometeors may briefly encounter regions of unsaturated air within the cloud due to entrainment, be exposed to unsaturated air outside the cloud due to detrainment, or fall from the cloud as precipitation. In any of these cases, hydrometeor growth stops, and a reduction in mass by evaporation or sublimation begins. Their fate then depends on their size, fall speed, the time of their exposure to the unsaturated air, and the relative humidity of that air. Hydrometeor size is also important: as previously detailed in Sec. 9.1.7, the greater fall speeds of larger hydrometeors increase their rate of evaporation or sublimation—referred to as the ventilation effect—beyond what would occur if they were stationary.

Some precipitating hydrometeors may completely evaporate or sublimate before reaching the ground. However, if they are large enough and/or if the unsaturated layer is relatively shallow or has a high relative humidity, they may lose only part of their mass. Any mass lost from the hydrometeors is added to the air as water vapor, with the associated latent heat of vaporization or sublimation being supplied by the air surrounding the hydrometeors. Consequently, this process both moistens and cools the unsaturated air through which the hydrometeors fall, sometimes even briefly bringing the air to saturation and creating a temporary fog at the ground, as discussed in Sec. 2.5 (see also Fig. 2.6). In extreme cases, latent cooling can produce negatively buoyant air, as described in the next section.

12.2.1 Downdrafts

i) Cooling through phase changes

Downdrafts often owe their existence, at least initially, to evaporating liquid hydrometeors or sublimating frozen hydrometeors.[4] Let us consider a small, incremental change in the phase of water or ice as it evaporates or sublimates in a parcel of clear air. We start with the First Law of Thermodynamics from (2.1):

$$\mathrm{d}q = L\,\mathrm{d}q_x = C_p\,\mathrm{d}T - \alpha\,\mathrm{d}p\,, \tag{12.1}$$

where $\mathrm{d}q$ is the heating/cooling experienced by the air in the parcel, L is the latent heat of vaporization or sublimation (depending on hydrometeor phase), $\mathrm{d}q_x$ is the change in the mass mixing ratio of the liquid or frozen

[4]Some downdrafts are also assisted by the overturning motions of eddies common in turbulent clouds.

hydrometeors, C_p is the specific heat of air at constant pressure, $\mathrm{d}T$ is the change in the parcel air temperature, α is the specific volume of the air (the reciprocal of the air density ρ), and $\mathrm{d}p$ is the change in the parcel pressure as it moves vertically.

Now let's assume that the change in pressure $\mathrm{d}p$ during the evaporation or sublimation is effectively zero, as we are considering an incremental change in hydrometeor mass over a short time. Then (12.1) can be simplified and rearranged to give

$$\mathrm{d}T = \frac{L}{C_p}\,\mathrm{d}q_x. \tag{12.2}$$

It is now clear that a decrease in hydrometeor mass due to evaporation or sublimation ($\mathrm{d}q_x < 0$) results in a proportional decrease in the temperature of the surrounding air. Because the latent heat of sublimation is greater than that for vaporization, the sublimation of a given mass of ice will cool the air somewhat more than the evaporation of the same mass of liquid drops.[5] Substituting values for L and C_p into (12.2), we find that for every gram of liquid drops evaporated per kilogram of air, the air temperature decreases by 2.5 °C. For the sublimation of ice, the decrease is a little greater, 2.8 °C.

This latent cooling of the air by evaporation or sublimation can make the air parcel negatively buoyant, being cooler than the undisturbed air outside it. The vertical acceleration a due to the buoyancy of the parcel can be written as:

$$a = \frac{T_{v,\mathrm{parcel}} - T_{v,\mathrm{env}}}{T_{v,\mathrm{env}}}\,g\,, \tag{12.3}$$

where $T_{v,\mathrm{parcel}}$ and $T_{v,\mathrm{env}}$ are the respective virtual temperatures (Sec. 2.3.6) of the parcel and the environment, both in units of kelvin, and g is the acceleration due to gravity. When the parcel virtual temperature is less than that of the environment,[6] the air parcel will experience a negative acceleration (i.e., accelerate downward).

---

[5]While the melting of frozen hydrometeors also cools the air, its role in driving downdrafts is comparatively minor. This is because the latent heat of fusion is an order of magnitude less than that for sublimation or vaporization. Note, however, that the latent heat of fusion is responsible for the difference between the cooling effects of sublimation and evaporation.

[6]During evaporation or sublimation of the hydrometeors in the parcel, the air is not only cooled but also moistened by the phase change from liquid or ice to water vapor. However, the contribution of this added water vapor to the parcel virtual temperature is small, and the latent cooling effect dominates.

The air parcel will begin to descend, warming adiabatically due to compression. However, this adiabatic warming may be partially or entirely offset by cooling from the evaporation or sublimation of hydrometeors. The extent of this cooling depends on the relative humidity within the parcel. When the latent cooling outweighs the compressional warming, the descending air parcel will have a virtual temperature less than its environment, making it negatively buoyant and accelerating downward according to (12.3).

ii) Hydrometeor loading

Falling precipitation can also initiate downdrafts, or add to their negative buoyancy, simply by virtue of its weight within an air parcel. This phenomenon is referred to as **hydrometeor loading**. To quantify its effects, we add a negative term $-q_t\, g$ to (12.3):

$$a = \frac{T_{v,\text{parcel}} - T_{v,\text{env}}}{T_{v,\text{env}}} g - q_t\, g = \left(\frac{T_{v,\text{parcel}} - T_{v,\text{env}}}{T_{v,\text{env}}} - q_t \right) g\,, \qquad (12.4)$$

where q_t is the total hydrometeor mass mixing ratio (unitless).

We see that a mass of hydrometeors within an air parcel "weighs it down." If the air parcel was already negatively buoyant due to its virtual temperature difference with its environment, hydrometeor loading will increase the downward acceleration of the parcel even more.

The effect of hydrometeor loading can potentially overcome any positive buoyancy that an air parcel might otherwise have. Thus, an air parcel in an updraft containing appreciable amounts of hydrometeor mass will not fully achieve the speed expected from its excess in virtual temperature relative to its environment. If the net buoyancy is negative due to hydrometeor loading, then an initially rising or stationary parcel of air will begin to descend as a downdraft.

We can use (12.4) to compare the contributions of hydrometeor loading and virtual temperature excess/deficit to the air parcel buoyancy. For example, taking $T_{v,\text{env}} \approx 0\ ^\circ\text{C}$, the loading effect due to each gram of hydrometeors per kilogram of air is equivalent to a 0.27 °C reduction in the parcel virtual temperature. In very heavy precipitation, the total hydrometeor mass can be on the order of several grams per kilogram of air or more, so the total loading effect on buoyancy could be equivalent to a full degree or more of cooling.[7]

[7]Note that the calculation using (12.4) must be performed with all virtual temperatures expressed in units of kelvin. The discussion here has converted the temperature reduction to degrees Celsius because it is more intuitive.

iii) Downbursts

We have just discussed how the sublimation or evaporation of hydrometeors, and the loading of hydrometeors, can induce negative accelerations of air parcels. These parcels will then form *downdrafts,* descending currents of air that can be located within, adjacent to, or below a cloud.

From (12.4) and using our earlier example where $T_{v,\mathrm{env}} \approx 0\,^{\circ}\mathrm{C}$, it can be shown that for every degree of latent cooling of an air parcel, or its equivalent due to hydrometeor loading, an adiabatic parcel of air experiences a downward acceleration of approximately 0.036 m s^{-2}. This value may seem very small, but if an air parcel starting at rest descends a distance z at constant acceleration a, the final downward velocity w is then

$$w = \sqrt{2az}. \tag{12.5}$$

Thus, for a one-kilometer descent at an acceleration rate of 0.036 m s^{-2}, the final downward velocity is a substantial 8.5 m s^{-1}. Because this expression assumes that the parcel does not mix with its environment and experiences no friction or other forces except gravity, it represents the theoretical maximum downward velocity.

Even so, much larger accelerations are possible, such as when the atmosphere from immediately below the cloud base to the ground has very low relative humidity, greatly enhancing cooling by evaporation of rain. Especially when compounded by hydrometeor loading due to heavy precipitation in thunderstorms, these downdrafts can become extreme, leading to a phenomenon called a **downburst**. A downburst is an area of strong winds near the ground produced by one or more convective downdrafts. Downbursts are capable of flattening trees and some structures on the ground and pose a severe hazard to aircraft during takeoff or landing.

12.2.2 Cold pools

Raindrops undergo evaporation within downdrafts, while frozen hydrometeors might first experience sublimation, then melting near the 0 °C isotherm, and finally evaporation (now as raindrops) all within a single downdraft, if it extends above and below the 0 °C isotherm. All these phase changes latently cool the air parcels containing the hydrometeors. Provided the air parcels remain unsaturated, these processes will continue to lower the temperature of the air within the downdraft. When this air reaches the ground, it will often be cooler than the surrounding air at the same level. We then refer to this latently cooled air as a **cold pool**.

Especially if the air into which hydrometeors are evaporating is initially very dry, the final temperature of the air within the downdraft as it reaches the ground might be significantly cooler than the surrounding

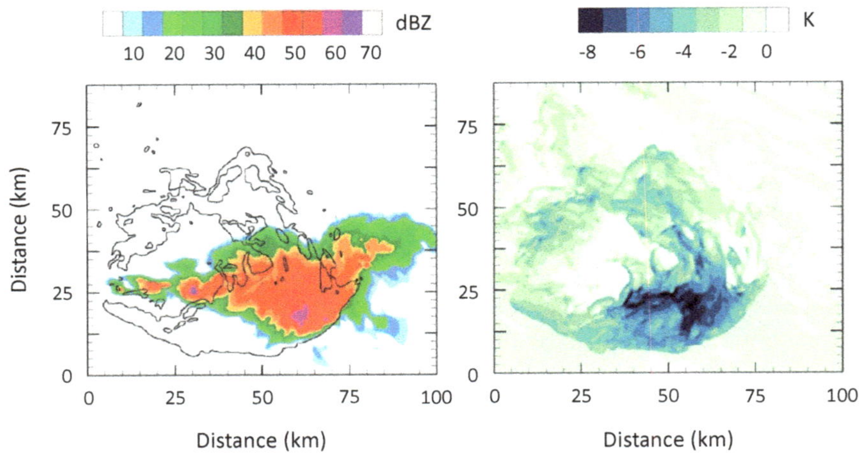

Fig. 12.7: Numerical simulation of a mesoscale convective system and its cold pool. *Left*: Derived lowest-level radar reflectivity (dBZ, shaded), with cold pool edge (black line) overlaid. *Right*: Potential temperature difference (*K*, shaded) between the cold pool and the surrounding air. *Adapted from Mallinson and Lasher-Trapp (2019). © American Meteorological Society. Used with permission.*

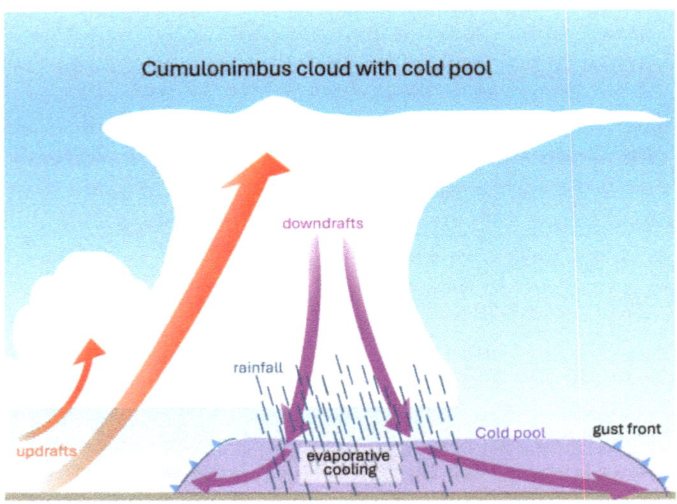

Fig. 12.8: Idealized depiction of a cumulonimbus cloud and the relative locations of the thunderstorm downdrafts, the rainfall, and the evaporative cooling of that rainfall that results in the formation of a cold pool at the ground that spreads outward in time. The gust front along the leading edge of the cold pool is marked by cold front symbols. In this example, the low-level winds are blowing from left to right, limiting the cold pool spread on the upwind side, and enhancing it on the downwind side.

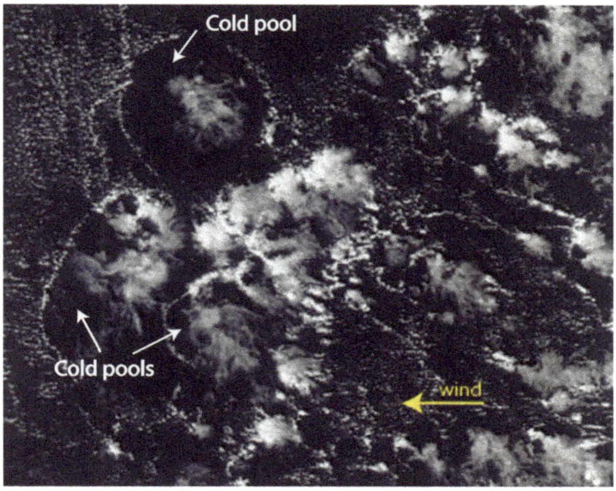

Fig. 12.9: Satellite image over the Caribbean Sea, showing clouds in various stages of evolution, including some that have produced cold pools. The largest cold pools are thelabeled regions mostly devoid of clouds on the left side of the image, each outlined by its prominent ring of clouds at its leading edge. The cold pools were likely created by the fuzzy clouds near their centers, which are in a state of decay after precipitating. The two largest labeled cold pools span approximately 100 km. *From Zuidema et al. (2017). (CC-BY)*

air, by several degrees or more, as shown in Fig. 12.7. In this case, the resulting cooler and denser air spreads outward, displacing the pre-existing warmer air. Air within a strong cold pool can be as much as 10 °C colder than the air outside of it.

An idealized illustration of the location of the cold pool relative to a cumulonimbus cloud is shown in Fig. 12.8. The advancing edge of the cold pool is called a **gust front**, due to the associated sharp change in wind direction and intensity that can be observed near the ground as the cold pool advances. You may have experienced a gust front as a short-lived, strong, cool wind preceding the arrival of a precipitating thunderstorm.

Similar to a cold front but on a much smaller scale, the advancing gust front can displace the warmer air in its path upward. This uplift can prolong the life of the parent storm or initiate new clouds or storms. Its ability to do so is largely dependent on the direction and strength of the low-level winds and the static stability of the low-level atmosphere surrounding the cold pool.

Behind the leading edge of the cold pool, the air near the surface is statically stable, suppressing new cloud formation until the cold pool dissipates. Fig. 12.9 shows an example of multiple cold pools over the Caribbean Ocean, with clear regions within the rings of slightly stronger clouds. The rings of clouds are being forced by gust fronts at the leading edge of the cold pools, while the cloud-free regions result from the stable air within the cold pools.

The evaporation or sublimation of hydrometeors near the ground is not only responsible for cold pools but also reduces the water mass reaching the ground as precipitation. In the next chapter, we address how this precipitation is characterized and quantified.

Questions for Review

1. What environmental factors determine the phase of precipitation (frozen versus liquid) reaching the ground?
2. What conditions favor melting of frozen hydrometeors as they fall into air warmer than 0 °C?
3. How can the local terrain influence hailstone sizes at the ground?
4. What causes a bright band on a radar image?
5. How do hydrometeors influence the buoyancy of the air parcel in which they are located?
6. How can hydrometeors initiate or enhance downdrafts?
7. What is a cold pool, and why is it important?

Homework problems and other resources for students and instructors can be found at bit.ly/4dYBwB5, or scan this code:

Further Reading

For more on melting [advanced]:

Lamb, D., and J. Verlinde, 2011: *Physics and Chemistry of Clouds*. Cambridge University Press

For more on the radar bright band:

Rauber, R. M., and S. W. Nesbitt, 2018: *Radar Meteorology. A First Course*. John Wiley & Sons

For more on the influence of hydrometeor phase changes on cold pools [advanced]:

Mallinson, H. M., and S. G. Lasher-Trapp, 2019: An investigation of hydrometeor latent cooling upon convective cold pool formation, sustainment, and properties. *Monthly Weather Review*, **147 (9)**, 3205–3222

CHAPTER 13

Characterizing and Quantifying Precipitation

In this final chapter, we focus on the precipitation that reaches the Earth's surface; that is, the end product of all the microphysical processes discussed in this book. We begin by describing the methods used to quantify surface precipitation, followed by an examination of the two main categories of precipitation and how their characteristics relate to the different clouds and microphysical processes that produce them. We also consider extreme precipitation events, using flash flooding as an example, to highlight how macroscale cloud properties, such as the direction and speed of their movement, influence the precipitation depth and duration at specific locations. The chapter concludes with a brief discussion on future research directions for extreme precipitation events.

13.1 Precipitation Accumulation

You are likely familiar with the concept of the accumulation of rainfall or snowfall on the ground. You may have had a simple rain gauge outside your home, which you checked after a storm passed to see how much rain fell. In colder regions or seasons, you might have used a ruler to measure the snow accumulation on the ground, hoping for enough snow to cancel school the next day.

If you compare these measurements with those taken by your friends across the region, you might be astonished by how different they are, even when the measurements are made in relatively close proximity. Precipitation is notoriously difficult to measure in a meaningful way.

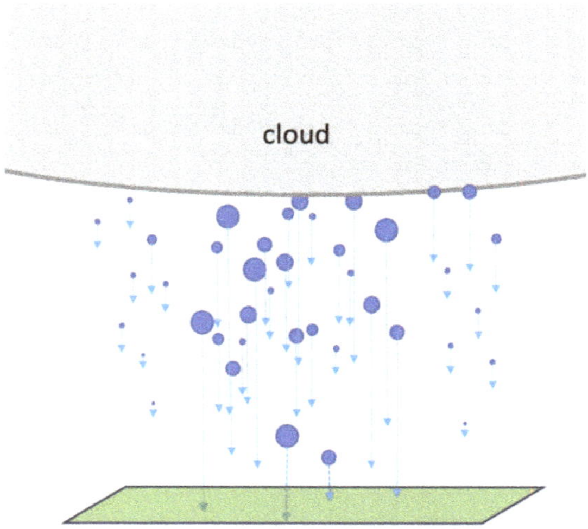

Fig. 13.1: Illustration of raindrops of different sizes (not drawn to scale) precipitating from a cloud toward the ground over an area shown by the green rectangle. The length of the downward arrow is scaled to the magnitude of the fall speed for each drop, except for those that will soon intercept the ground. Some of the smallest raindrops will likely evaporate before reaching the ground, thus contributing nothing to the accumulation.

The precipitation produced across a broad cloud at any instant can be quite variable, as suggested by the simple sketch shown in Fig. 13.1. We have seen in this book that clouds are not homogeneous across their volumes, so it should not be surprising that the locations within the cloud that produce precipitation can be equally variable.

Additionally, the rainfall pattern depicted above would neither remain constant over time nor simply be advected (passively move along) with the cloud. We have observed that both clouds and the precipitation they generate evolve over time, leading to varying amounts of rainfall even in regions that were directly beneath the cloud at some point.

The rate of movement of a cloud over a particular site will also affect the amount of precipitation that falls there. Consider two clouds having the same horizontal extent and producing the same amount of precipitation but moving at different speeds: the slower-moving cloud will produce greater accumulations on the ground than the faster-moving cloud, because the latter cloud is directly overhead for a shorter duration. Additionally, individual clouds have finite extents, so as they move, some locations will never be directly beneath them and thus will not be exposed to their precipitation.

We quantify the **precipitation accumulation** P as a *depth* of rainfall or snowfall on the ground over some period of time, typically expressed in units of millimeters, but in the United States, it is also expressed in inches. Snowfall can be reported as the actual snowfall depth or, if it is collected in a receptacle and melted or weighed, as the **snow water equivalent** (SWE).[1]

The meteorologist is fully aware that SWE is generally much smaller than the "actual" snowfall accumulation measured on the ground during a snowfall event. Weather forecasters predict the unmelted snowfall accumulation by converting the SWE (predicted in numerical weather prediction models) to an actual snowfall depth using a **snow-to-liquid ratio**. The assumed snow-to-liquid ratio is based on cloud temperatures, the humidity of the air ingested by the clouds, and records of past events in a given region. The typical range is from around 3:1 to 30:1 in the U.S., but it can vary widely depending on temperature, humidity, and other conditions. A ratio of 3:1 represents very dense, wet snow, while 30:1—that is, 30 mm of snow depth for every 1 mm of liquid water—represents exceptionally fluffy, powdery snow.

Whereas SWE is most useful for hydrological and agricultural purposes, actual snowfall depth is often more relevant to society, particularly for transportation issues and winter sports enthusiasts. Consequently, weather forecasters and the media typically report the actual snowfall depth.

Although graupel and hail can both reach the surface, accumulations are not reported separately for these hydrometeor types, though graupel occasionally makes up a significant portion of the total reported snow depth. Hailstones are usually too sparse to be measured as an accumulation, except in the rare reports of heavy showers of small hailstones that can pile up like snow on the ground.

The rainfall or snowfall accumulation can be reported as the difference between the measured depths of precipitation at the start of the event (d_i) and after it has finished (d_f):

$$P = d_f - d_i, \tag{13.1}$$

where each depth d, and thus P, is typically recorded in millimeters (although the U.S. National Weather Service reports P in inches). The initial depth d_i is, of course, zero if the measurement device is empty at the start of the precipitation or if there was no snow already on the ground before the event. For climatological purposes, the accumulation may be recorded as a 6-hour, 12-hour, or daily amount.

[1]SWE is commonly pronounced by the hydrology and snow science communities as "swee."

13.1.1 Rain gauges

A **rain gauge** measures the accumulation of rain in units of depth at regular time intervals. Measurement intervals for some rain gauges can be as short as one minute or even less. However, for many operational gauges around the world, accumulations are reported only hourly or sometimes even daily in remote regions.

The simplest rain gauges are nothing more than open containers that capture falling precipitation. If the container is cylindrical with straight vertical sides, the accumulated depth d can be measured directly with a ruler, similar to a cylindrical glass rain gauge with an etched scale that you may have outside your home. Otherwise, d is computed as the volume V of collected water divided by the area A of the top opening. These gauges require manual emptying after each event.

Some automated operational gauges use a similar principle of measurement: a *weighing gauge* measures the weight—and thus volume—of water collected in a bucket inside the gauge, recording it as a function of time (Fig. 13.2, top right). The bucket (colored blue in this image) is suspended by wires that vibrate at different frequencies as it becomes heavier; the change in frequency can be translated into an accumulation. They have been shown to have better accuracy than other gauges but require manual emptying of the bucket, so they must be located in accessible locations.

The most commonly used rain gauge for operational applications is called a *tipping bucket rain gauge*; its inner workings are shown in the bottom right image of Fig. 13.2. In these gauges, the "bucket" is the metal receptacle shown in the center of the image, divided into two halves by a central barrier. The bucket can pivot left or right about its center point. Rainfall entering the gauge falls into the half of the bucket that is tilted upward. When 0.254 mm (0.01 inches) of rain accumulates, its weight tips the bucket downward on that side, simultaneously triggering an electrical signal to record the rainfall and dumping the water. The other half tilts upward, ready to collect the next increment, repeating the process until the rain stops.

13.1.2 Snow measurements

Some of the devices already discussed for measuring rainfall can also be used for measuring the melted-equivalent snowfall (SWE). For example, many tipping bucket rain gauges contain thermostatically controlled heaters that automatically melt snow falling into the collection funnel, so that its water content can be measured in the usual way. Some weighing gauges are either heated or partly filled with non-evaporating antifreeze so that the collected snow is melted into water.

Fig. 13.2: *Left:* Side view of a weighing bucket gauge mounted on a portable trailer. The black cylinder on top is the gauge opening for precipitation to enter. The array of freely swinging triangular metal pieces surrounding the exterior of the gauge, called an Alter wind shield, helps to damp winds around the gauge opening that could blow precipitation away from it. *Right, top:* After taking off the silver exterior cover of the gauge shown to the left, the actual bucket (blue) inside the bottom of the gauge can be viewed, along with the wires that record changes in the bucket weight. *Right, bottom:* The interior workings of a tipping bucket rain gauge, with the collection funnel removed. The bucket in this case refers to the metal chute in the center with a barrier that divides it into halves and pivots around its center axis. The coiled tubing below the bucket protects the sensor wires that record each tip and the electrical lead to the rainfall recorder. *Photos left and top right by S. Lasher-Trapp; photo bottom right from WikimediaCommons.*

For measuring actual snowfall depth, different techniques are required. One method consists of a simple *snow stick*, which is essentially a long, graduated pole or ruler installed vertically—sometimes staked directly into the ground—in open, flat areas to avoid interference from snow drifts or nearby obstructions. The total snow depth is measured by visually reading the snow level against the graduated markings on the stick. Accumulated depth since the previous measurement is then just the difference between the two readings. Snow sticks are widely used by weather observers and ski resorts to simply and reliably monitor snow accumulation.

Another simple, low-tech device is called a *snowboard*. It consists of a flat board painted white so as to minimize the melting of snow by sunlight. It is placed in an open area away from trees and structures that might shelter it from the snow (and thus underestimate the snowfall). The board must be cleared off before the snow event begins, and the depth of snow on it is measured with a ruler set on its edge on top of the board. The snow depth should be measured as soon as the snowfall ends because the longer the snow sits on the snowboard, the more it compacts over time (as also occurs with snow that lies on the ground).

A few automated methods are sometimes used in mountainous areas that are less accessible, but they can be much more expensive and possibly less reliable. For example, *acoustic sensors* can be used to measure new snowfall by sensing the change in the overall snow depth as a function of time. These instruments essentially send ultrasonic pulses downward toward the ground, and they are reflected back to the sensor by the snow surface. As the snow deepens, the time between the emitted pulse and the received signal decreases, which can be converted to increases in the height of the snowpack surface. The accuracy of these measurements can be influenced by local wind speeds, as well as by the air temperature and relative humidity.

Optical sensors operate under somewhat similar principles. The instrument directs a laser downward toward the ground, and by measuring the time required for the light to reflect back to the sensor, the distance between the snow surface and the sensor can be calculated. The accuracy of these sensors does not suffer from the weather conditions that affect acoustic sensors, but they are much more expensive.

13.1.3 Challenges of precipitation measurements

Perhaps surprisingly, *accurate* measurements of precipitation accumulations present a significant technical challenge in operational meteorology. The problem arises because all common precipitation gauge designs are subject to difficult-to-eliminate errors and biases, especially under windy conditions or when the precipitation is light and/or frozen.

For example, when precipitation is light, it may wet the collection funnel, but a significant fraction might evaporate without ever flowing deeper into the gauge where it can be measured. Winds blowing around the gauge can create artificial air circulations that may even prevent precipitation from entering the gauge opening. This issue is especially significant for frozen precipitation, which falls more slowly than rain and is more susceptible to these circulations.

Various types of *wind screens* have been developed over the years to dampen such circulations. The array of freely-swinging triangular metal pieces surrounding the gauge in Fig. 13.2, known as an Alter wind screen, was designed to address this problem. It provides some help, but more recent advances have altered its shape and incorporated additional external fences around the gauge to mitigate artificial air motions. Strong winds during snowstorms can also cause snow drifts, which are regions of locally enhanced snowfall. Gauges located within snow drifts will measure inflated overall snowfall accumulation, so a series of *snow fences* is often installed around the gauge to address this issue.

Another concern regarding gauge measurements is representativeness: that is, the reality that the collection area of a gauge is minuscule relative to the scale of the precipitation-producing cloud overhead, capturing only a tiny fraction of the potentially complex pattern of precipitation reaching the ground. Consequently, any practical number of sparsely distributed gauges is typically inadequate for accurately capturing the spatial variability of precipitation across a cloud. Additional data sources, such as weather radar, are usually essential to provide a more comprehensive picture, albeit with their own limitations on accuracy.

To increase the spatial coverage of rainfall and snowfall measurements across the United States, a national community effort has been cultivated over the last two decades: the Community Collaborative Rain, Hail, and Snow Network (CoCoRaHS; cocorahs.org), a nonprofit organization sponsored by the National Oceanic and Atmospheric Administration (NOAA) and the National Science Foundation (NSF). Anyone is free to join, receive a rain gauge, and be trained on how to properly mount the rain gauge to attain the most accurate measurements. Volunteers are also instructed on how to make snowboards for measuring snow accumulation and how to melt the snow collected in their gauges for reporting SWE accumulation. These measurements are then submitted online or through the CoCoRaHS app and are made immediately available for use.

These data have proven invaluable in augmenting the available data for the meteorological, hydrological, agricultural, and engineering communities, among others. CoCoRaHS observations are now collected in all 50 states of the U.S., as well as in Canada and the Bahamas, with plans for expansion to additional countries worldwide.

13.2 Precipitation Rate

Surface precipitation occurs when hydrometeors, whether in liquid or frozen form, survive evaporation or sublimation to reach the ground. Since not all falling hydrometeors have the same size, they do not fall at the same speed, introducing variability in the time required for them to reach the ground. Within downdrafts inside or beneath the cloud, hydrometeors will approach the ground more quickly than their individual fall speeds.

When evaluating how intensely clouds and storms can precipitate, we utilize the **precipitation rate,**[2] R, defined as the downward flux of water mass through or onto a specified horizontal reference plane. As shown by the green square in Fig. 13.1, this plane is often taken to be ground level, although it does not have to be.

While R can be defined, computed, and/or measured as a mass of water per unit time per unit horizontal area, the precipitation rate is more commonly expressed as an equivalent *depth per unit time*, usually in millimeters per hour. Such a convention is convenient because we already use precipitation gauges to express accumulation as a depth. The mean precipitation rate $\overline{R}$ over any given time interval can then be obtained from precipitation gauges as

$$\overline{R} = \frac{P_2 - P_1}{\Delta t} \tag{13.2}$$

where P_1 is the accumulation at some initial time, P_2 is the accumulation at a later time, and Δt is the difference in those times. A true instantaneous rain rate would require $\Delta t \rightarrow 0$, in which case

$$R \equiv \frac{\mathrm{d}P}{\mathrm{d}t}, \tag{13.3}$$

but real-world rain gauges rarely have the resolution to even approach this idealization.

Rather, the interval over which the rate is actually computed is often at the finest time resolution that the gauge records new values of accumulation, which may be one minute, five minutes, or more. For a tipping bucket-type gauge, in fact, the optimal way to obtain R at highest possible resolution is as the fixed rainfall increment ΔP corresponding to a single tip—typically 0.254 mm (0.01 inch)—divided by the observed time Δt since the previous tip. The latter can range from an hour or more in extremely light precipitation down to a minute or less in an extremely heavy downpour.

Because the precipitation rate R can be expressed as the time derivative of the accumulation P, the accumulation P can in turn be expressed as a time integral of the precipitation rate R:

$$P = \int_{t_i}^{t_f} R \, dt \tag{13.4}$$

---

[2]It is common to denote precipitation rate with R. We have often used R in previous chapters to represent the radius of a given hydrometeor. To prevent confusion in this chapter, we will use the hydrometeor diameter D to represent its size.

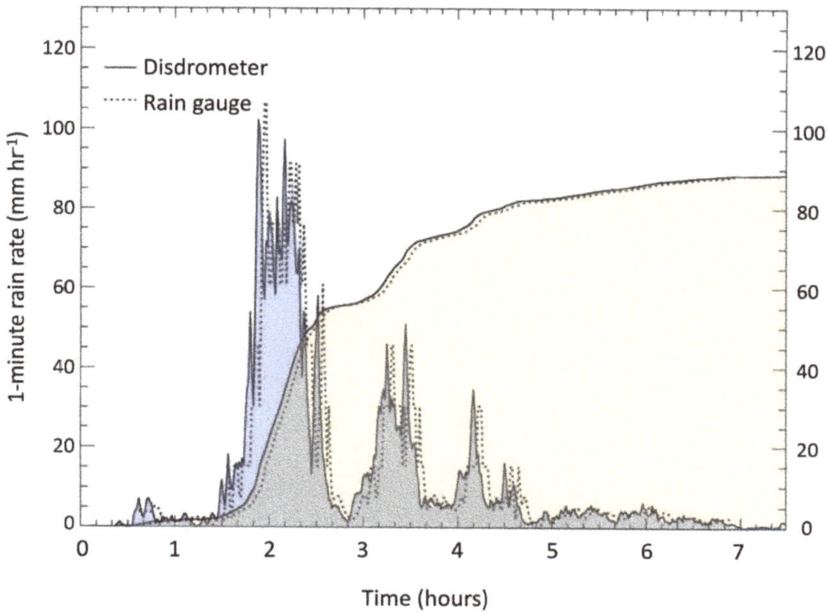

Fig. 13.3: Comparison of 1-minute rainfall rates (filled blue line and accompanying jagged dashed line, corresponding to the scale on the left) and accumulated rainfall (filled yellow line and accompanying smooth dashed line, corresponding to the scale on the right) by a disdrometer and a tipping bucket rain gauge. To calculate the rainfall rate and accumulation to compare to the tipping bucket measurements, the raindrop size distribution would be used at one-minute intervals with equations (13.4) and (13.10). The rain rate measurements are slightly offset from one another due to the use of different clocks on the instruments. *Adapted from Wang et al. (2008). © American Meteorological Society. Used with permission.*

where t_i is the initial time when precipitation began at a given location, t_f is the final time when precipitation was recorded at that location, and R is the rain rate, melted-equivalent snowfall rate, or actual snowfall rate.

Rarely are precipitation rates even approximately constant over an entire rain event, so R is a function of time. For example, Fig. 13.3 shows that R can increase or decrease by a factor of two or more over the course of a minute. This variability in R is, of course, lost if only the final accumulation P is reported, such as the single value at the end of the time series.

For rain rates, the range of R measured at the ground spans values from a few millimeters or less per hour (what meteorologists would label "drizzle") to 25 millimeters per hour (approximately 1 inch per hour; associated with convective clouds) and, in rare cases, up to several hundred millimeters per hour (several inches per hour; observed only in the

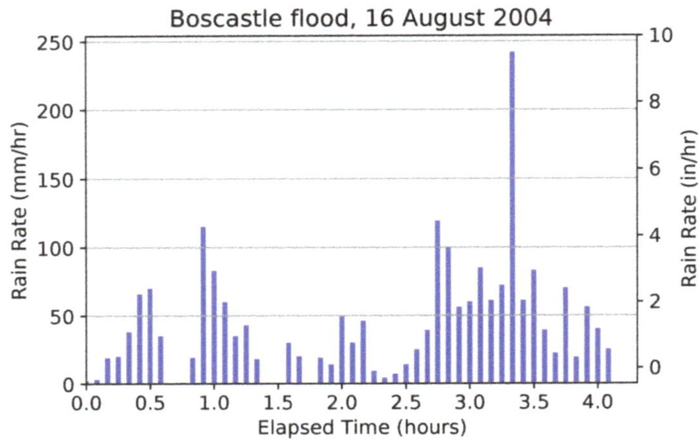

Fig. 13.4: Measured rainfall rates over 5-minute periods at a single weather station in Southwest England. A succession of storms moved over the area and caused flash flooding. Values of the corresponding rainfall rate in inches per hour can be read from the scale on the right-hand side of the plot. *Data from Burt (2005).*

strongest storms). Fortunately, such extreme rates rarely last longer than a few minutes.

A real-world example of extreme rain rates is shown in Fig. 13.4, where 5-minute rain rates are plotted over a period of four hours during a flash flooding event. Several instances of rainfall rates exceeding 100 mm hr^{-1} (roughly 4 inches per hour) occurred, and at one time interval, an intense rainfall rate of nearly 250 mm hr^{-1} ($\sim$10 inches per hour) was measured!

Snowfall rates, like accumulations, may be reported as the rate of unmelted snow accumulation on the ground, which is typically the primary concern for the public, or as the **liquid water equivalent** (LWE) **rate**, which is more pertinent for hydrological and agricultural purposes in determining the rate of water reaching the ground. While unmelted snowfall rates can be as high as 150 mm hr^{-1} (about 6 inches per hour) during lake-effect snowstorms or in mountainous areas, their LWE rates may be only 7.5 to 15 mm hr^{-1}. For more moderate snowfall rates, the LWE rates are typically 1 to 5 mm hr^{-1} or less. If, however, some of the crystals or aggregates are rimed, the LWE rates will be higher, 10 mm hr^{-1} or more. These LWE rate maxima, however, are still much smaller than the extreme rates recorded during heavy rainfall.

13.2.1 Rain rate from raindrop size distributions

In addition to rain gauges that simply measure accumulated volume, instruments called *disdrometers* can measure the size distribution of rain-

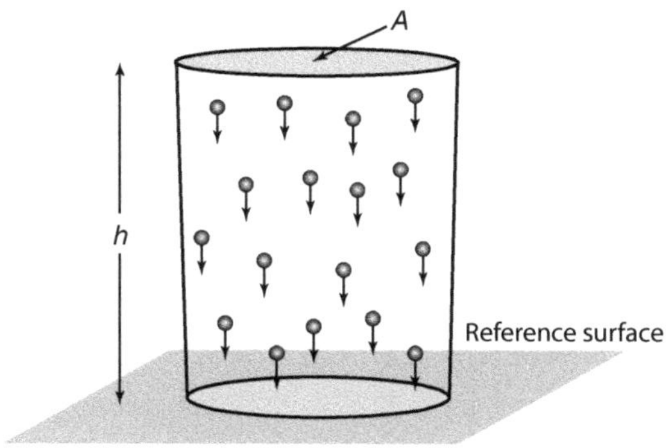

Fig. 13.5: A cylindrical volume of height h and horizontal area A encompassing a hypothetical ensemble of raindrops of identical diameters and, thus, identical fall speeds V_t. The cylinder falls with the drops through the horizontal reference surface in time $\Delta t = h/V_t$.

drops falling to the ground, which can in turn be used to calculate the rain rate. An example of such calculations can be seen in Fig. 13.3. Numerical weather forecast models also use raindrop size distributions to calculate a rain rate, which can then be used with (13.4) to predict an accumulation over a given period of time.

For simplicity, we first consider the very idealized case that all raindrops have the same diameter D and therefore the same fall speed V_t. If there are N of these drops per unit volume of air, then the volume of water per unit volume of air, called the volume fraction of water f_v, is simply

$$f_v = N\frac{\pi}{6}D^3 \,. \tag{13.5}$$

Now imagine a cylindrical volume in space with vertical dimension h and horizontal area A (Fig. 13.5). Its volume is then hA, and the total volume of rainwater it encloses is

$$v = hAf_v = hAN\frac{\pi}{6}D^3 \,. \tag{13.6}$$

If we now allow the cylinder enclosing the drops to follow them downward at their shared terminal velocity V_t, then the time required for it to pass through our reference surface is

$$\Delta t = \frac{h}{V_t} \,. \tag{13.7}$$

It follows that the volume of rainwater per unit time per unit area can be reduced to a depth per unit time, thus equivalent to the precipitation rate R, given by

$$R = \frac{v}{A\Delta t} = \frac{hAN\pi D^3 V_t}{6hA} \,, \tag{13.8}$$

which can be simplified further to

$$R = N\frac{\pi}{6}D^3 V_t. \tag{13.9}$$

In words, the contribution of raindrops of size D to the precipitation rate R can be written as

$$\begin{aligned} R =& [\text{number of raindrops per unit volume of air}] \\ &\times [\text{volume of each raindrop}] \\ &\times [\text{fall speed}]. \end{aligned}$$

We now need to generalize our expression (13.9) to the real-world case where precipitation consists of a wide range of drop sizes with a distribution given by n_D, where $n_D\, dD$ is the number of raindrops per unit volume of air with diameters between D and $D + dD$, as discussed in Sec. 4.6.

Recall that the units of n_D are number per unit volume of air (m^{-3}) per unit raindrop diameter (e.g., mm^{-1}). Multiplying by the differential dD converts n_D to the number concentration of raindrops (m^{-3}) having diameters between D and $D + dD$. We can directly substitute $n_D\, dD$ for N in (13.9) and then integrate to obtain the total precipitation rate contributed by drops of all sizes:

$$R = \frac{\pi}{6}\int_0^\infty n_D D^3 V_t \, dD\,, \tag{13.10}$$

where both n_D and V_t are functions of D.

In Sec. 4.6.4, raindrop sizes were modeled as an exponential distribution that is shown again here:

$$n_D = N_0\, e^{-\Lambda D}\,, \tag{13.11}$$

where n_D is the number of raindrops per unit volume of air per unit diameter, N_0 is the hypothetical value at diameter $D = 0$, and Λ expresses the rate of decrease in n_D as D increases. When plotted on a logarithmic y-axis, N_0 represents the y-intercept and $-\Lambda$ is the slope. If (13.11) is substituted into (13.10), then the rain rate can be computed from just the two parameters N_0 and Λ as follows:

$$R = \frac{N_0 \pi}{6}\int_0^\infty e^{-\Lambda D} D^3 V_t \; dD\,. \tag{13.12}$$

Based in part on the pioneering measurements of Marshall and Palmer, it was shown that heavier rainfall rates tend to be associated with larger raindrops; that is, Λ decreases (the slope flattens) to increase the overall proportion of large drops. If we examine the expression *inside* the integral

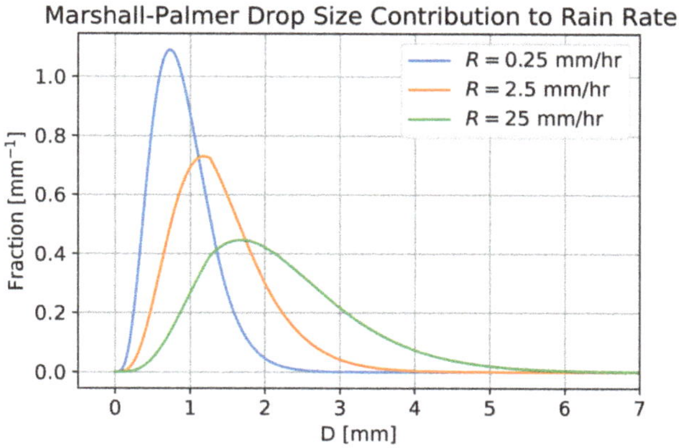

Fig. 13.6: The relative contribution of different drop sizes to three different total rain rates (expressed as a fraction of the total), assuming an ideal Marshall-Palmer drop size distribution and the fall speed relationships (4.13)–(4.15). For any chosen range of drop diameters, $[D_1, D_2]$, the area under the curve within that range represents the fraction of the total rain rate contributed by drops of those sizes.

in (13.12) and determine how its magnitude varies with D, we can identify the range of raindrop sizes that contribute the most to the total rain rate, taking into account the number, size, and fall speeds of the drops.

Using the Marshall-Palmer values for N_0 and Λ from (4.9) and (4.10), Fig. 13.6 presents the raindrop sizes contributing the most to selected rain rates that range from extremely light to moderately heavy. In general, as the rain rate R increases, so do the sizes of the raindrops contributing to it.

For example, for $R = 0.25$ mm hr^{-1} (about 0.01 inch per hour), the bulk of the total rainfall is contributed by raindrops smaller than 1.5 mm in diameter, with a peak contribution near 0.8 mm. For $R = 2.5$ mm hr^{-1}, on the threshold for "moderate" rain rates, the peak contribution increases to 1.2 mm in diameter, with significant additional contributions from raindrops up to around 3 mm in diameter. Finally, for the relatively heavy rain rate $R = 25$ mm hr^{-1}, the peak contribution comes from raindrops near 1.8 mm in diameter, but those of 4 mm in diameter and greater are also significant contributors. For the lighter rain rates, raindrops of these large sizes are essentially absent. Although not all raindrop size distributions are well approximated by a Marshall-Palmer distribution, these trends generally hold.

13.2.2 Snowfall and rainfall rates from aggregate size distributions

In contrast to raindrops, the size distributions of ice crystals and aggregates contributing to snowfall are measured less frequently. Although some (costly) *2D video disdrometers* can count and size the images of the crystals and aggregates as they fall, the true maximum dimensions of the aggregates can be difficult to establish, as it depends on the viewing angle of the instrument relative to an asymmetrical aggregate as seen, for example, in Fig. 11.1. In earlier times, crystals and aggregates were observed during natural snowfall by collecting them on glass slides coated with a chemical solution called *Formvar* to preserve their shapes until they could be counted and sized under a microscope. Given the scarcity of observations of aggregate sizes during natural snowfalls, and the uncertainties of any data that are collected, it is uncommon to use them to calculate a snowfall rate.

However, numerical models must assume that aggregates follow a specific size distribution. As noted in Sec. 4.6.4, the size distributions measured within a snowfall tend to be exponential, i.e., having more particles at the smaller sizes, with far fewer aggregates (falling off exponentially) at larger sizes. Therefore, most numerical models use exponential distributions to represent the range of aggregate sizes in their computations.

Despite limited observations, some studies have suggested that larger aggregates are associated with higher actual (unmelted) snowfall rates. This is sometimes also true for their associated LWE (liquid water equivalent) rates, although the LWE rates can be more affected by the presence of any rime on the crystals or aggregates (increasing their bulk densities) than by aggregate size.

In colder climate zones, aggregation is responsible for most of the rainfall that reaches the surface. When the air near the ground is warmer than 0 °C, melted aggregates may result in rainfall, as shown in the rightmost microphysical pathway of Fig. 9.1. Large single ice crystals grown by diffusion alone, when melted, result in drizzle drops of less than 1-mm diameter, unless they were rimed. On the other hand, typical aggregate sizes when melted can produce small raindrops, up to a few millimeters in diameter. The aggregate sizes required to produce large raindrops are very rare—unless they are rimed—as they tend to break up before melting. Aggregation and subsequent melting typically produce only light or moderate rainfall rates ($\lesssim 10$ mm hr^{-1}).

In numerical weather prediction models, snowfall rates are commonly calculated as the LWE rates, requiring an assumption about the bulk density of the aggregates. Subsequently, a typical rain-to-snow ratio is applied to convert the LWE rate to the actual snowfall rate.

13.3 Measuring Hail on the Ground

In contrast to rainfall and snowfall, the surface accumulation and precipitation rate of graupel or hail alone are not standard measurements. Graupel, regardless of whether or not it melts upon reaching the ground, can be collected by rain and snow gauges but is typically combined with the total accumulated rainfall or snowfall. Hailstones, on the other hand, are usually not produced in sufficient numbers to build up a measurable accumulation at the ground, except for rare events where accumulations of small hailstones sometimes occur.

However, the *size* of hailstones reaching the ground is of significant societal importance. In this section, we briefly shift from general precipitation measurement to focus upon how hailstone sizes are determined. Along with their fall speeds, hailstone sizes regulate the potential damage to rooftops, vehicles, crops, and livestock. Several methods are currently in use for ground-based measurements.

Reports of maximum hailstone size are commonly communicated to National Weather Service offices across the United States using a ruler or reference objects like coins (dime, nickel, quarter), sports balls (golf ball, tennis ball, baseball, softball), or vegetables or fruit (pea, lime, grapefruit). After the reports are quality-checked for plausibility—for example, whether a storm was actually over that location—they are entered into an official national weather database. However, such hail reports tend to cluster around more highly populated areas, leading to spatial biases in the data.

To improve the spatial density of hail reports, particularly in less populated areas, the volunteers of the CoCoRaHS effort in the U.S. (see Sec. 13.1.3) are also trained to make hail reports using **hail pads** (Fig. 13.7). This inexpensive, historically labor-intensive method has been used since the 1960s to measure the size distributions of hailstones falling to the ground. Hail pads can simply consist of aluminum foil-wrapped squares of foam that can be placed flat upon the ground before a hailfall event. When a hailstone impacts the hail pad, it creates a dent of equal size. By counting and measuring the sizes of all the dents on the hail pad, and knowing the area of the hail pad, a size distribution of hailstones can be derived.

While not necessarily addressing the paucity of hail reports in more rural areas, so-called *crowdsourced data* can augment the overall hail report database and indicate variability in hailstone sizes at smaller spatial scales. In the United States, a recent weather app called mPING[3] has been made freely available to facilitate hail reporting for the general public. Similar apps are being used in other countries, and together they are contributing to the collection of massive amounts of data. Users simply enter the time

---

[3]See mping.nssl.noaa.gov for more information.

Fig. 13.7: A hailpad after a hailfall event showing dents created by a variety of hailstone sizes. *Photo courtesy of cocorahs.org.*

of their report into the app and select the maximum size of the hailstones, choosing equivalent sizes like peas, marbles, coins, golf balls, etc. The app then automatically records the report location from their cell phone. While not providing a size distribution, these reports have been invaluable for detecting the largest hailstone sizes produced by a given storm, which is useful for researchers and insurance agencies to determine damage and loss due to hailfall events.

Scientists worldwide are beginning to use reports from such apps or even those posted on social media outlets to add to hail databases. For these crowdsourced data, scientists must manually verify reports by cross-referencing them with storm tracks from radar and official hail reports.

Another method of measuring hailstone distributions on the ground, developed in Switzerland, involves the use of drones. Immediately after a hail event concludes, these drones fly over the affected area to photograph the hailstones. Image recognition software is then utilized to characterize the size distributions of hailstones in the captured images. However, a major challenge of this method is the need for an accurate forecast to position drones in advance of a hailfall event. The drones must be deployed immediately after the event to photograph the stones before melting begins.

The prediction of precise hailfall locations and timing remains difficult, complicating timely drone deployment.

13.4 Precipitation Efficiency

Clouds ingest water vapor and may convert some of it into precipitation that reaches the Earth's surface, but never all of it. This conversion is inherently imperfect for several reasons:

(i) As discussed in Chapter 3, the entrainment and mixing of dry air into the cloud can lower the relative humidity below saturation, causing some hydrometeors to evaporate or sublimate.

(ii) Once the processes driving cloud formation—such as large-scale cooling or updrafts—cease, smaller hydrometeors (e.g., cloud droplets or ice crystals) that have not sufficiently grown through collision-coalescence, aggregation, or riming will evaporate or sublimate before reaching the ground.

(iii) Not all ingested water vapor can be converted to liquid or ice through condensation or deposition. Even in the coldest tropospheric clouds, the saturation vapor pressure over ice is still nonzero ($e_{si} \sim 1$ Pa at -60 °C), meaning that at least a small amount of water vapor always remains unavailable for ice formation.

(iv) Once precipitation falls outside the cloud boundaries, it loses mass through evaporation or sublimation in the unsaturated air encountered on its way to the ground.

We quantify how much of the water vapor ingested by the cloud is converted into precipitation by its **precipitation efficiency** (PE). It can be defined in numerous ways, but most commonly as:

$$\mathrm{PE} = \frac{M_p}{M_v} \times 100\% \tag{13.13}$$

where M_p is the mass of precipitation reaching the ground, and M_v is the mass of water vapor ingested by the cloud. Both M_p and M_v should be summed over the cloud lifetime.[4] This ratio, which can be thought of as "precipitation out" to "water vapor in" over the lifetime of the cloud, is commonly expressed as a percentage.

[4]Note that because precipitation on the ground is time-lagged from the ingestion of water vapor by the cloud, it is always safest to use the cloud lifetime totals. Otherwise, initially all clouds have PE=0%, and at the end of their lifetimes, their PE values may approach ∞!

The PE of clouds can range from 0%—that is, the cloud *never* produces any surface precipitation—to values over 90%. Even within the same type of cloud, such as stratocumulus clouds or cumulonimbus clouds, PE values can vary. Cloud characteristics that can affect the PE include the amount of CCN and/or INP ingested by the cloud, the cloud liquid water or ice water content—also affected by entrainment—and the duration of the cloud's lifetime.

The PE is difficult to determine from observations. Assumptions must be made to estimate the amount of water vapor ingested by the cloud during its lifetime. In addition, as discussed earlier in this chapter, precipitation reaching the ground is notoriously variable, even across short distances between two different gauges. Observational estimates typically suggest values of 20–30% for thunderstorms, but significant deviations from these values are possible due to these difficulties, along with likely variability among storms in their actual PE.

Smaller precipitating clouds may actually have greater PE than large thunderstorms, even though their precipitation rates are far less than those typical of thunderstorms. PE is an expression of how efficiently a cloud or storm converts ingested water vapor to precipitation on the ground; it does not indicate the intensity of the precipitation, nor the fact that a thunderstorm ingests far more water vapor during its lifetime than, for example, a small cumulus cloud.

An important characteristic of the cloud environment that can greatly affect the cloud PE is *vertical wind shear*, because of its possible influence on evaporation or sublimation. Consider first the case of no shear, as depicted by Fig. 13.8a. Precipitation formed within the cloud tends to fall through the entire depth of the cloud below it before exiting the base. Not only does this extend the time hydrometeors have to grow, it also greatly reduces their exposure to unsaturated air before they reach the ground.

On the other hand, when there is significant wind shear—meaning the winds near the cloud base are weaker or differ in direction compared to those near the cloud top—the cloud will lean over with height, as shown in Fig. 13.8b.

Any precipitation produced near the upper levels of the cloud falls a greater distance through unsaturated air and thus has more time to evaporate. Also, condensed water at lower levels within the cloud is less likely to be collected by hydrometeors falling from higher levels. For both reasons, the PE is generally smaller in clouds formed in environments with strong vertical wind shear, all else being equal.

However, despite the tendency of strong vertical wind shear to reduce the PE in clouds, high rain rates can still be produced by clouds in these environments. As stated earlier, a lower PE does not necessarily preclude the production of high precipitation rates.

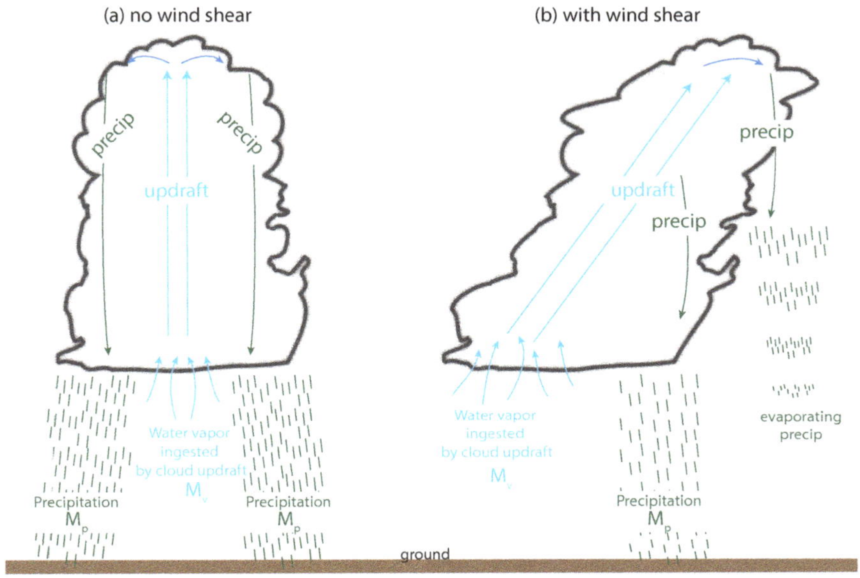

Fig. 13.8: Illustrations of the precipitation efficiency of a cloud in (a) an environment lacking vertical wind shear and (b) in an environment having vertical wind shear (in this example with stronger winds in the upper half of the cloud blowing toward the right of the diagram). Given the same water vapor mass M_v ingested by both clouds, the PE would be higher in (a) because more of the precipitation is reaching the ground than in (b).

For example, supercell thunderstorms form in environments with substantial vertical wind shear. In numerical simulations of supercell thunderstorms, as the vertical wind shear in the environment increases, so do their updraft speeds. This allows the storm to ingest more water vapor per unit time and potentially create more overall precipitation. However, the stronger updraft also carries more hydrometeor mass high into the anvil region of the storm, where it is unlikely to survive sublimation and evaporation before reaching the ground. Therefore, even though stronger storms in environments with greater wind shear may produce more precipitation internally, a reduced PE means that they will not necessarily yield more precipitation on the ground.

13.5 Convective vs. Stratiform Precipitation

In Chapter 1, the general differences between convective (i.e., cumuliform) and stratiform clouds were noted, and similar characterizations of the precipitation they may produce are also useful.

13.5.1 Qualitative definitions

Convective precipitation tends to consist of localized, shorter-term episodes of higher rainfall rates, sometimes 25 mm hr^{-1} or greater, from convective clouds. Convective precipitation often includes large raindrops and possibly graupel and/or hail. In cold seasons, even relatively weak convective updrafts can produce localized, very high snowfall rates, possibly containing rimed aggregates. Convective precipitation is associated with localized stronger radar echoes, like those shaded red in Fig. 13.9.

Stratiform precipitation, falling from stratiform clouds, tends to be widespread and continuous, with light to moderate rainfall rates, typically around 2.5 mm hr^{-1}, but possibly up to 10 mm hr^{-1} in the deepest, wettest nimbostratus clouds. Stratiform precipitation consists of drizzle or small raindrops. In cold seasons, it consists of light to moderate snowfall containing individual ice crystals and/or aggregates, with minimal rime. Stratiform precipitation is associated with broader, weaker radar echoes, similar to the more diffuse areas shaded yellow, purple, and blue in Fig. 13.9.

A large cloud complex, such as a *mesoscale convective system* (MCS), can contain both types of precipitation because the parent cloud includes both convective and stratiform elements. The radar image in Fig. 13.9 depicts a quasi-linear type of MCS, often called a *bow echo* due to its curved shape on radar or a *squall line* owing to its common association with strong and potentially damaging winds. The leading edge features an intense line of convection, with high radar reflectivity values indicating heavy convective precipitation. The individual convective elements are often referred to as *cells*, defined as a local maximum in radar reflectivity within the system, as can be easily seen as separate entities within the convective line at the bottom of the radar image. Hydrometeors within the strong convective updrafts can be transported to the top and behind the convective line into the stratiform region. Here, they fall within weaker updrafts or downdrafts, producing lighter stratiform precipitation with a broader, weaker radar echo. A city in the storm's path would first experience heavy rain (potentially containing some graupel or hail as well), followed by a longer period of light rain or drizzle.

13.5.2 The roles of different collection processes

Let us now revisit the different collection processes that can be active in creating convective and stratiform precipitation: collision-coalescence, aggregation, and riming. A comparison of the key aspects of these processes in convective and stratiform clouds can provide more insight into the reasons for the different characteristics of their respective precipitation.

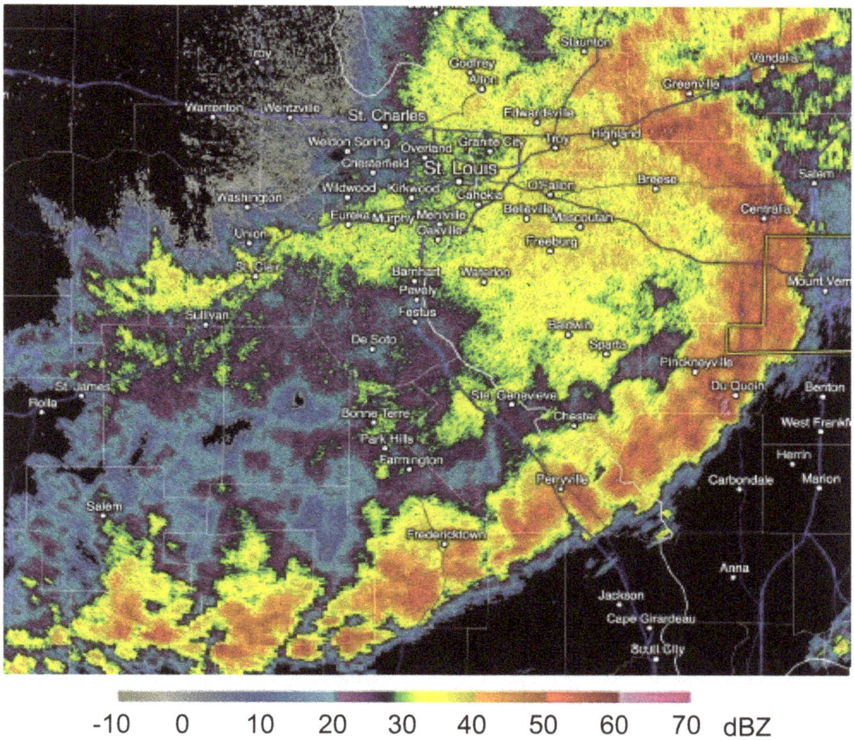

Fig. 13.9: Example of a mesoscale convective system as seen by a weather radar based near St. Louis, Missouri, with color shading according to the radar reflectivity factor in dBZ, as shown at the bottom of the figure. The system is moving toward the lower right of the figure. Along the leading line of the system, convective precipitation is falling, with radar reflectivities exceeding 50 dBZ, while the reflectivities in the trailing stratiform precipitation behind it are often less than 35 dBZ. *Courtesy of R. J.Trapp.*

For reference, recall the continuous collection model introduced in Sec. 10.2.2, which provides a basic foundation for understanding the growth of hydrometeors via collision-coalescence, aggregation, or riming, assuming spherical particles for simplicity:

$$\frac{\mathrm{d}M}{\mathrm{d}t} = \pi \,(R+r)^2 \Delta V_t E L. \tag{13.14}$$

The rate of mass increase $\mathrm{d}M/\mathrm{d}t$ of the collector particle depends on the liquid water content, L, (or the ice water content, I, for aggregation), the collection/aggregation/riming efficiency, E, the radii of the collector and collected hydrometeors, R and r, and the difference in their fall speeds, ΔV_t. The mass growth rates are therefore maximized when these individual factors are at their largest possible values.

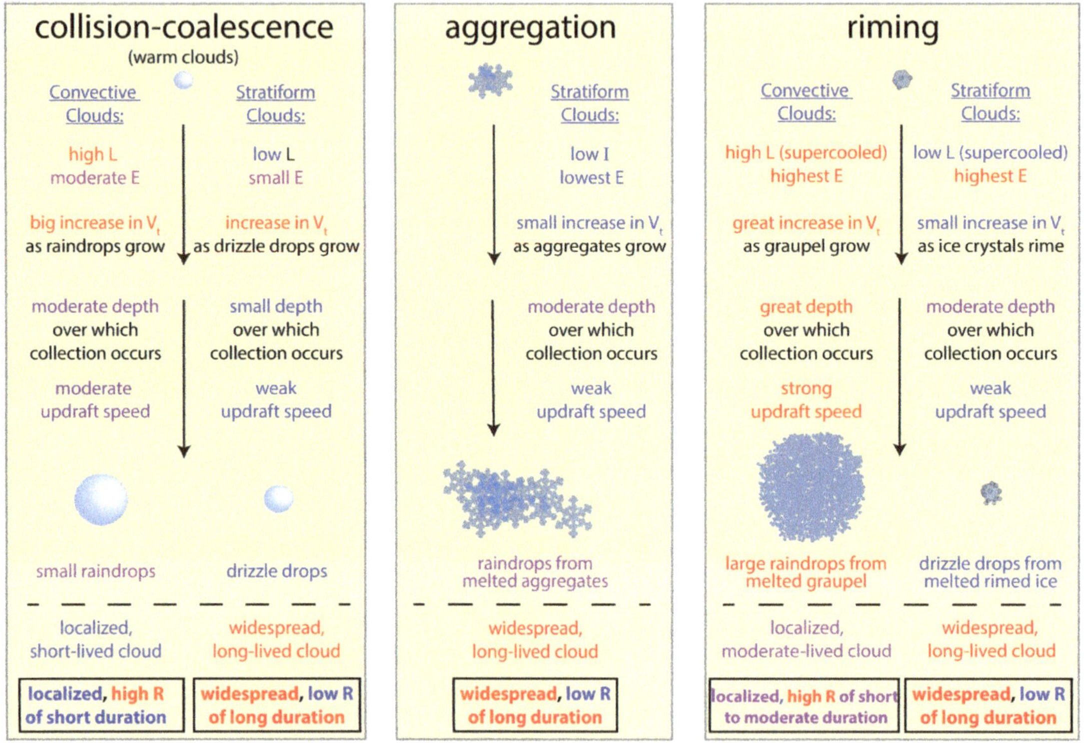

Fig. 13.10: Schematic overview of the relative importance of different factors influencing collection processes in convective and stratiform clouds, as well as the resulting maximum surface precipitation rates and durations. Font color indicates the qualitative magnitude of each aspect for each type of cloud, with blue, purple, and red denoting small, moderate, and large values, respectively. Below the dashed line, the noted horizontal cloud extent and cloud lifetime also influence the character of the precipitation that is listed at the very bottom. "Low *R*" and "high *R*" denote values less than or greater than 25 mm hr^{-1}, respectively. "Short duration" implies time scales of minutes, "moderate duration" implies 30 to 60 minutes, and "long duration" implies one or more hours.

In Fig. 13.10, the relative importance of these factors is summarized for each type of collection process in either convective or stratiform clouds, culminating in the typical maximum sizes of raindrops, or the liquid-equivalent raindrop size for frozen hydrometeors. To simplify this discussion, we neglect entrainment—which could reduce or even prevent precipitation production within the clouds—as well as the evaporation or sublimation of precipitation as it falls from the cloud to the ground.

Convective clouds primarily create precipitation through collision-coalescence (warm clouds[5]) and/or riming (mixed-phase clouds). Aggregation within cold cumuliform clouds is not addressed here, as it is rarely as important as riming for creating surface precipitation.

Stratiform clouds, on the other hand, might produce some precipitation through collision-coalescence (warm clouds), but otherwise their precipitation will predominantly be created by aggregation (cold or mixed-phase clouds). Riming is generally of secondary importance in these clouds, due to the limited amounts of supercooled water available for collection, but it is common to witness lightly rimed crystals and aggregates in stratiform precipitation.

Keeping (13.14) in mind and working our way downward in Fig. 13.10, we know that, in general, convective clouds have much greater liquid water content than stratiform clouds. This enhances both the collision-coalescence and riming processes in warm and mixed-phase convective clouds, respectively, compared to their stratiform counterparts. Cold stratiform clouds also contain minimal ice water content for aggregation, slowing that process compared to the others.

Riming efficiencies are generally highest in both convective and stratiform clouds, with moderate collection efficiencies possible for collision-coalescence in convective clouds but smaller values in stratiform clouds, where drops typically are not as large. Aggregation efficiencies, although known with much less certainty than the other efficiencies, are believed to be the smallest for any of the collection processes, again slowing that process compared to the others.

The mass growth rate of a hydrometeor through the collection process is also a function of its fall speed, which changes as its mass increases. If we temporarily ignore any variations in the fall speeds of the smaller collected hydrometeors, we can envision that as the fall speed of the collector drop, aggregate, or rimer increases with growth (i.e., ΔV_t increases), its mass growth rate will also increase. However, this acceleration in the col-

---

[5]For simplicity, here we are considering collision-coalescence only in warm clouds, although it can also contribute to precipitation in mixed-phase clouds.

lection process is much smaller for aggregates because their fall speeds do not increase as significantly with size compared with raindrops or rimers (graupel particles or hailstones). Consequently, the mass growth rate of an aggregate will be slower over time compared to a raindrop or a riming hydrometeor.

Implicit in (13.14) is that the final mass of the collector particle depends significantly on the *total time* available for it to grow by collection before it falls out of the cloud. In this context, a deeper cloud allows for more growth time. As described in Fig. 13.10, both warm convective and warm stratiform clouds have small to moderate depths, putting them at a disadvantage in this regard. Cold and mixed-phase stratiform clouds are similarly disadvantaged, except in the deepest cases of nimbostratus clouds. Mixed-phase convective clouds, which undergo the riming process, tend to have the greatest depths, allowing their hydrometeors to grow much larger than in any other cloud or through any other collection process.

Moreover, the maximum collector particle size that can be attained by any of the collection processes is also partly a function of the maximum updraft speed within the cloud. The collector particle begins its final descent through the cloud once its fall speed exceeds the updraft speed, but can grow to larger sizes before this occurs if the updraft is very strong. Mixed-phase convective clouds have much greater updraft speeds than stratiform clouds, and thus allow some hydrometeors to grow to much larger sizes within the cloud before they become too heavy and fall from it.

Combining all these factors, we see that the column in Fig. 13.10 having the greatest number of favorable aspects for collection growth (listed in red font) is that for the mixed-phase convective clouds where riming is occurring. Large graupel particles or even small hailstones, when melted before reaching the ground, can produce the largest raindrops. Melted aggregates produced in cold stratiform clouds,[6] and the raindrops produced by warm convective clouds, would tend to be of more moderate sizes. Most warm stratiform clouds, or the light riming possible in mixed-phase stratiform clouds, would produce much smaller drizzle drops, as would the aggregation process in cold stratiform clouds if any of the contributing factors were very small.

Now let us consider the influence of the typical maximum hydrometeor sizes produced by these collection processes acting within convective

[6]Note that the light riming of aggregates in a mixed-phase cloud is not explicitly considered in Fig. 13.10, but could increase the sizes of raindrops when they melt. However, the raindrops would still not typically be as large as those produced by melting graupel.

or stratiform clouds upon their liquid-water equivalent surface precipitation rates. Using a simple equation like (13.9) for reference, the rain rate R is proportional to the third power of the drop diameter D and to its fall speed V_t (which is larger for larger D). This makes sense, given our discussion in previous chapters that larger rain rates tend to be dominated by larger raindrops in the size distribution.

From the preceding discussion on the relative sizes of raindrops produced by the different collection processes within convective and stratiform clouds, we have estimated that the largest and thus fastest-falling raindrops will be produced by deep, mixed-phase convective clouds where riming is the dominant collection process. Shallow, warm convective clouds can also produce moderate-sized raindrops with moderate fall speeds from the collision-coalescence process. It thus follows that *the rain rates of convective clouds can be very high.* In contrast, the raindrops produced by any of the collection processes acting within stratiform clouds are much smaller, often drizzle-sized, with smaller fall speeds. Thus, *low rain rates are typical of most stratiform clouds.*

13.5.3 The influence of cloud horizontal extent and lifetime

The spatial and temporal extents of the surface precipitation produced by stratiform and convective clouds also follow from their respective horizontal dimensions and lifetimes as listed beneath the horizontal dashed lines in Fig. 13.10. The contribution by stratiform clouds to the Earth's overall precipitation remains substantial because their lower precipitation rates are compensated by their much greater horizontal extent and longer duration. This is particularly true in regions and seasons where convective precipitation is less common.

On the other hand, the surface precipitation produced by individual convective clouds tends to be localized and of much shorter duration, because of the much smaller horizontal dimensions and shorter lifetimes of convective clouds. However, their high precipitation rates can still yield substantial accumulations at the ground, particularly for deep mixed-phase convective clouds.

The horizontal extent and duration of precipitation produced by convective clouds are enhanced greatly whenever they exist as large clusters, such as in mesoscale convective systems. Within these systems, multiple generations of convective cells can grow and decay over time. The overall lifetime and horizontal extent of the system then become much greater than those of the individual convective cells that they contain. Likewise, the areal extent and duration of the precipitation are also greatly increased. In the next section, we examine how such a system can produce extreme amounts of precipitation, leading to very hazardous conditions.

13.6 Flash flooding

It is of interest to meteorologists—and to the general public—to understand the conditions under which deep convective storms may produce extreme amounts of rainfall on the ground for an extended period of time, giving rise to **flash flooding**. While flash flooding can result from factors such as terrain or other conditions affecting the runoff of precipitation across the ground and into nearby streams and rivers, our focus here is on those factors related specifically to the storms or storm systems themselves.

From (13.4), the accumulation of rainfall, P, over any area depends upon the rainfall rate, R, and the duration over which the rainfall occurs ($t_f - t_i = \Delta t$). Thus, the greatest rainfall accumulations will occur where either R or Δt, and especially both factors combined, are greatest. As we saw in the previous section, convective clouds are very capable of producing high values of R, but the rainfall duration Δt tends to be short or at most moderate for larger convective systems. To understand when the accumulation of precipitation might be unusually large over a given location from convective clouds, we need to consider how the duration of convective precipitation might be increased.

Two factors maximize the duration of rainfall: the horizontal extent of the storm and the rate at which it moves. The duration of precipitation Δt experienced at a given location (e.g., the town where you live) from a given storm can be expressed in terms of the magnitude of the storm motion vector $|\vec{C}|$ and the spatial extent L of the precipitation reaching the ground along the direction of $\vec{C}$:

$$\Delta t = \frac{L}{|\vec{C}|}. \tag{13.15}$$

Some hypothetical examples of two different MCSs with different storm motions are shown in Fig. 13.11. These examples illustrate how the alignment of the storm motion and its precipitating area influences the rainfall at a given location. If the *narrowest* area of convective precipitation is aligned with the motion of the storm system, the precipitation event can be quite short, lasting only a few minutes (far left column). However, as the length of the convective precipitation becomes more aligned with the storm motion vector (moving from the left toward the right columns in the same figure), the duration of the heavy precipitation increases. In the rightmost column, an extremely dangerous alignment can occur where the *broadest* area of convective precipitation is aligned with the storm motion vector, causing the precipitation to continue at a particular location for a significantly longer Δt.

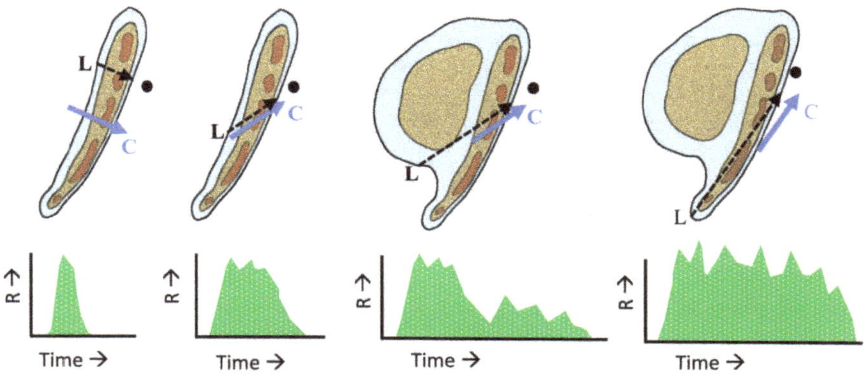

Fig. 13.11: *Top row:* hypothetical radar echoes from storm systems of different shapes that show locations of convective precipitation (shaded red) and stratiform precipitation (shaded yellow and blue). The length of the precipitation (L; dashed black arrow) along the direction of the storm motion vector (C; blue arrow) determines the duration of rainfall at a given point (black dot). *Bottom row:* accompanying hypothetical rain rate R that would be experienced at the location of the black dot as a function of time, based upon the passage of the storm system along its motion vector C. *Adapted from Doswell et al. (1996).*

The worst scenario is a phenomenon called *training*, where new storms embedded within multi-storm systems are continually regenerated at one end of the system and then move along the major axis of that system, producing frequent episodes of heavy rainfall over the same area for a prolonged time. This could occur, for example, in the case shown at the far right of Fig. 13.11, if new storms are continuously generated at the bottom of the convective line and subsequently move along the line toward the black dot. It is important to understand that not every storm that produces very high rain rates will result in flash flooding; rather, it is the *long duration* over which a given location experiences such high rain rates that leads to flash flooding. The training of these storms thus increases the duration of high rain rates. When storms are training within a larger convective system like this, and the radar indicates that R is high within them, flash flooding is likely.

Sometimes flash flood events occur because storms regenerate over a given area with a nearly negligible storm motion vector, i.e., hardly any movement of the storms at all. For example, many flash flood events occur near high terrain, where strong winds impinging upon its leading edge can "lock" a succession of storms with high R into place over a given area. Such a scenario enhances the duration of convective precipitation over those areas, often with disastrous consequences.

13.7 A Brief Note on Future Research

Extreme precipitation events, such as the flash flooding described in the previous section, as well as extreme snowfall or hailfall events, result not only from (i) the microphysical processes within the storm, but also (ii) the dynamical factors of the storms themselves, and (iii) the thermodynamics and dynamics of the larger-scale environments in which they occur. Currently, extreme precipitation events appear to be increasing in frequency, possibly due to the changing, warming climate, but this is an area of intense research with little certainty. Considering any of these factors (i-iii) *alone* will not be particularly useful in advancing our progress in predicting these events; understanding the interactions among these different factors is critical. Throughout this textbook, this author has aimed to emphasize the importance of these interactions.

The remaining gaps in our understanding of precipitation formation, particularly extreme precipitation events, offer challenging opportunities for current and future scientists. Some of these gaps have been highlighted herein, while others await discovery. This author has found great enjoyment and value in addressing even small gaps in our knowledge of clouds and precipitation, and she hopes this textbook will inspire you to explore these and other challenges with curiosity and rigor, contributing to the ongoing advancement of atmospheric science.

Questions for Review

1. In what ways is precipitation accumulation quantified and measured? What are the limitations and challenges associated with such measurements?
2. How can hydrometeor size distributions be converted to a precipitation rate?
3. What is meant by the precipitation efficiency of a cloud?
4. How do stratiform and convective precipitation differ?
5. What different characteristics of stratiform and convective clouds influence the size of the precipitating hydrometeors produced by collision-coalescence, aggregation, and/or riming?
6. What different characteristics of stratiform and convective clouds influence the extent and duration of their precipitation?
7. How are flash flooding conditions produced?

Homework problems and other resources for students and instructors can be found at bit.ly/4dYBwB5, or scan this code:

Further Reading

For more about rain gauges:

Kidd, C., A. Becker, G. J. Huffman, C. L. Muller, P. Joe, G. Skofronick-Jackson, and D. B. Kirschbaum, 2017: So, how much of the Earth's surface is covered by rain gauges? *Bulletin of the American Meteorological Society*, **98 (1)**, 69–78

For more about the challenges of measuring snowfall:

Rasmussen, R., and Coauthors, 2012: How well are we measuring snow: The NOAA/FAA/NCAR winter precipitation test bed. *Bulletin of the American Meteorological Society*, **93 (6)**, 811–829

For more on convective versus stratiform precipitation [advanced]:

Houze, R. A., Jr., 2014: *Cloud Dynamics*. 2nd ed., Academic Press

Symbols and Abbreviations

Symbol	Description	Page
a	half-length along the a-axis used in describing some non-spherical ice particles	217
a, b	coefficients used in a mass-diameter relationship	77
a, b, c, d	coefficients appearing in the Meyers relationships for ice particle nucleation	184
a, b, c, d	coefficients appearing in the Demott relationship for N_{INP}	
a, b	curvature and solute effect coefficients in the Köhler equation	149
A, β	coefficents appearing in the Fletcher curve for ice particle nucleation	184
A_a, A_i	cross-sectional area of aggregates, ice crystals	272
c	speed of light	113
c	half-length along the c-axis used in describing some non-spherical ice particles	217
C, k	coefficients appearing in the dependence of N_{CCN} on supersaturation	176
$\vec{C}$	storm motion vector	340
C_l	specific heat capacity of liquid water	203
C_p	specific heat capacity of air at constant pressure	24
C_0	capacitance shape parameter	215
D	diameter; e.g., of a hydrometeor	58
D_v	diffusivity of water vapor in air	196
e	vapor pressure	25
e_s	saturation vapor pressure with respect to liquid water	25, 130
e_s'	modified saturation vapor pressure with respect to water with a dissolved solute	146
e_{si}	saturation vapor pressure with respect to ice	25, 131
$e_{s,r}$	saturation vapor pressure with respect to pure liquid water accounting for curvature	136
$e_{s,r}'$	saturation vapor pressure accounting for both curvature and dissolved solute in a droplet	148
E	collection efficiency	237
E_a	aggregation efficiency	270
E_c	collision efficiency	235
E_c'	coalescence efficiency	237
E_r	riming efficiency	275

Symbol	Description	Page
F_D	drag force; e.g., on a hydrometeor	78
F_g	gravitational force; e.g., on a hydrometeor	78
F_v	flux of water vapor	196
g	acceleration due to gravity	24
G	Gibbs free energy	132
G, G_i	diffusional growth rate factors for liquid and frozen hydrometeors	204, 219
h	height; e.g., one dimension of an ice crystal	75
i	van't Hoff factor	147
I	ice water content	270
k	Boltzmann constant	133
K	coefficient of thermal conductivity	203
K	collection kernel in collision-coalescence	238
L	cloud liquid water content (mass per unit volume)	42
L_A	adiabatic cloud liquid water content	42
L_v , L_f, L_s	latent heats of vaporization, fusion (melting), sublimation	25, 129
m	mass; e.g., of hydrometeor	
M	mass of a frozen hydrometeor or collector drop	77, 243
M_s, M_w	molecular weight of solute, water	147
N	total number concentration of a particle type	58
$N_{AP,0.5}$	number concentration of aerosol particles with diameters larger than 0.5 μm	186
N_{CCN}	number concentration of activated CCN	176
N_i	number concentration of ice crystals	184
N_{INP}	number concentration of INP	185
N_{max}	concentration of cloud droplets activated at supersaturation s_{max}	180
N_0	overall amplitude coefficient in $n(D)$	71
n	normalized number concentration	69
n	water molecules per unit volume	133
$n(D)$	particle/hydrometeor size distribution function	71
n'	solution molecules per unit volume	148
n_D	number concentration in a size bin	68
n_i	molar density of ice	139
n_w, n_s	moles of water, solute in a droplet solution	146
p	pressure	24
P	precipitation accumulation	316
q	heat energy per unit mass added to an air parcel	24
Q_1, Q_2	coefficients in an expression for dS/dt	179
q_c, q_i	cloud liquid water, cloud ice mixing ratios	30
q_t	total hydrometeor mixing ratio	310
q_v	water vapor mixing ratio	27
q_{vs}	saturation mixing ratio	28
r	droplet radius	58
r^*	critical radius for nucleating or activating a cloud droplet	134, 154
R	precipitation rate	119
R	drop radius	194
R_v	gas constant for water vapor	25
s, s_i	supersaturation with respect to liquid water, ice	29
s_{max}	maximum supersaturation in an updraft	180
S	saturation ratio	28

Symbol	Description	Page
S^*	critical saturation ratio	154
t	elapsed time	113
T	temperature (absolute, except where noted)	24
T_0	reference temperature (e.g., 273.15 K)	139
T_s	amount of supercooling; i.e., $T_0 - T$ in (6.10)	139
T_d	dewpoint temperature	26
T_v	virtual temperature	28
T_w, T_{wi}	wet-bulb temperature, ice-bulb temperature	39
V	volume	147
V_{swept}	sweepout volume per unit time for a falling drop	234
V_t	terminal fall speed	78
w	updraft speed	89
w_{max}	maximum updraft speed	180
x	radial distance from a droplet center	195
y	collision radius	235
z	height above ground level or sea level	24
Z	radar reflectivity factor	117
Z_{DR}	radar depolarization ratio	121
Z_h, Z_v	radar reflectivities for horizontal and vertical polarization	121
α	specific volume	24
α, β	volume and surface area correction factors for an ice crystal relative to a sphere	136
ΔT_d	dewpoint depression	26
Γ_d	dry adiabatic lapse rate	24
Γ_s	saturated adiabatic lapse rate	32
Λ	slope parameter in gamma and exponential size distributions	71
μ	shape parameter in gamma size distribution	71
ρ	air density	79
ρ'	density of a solution droplet	147
ρ_0	reference air density at sea level	79
ρ_v	vapor density	26
ρ_l, ρ_i	density of liquid water, pure ice	69, 74
ρ_x	bulk density of a frozen hydrometeor of type x	75
ρ_z	density of dry air at altitude z	43
σ_{lv}	surface energy of the liquid-vapor interface	132
σ_{iv}	surface energy of the ice-vapor interface	132
AP	aerosol particles	59
CAPE	convective available potential energy	52
CCL	convective condensation level	52
CCN	cloud condensation nuclei	156
EL	equilibrium level for a buoyant parcel	52
INP	ice nucleating particles	143
LFC	level of free convection	35
LCL	lifting condensation level	34
LWE	liquid water equivalent, used with a snowfall rate	324
RH, RH_i	relative humidity w/ respect to liquid water, ice	26
SWE	snow water equivalent, used with an accumulation	317
WBF	Wegener-Bergeron-Findeisen process	225

Credits and Permissions

For those figures attributed to other published sources, see the Bibliography (p. 355) for complete citations.

Cover	Photo by G. Petty.
Fig. 1.1	Courtesy of NASA Visible Earth, visibleearth.nasa.gov
Fig. 1.2	Photos by S. Lasher-Trapp.
Fig. 1.3	Figure by S. Lasher-Trapp.
Fig. 1.4	Photos by S. Lasher-Trapp.
Fig. 1.5	Photos by S. Lasher-Trapp.
Fig. 1.6	Figure by S. Lasher-Trapp.
Fig. 1.7	Photos by S. Lasher-Trapp.
Fig. 1.8	Photo by S. Lasher-Trapp.
Fig. 1.9	Photo by S. Lasher-Trapp.
Fig. 2.1	Figure by S. Lasher-Trapp.
Fig. 2.2	Figure by S. Lasher-Trapp.
Fig. 2.3	Figure by S. Lasher-Trapp.
Fig. 2.4	Figure by S. Lasher-Trapp.
Fig. 2.5	Figure by G. Petty.
Fig. 2.6	Figure by S. Lasher-Trapp.
Fig. 3.1	Figure by S. Lasher-Trapp.
Fig. 3.2	Figure by S. Lasher-Trapp.
Fig. 3.3	Figure by S. Lasher-Trapp.
Fig. 3.4	Figures by S. Lasher-Trapp. Data from the Small Cumulus Microphysics Study (SCMS95). DOI : doi.org/10.5065/D6B856F0
Fig. 3.5	Photo by S. Lasher-Trapp.
Fig. 3.6	Modified from Cooper et al. (2013). © American Meteorological Society. Used with permission.
Fig. 3.7	Left: Modified from Blyth et al. (1988). © American Meteorological Society. Used with permission. Right: Image created with software of the former David Ebert Visualization Group at Purdue University.

Fig. 3.8 Modified from Blyth et al. (2005). ©1974 John Wiley and Sons, used with permission.

Fig. 3.9 Figure by S. Lasher-Trapp.

Fig. 3.10 Figures by S. Lasher-Trapp. Data from the Small Cumulus Microphysics Study (SCMS95). DOI: doi.org/10.5065/D6B856F0

Fig. 3.11 Modified from Lasher-Trapp et al. (2021). © American Meteorological Society. Used with permission.

Fig. 4.1 After McDonald (1958).

Fig. 4.2 Left and center images from the Alliance Icing Research Study II (AIRS-II). DOI: doi.org/10.5065/D68C9TK5. Right: Photo by S. Lasher-Trapp.

Fig. 4.3 Collage courtesy of Dr. Justin Stachnik. Images from the Alliance Icing Research Study II (AIRS-II). DOI: doi.org/10.5065/D68C9TK5.

Fig. 4.4 Photo by S. Lasher-Trapp.

Fig. 4.5 Top image from Praz et al. (2017), distributed under the Creative Commons Attribution 3.0 License (CC-BY). Bottom image courtesy of Dr. Sandra Yuter, Dept. of Marine, Earth and Atmospheric Sciences, North Carolina State University.

Fig. 4.6 Top: Photo by S. Lasher-Trapp. Bottom: Photo courtesy of the Aberdeen SD National Weather Service office.

Fig. 4.7 Figure by S. Lasher-Trapp.

Fig. 4.8 Figure by S. Lasher-Trapp.

Fig. 4.9 Figure by S. Lasher-Trapp.

Fig. 4.10 Modified from Marshall and Palmer (1948). Published 1948 by the American Meteorological Society.

Fig. 4.11 Modified from Brandes et al. (2007). © American Meteorological Society. Used with permission.

Fig. 4.12 Figure by S. Lasher-Trapp.

Fig. 4.13 Figure by S. Lasher-Trapp.

Fig. 4.14 Figure by S. Lasher-Trapp.

Fig. 4.15 Modified from Beard et al. (1989). © American Meteorological Society. Used with permission.

Fig. 4.16 Reprinted from Stout et al. (2024). © Jennifer R. Stout, Christopher D. Westbrook, Thorwald H. M. Stein, and Mark W. McCorquodale, 2024. This work is distributed under the Creative Commons Attribution 4.0 License (CC-BY).

Fig. 4.17 Top: After Kajikawa (1972). Bottom: After Jayaweera and Cottis (1969).

Fig. 4.18 After Sasyo and Matsuo (1980).

Fig. 4.19 Modified from Böhm (1989). © American Meteorological Society. Used with permission..

Fig. 4.20 Reprinted from Chueh, C.-C., P. K. Wang, and T. Hashino, 2018: Numerical study of motion of falling conical graupel. *Atmospheric Research*, **199**, 82–92, ©2018, with permission from Elsevier.

Fig. 4.21 Reprinted from Knight and Knight (2001). © American Meteorological Society. Used with permission.

Fig. 4.22 Modified from Lasher-Trapp et al. (2001). © American Meteorological Society. Used with permission.

Fig. 5.1 Modified from Connolly et al. (2012). This work is distributed under the Creative Commons Attribution 3.0 License (CC-BY). Manchester Cloud Chamber website: data.cas.manchester.ac.uk/micc/micc.htm

Fig. 5.2 Photo by Peter Stein/stock.adobe.com. Sphinx Observatory website: www.hfsjg.ch/en/jungfraujoch/infrastructure/sphinx-laboratories/

Fig. 5.3 Top: Photo from the National Science Foundation Multimedia Gallery. Center: Photo by S. Lasher-Trapp. Bottom: Photo courtesy of Dr. Jeffrey French, Dept. of Atmospheric Science, University of Wyoming. C-130 aircraft DOI: doi.org/10.5065/D6WM1BG0. HIAPER aircraft DOI: doi.org/10.5065/D6DR2SJP. King Air website :www.uwyo.edu/atsc/uwka/king-air.html

Fig. 5.4 Top image from the Ice in Cumulus Experiment-Tropical (ICE-T) field campaign DOI: doi.org/10.5065/D68C9TMM. Bottom: Figure by S. Lasher-Trapp.

Fig. 5.5 Photo courtesy of Tom A. Warner, formerly at the South Dakota School of Mines and Technology.

Fig. 5.6 Figure by G. Petty.

Fig. 5.7 Photo by S. Lasher-Trapp.

Fig. 5.8 Modified from Baumgardner (1989).

Fig. 5.9 Images courtesy of Dr. Alexandria Johnson, Dept. of Earth, Atmospheric and Planetary Sciences, Purdue University. Data from Ice in Cumulus Experiment-Tropical (ICE-T): doi.org/10.5065/D6B27SND

Fig. 5.10 Images courtesy of Dr. Emma Järvinen and Dr. Martin Schnaiter, Karlsruhe Institute of Technology.

Fig. 5.11 Figure by S. Lasher-Trapp.

Fig. 5.12 Panels (a) and (b): Images courtesy of Tobias Ross, Dept. of Climate, Meteorology and Atmospheric Sciences, University of Illinois Urbana-Champaign. Panel (c): Image courtesy of Dr. David Leon, Dept. of Atmospheric Science, University of Wyoming. Radar data in (a) and (b) from joint operations of the Remote sensing of Electrification, Lightning, And Mesoscale/microscale Processes with Adaptive Ground Observations (RELAMPAGO) DOI: doi.org/10.26023/DA1G-MCNC-YB0F, and Clouds, Aerosols, and Complex Terrain Interactions (CACTI; DOI: doi.org/10.5439/1617690) field campaigns. Radar data in (c) from the Ice in Cumulus Experiment-Tropical (ICE-T) field campaign (flights.uwyo.edu/uwka/wcr/projects/icet11/).

Fig. 5.13 Top: Photo courtesy of the U.S. National Weather Service. Bottom: Photo courtesy of Eddie Wolff, Department of Climate, Meteorology, and Atmospheric Sciences, University of Illinois Urbana-Champaign.

Fig. 5.14 Reprinted from Ryzhkov et al. (2005). © American Meteorological Society. Used with permission.

Fig. 5.15 Images courtesy of Dr. J. Vivekanandan, National Center for Atmospheric Research.

Fig. 6.1 After Curry and Webster (1998).

Fig. 6.2 After Curry and Webster (1998).

Fig. 6.3 Figure by G. Petty.

Fig. 6.4 Figure by G. Petty.

Fig. 6.5 Figure by S. Lasher-Trapp.

Fig. 6.6 Figure by S. Lasher-Trapp.

Fig. 6.7	Figure by S. Lasher-Trapp.
Fig. 7.1	Figure by G. Petty.
Fig. 7.2	Figure by S. Lasher-Trapp.
Fig. 7.3	Figure by G. Petty.
Fig. 7.4	Figure by G. Petty.
Fig. 7.5	Figure by S. Lasher-Trapp.
Fig. 7.6	Modified from Kanji et al. (2017). © American Meteorological Society. Used with permission.
Fig. 7.7	Figure by S. Lasher-Trapp.
Fig. 8.1	Figure by S. Lasher-Trapp.
Fig. 8.2	Reprinted from Wang et al. (2021), distributed under the Creative Commons Attribution 4.0 License (CC-BY).
Fig. 8.3	Modified from Lasher-Trapp and Stachnik (2007) and Lasher-Trapp et al. (2008). Both sources © American Meteorological Society. Used with permission.
Fig. 8.4	Reprinted from Lasher-Trapp and Stachnik (2007). © American Meteorological Society. Used with permission.
Fig. 8.5	Modified from Skeie et al. (2011), distributed under the Creative Commons Attribution 3.0 License (CC-BY-SA).
Fig. 8.6	After Wallace and Hobbs (2006).
Fig. 8.7	Figures by S. Lasher-Trapp; data courtesy of Dr. Jim Hudson. Data from the Alliance Icing Research Study-II (AIRS-II) field campaign DOI doi.org/10.5065/D60V8B32
Fig. 8.8	After Twomey (1959).
Fig. 8.9	Modified from Hudson and Yum (2001). © American Meteorological Society. Used with permission.
Fig. 8.10	Reprinted from Kanji et al. (2017). © American Meteorological Society. Used with permission.
Fig. 8.11	Reprinted from Murray et al. (2012) with permission from The Royal Society of Chemistry. pubs.rsc.org/en/content/articlehtml/2012/cs/c2cs35200a.
Fig. 8.12	Reprinted from DeMott et al. (2010). Used with permission from Proceedings of the National Academy of Sciences (PNAS).
Fig. 8.13	Reprinted from DeMott et al. (2010). Used with permission from Proceedings of the National Academy of Sciences (PNAS).
Fig. 8.15	Modified from Lasher-Trapp et al. (2016). © American Meteorological Society. Used with permission.
Fig. 9.1	After Lamb (2001).
Fig. 9.2	Figure by G. Petty.
Fig. 9.3	Figure by G. Petty.
Fig. 9.4	Figure by G. Petty.
Fig. 9.5	Figure by G. Petty.
Fig. 9.6	Figure by S. Lasher-Trapp.
Fig. 9.7	After Rogers and Yau (1989).

Fig. 9.8 Reprinted from Kikuchi, K., T. Kameda, K. Higuchi, A. Yamashita, and Coauthors, 2013: A global classification of snow crystals, ice crystals, and solid precipitation based on observations from middle latitudes to polar regions. *Atmospheric Research*, **132**, 460–472, ©2013, with permission from Elsevier.

Fig. 9.9 Figure by S. Lasher-Trapp.

Fig. 9.10 Photos courtesy of Pomona Carrington-Hoekstra, Dept. of Integrative Biology and Mathematics, University of Illinois Urbana-Champaign.

Fig. 9.11 Reprinted from Hueholt et al. (2022). © American Meteorological Society. Used with permission.

Fig. 9.12 Modified from Hueholt et al. (2022). © American Meteorological Society. Used with permission.

Fig. 9.13 Figure by S. Lasher-Trapp.

Fig. 9.14 Figure by S. Lasher-Trapp.

Fig. 9.15 Reprinted from Harrington et al. (2013). © American Meteorological Society. Used with permission.

Fig. 9.16 Figure by S. Lasher-Trapp.

Fig. 9.17 Figure by S. Lasher-Trapp.

Fig. 9.18 Figure by G. Petty.

Fig. 9.19 Figure by S. Lasher-Trapp.

Fig. 9.20 Photo by H. Raab, released under CC-BY-SA.

Fig. 10.1 Modified from Beard (1976). Published 1976 by the American Meteorological Society.

Fig. 10.2 Figure by S. Lasher-Trapp.

Fig. 10.3 Figures by S. Lasher-Trapp based on data from Klett and Davis (1973) and Beard and Ochs (1984).

Fig. 10.4 Figure by G. Petty.

Fig. 10.5 Figures by S. Lasher-Trapp.

Fig. 10.6 After Young (1975).

Fig. 10.7 After Young (1975)

Fig. 10.8 Figure by S. Lasher-Trapp.

Fig. 10.9 Modified from Young (1993), *Microphysical Processes in Clouds*. © Oxford Publishing Limited. Reproduced with permission of the Licensor through PLSclear.

Fig. 10.10 From the cover of *Bull. Amer. Meteor. Soc.*, Dec. 2007. ©American Meteorological Society. Reproduced with permission.

Fig. 10.11 Figures by S. Lasher-Trapp.

Fig. 10.12 Figure by S. Lasher-Trapp.

Fig. 10.13 Figure by S. Lasher-Trapp.

Fig. 10.14 Reprinted from Ayala et al. (2008). © Deutsche Physikalische Gesellschaft, used by permission of the authors and IOP Publishing CC-BY-NC-SA.

Fig. 10.15 Figure by S. Lasher-Trapp.

Fig. 10.16 Figure by S. Lasher-Trapp.

Fig. 11.1 Reprinted from Jiang et al. (2019). © American Meteorological Society. Used with permission.

Fig. 11.2 Figure by S. Lasher-Trapp.

Fig. 11.3 Modified from Hobbs et al. (1974). ©1974 John Wiley and Sons, used with permission.

Fig. 11.4 Image courtesy of Dr. Sandra Yuter, Dept. of Marine, Earth and Atmospheric Sciences, North Carolina State University.

Fig. 11.5 Modified from Connolly et al. (2012), distributed under the Creative Commons Attribution 3.0 License (CC-BY).

Fig. 11.6 Figure by S. Lasher-Trapp.

Fig. 11.7 Photo by S. Lasher-Trapp.

Fig. 11.8 Figure by S. Lasher-Trapp based on data from Wang and Ji (2000).

Fig. 11.9 Figure by S. Lasher-Trapp based on data from from Khain et al. (2001).

Fig. 11.10 Photo courtesy of Alyssa Pappas, Dept. of Atmospheric Sciences, University of Illinois Urbana-Champaign.

Fig. 11.11 Figure by S. Lasher-Trapp.

Fig. 11.12 Photo from the University Corporation for Atmospheric Research (UCAR) via opensky.ucar.edu.

Fig. 11.13 Figure by S. Lasher-Trapp after Young (1993)

Fig. 11.14 Figure by S. Lasher-Trapp.

Fig. 11.15 Photo by Petr Brož, released via WikimediaCommons under the Creative Commons Attribution 3.0 License (CC-BY)

Fig. 11.17 Modified from Knight and Knight (1970). Published 1970 by the American Meteorological Society.

Fig. 11.16 Figure by S. Lasher-Trapp after Browning and Foote (1976) and Young (1977).

Fig. 12.1 Figure by S. Lasher-Trapp.

Fig. 12.2 Figure by G. Petty.

Fig. 12.3 Figure by S. Lasher-Trapp.

Fig. 12.4 Modified from Rasmussen and Pruppacher (1982). © American Meteorological Society. Used with permission.

Fig. 12.5 Calculations and plots courtesy of Marley Majetic and Antonio Elizondo, Dept. of Atmospheric Sciences, University of Illinois Urbana-Champaign. Data are from the CMIP5 project. We acknowledge the World Climate research Programme's Working Group on Coupled Modelling, which is responsible for CMIP, and we thank the climate modeling groups (MIROC, NCAR, NOAA GFDL) for producing and making available their model output. For CMIP the U.S. Department of Energy's Program for Climate Model Diagnosis and Intercomparison provides coordinating support and led development of software infrastructure in partnership with the Global Organization for Earth System Science Portals. Data website: pcmdi.llnl.gov/mips/cmip5/

Fig. 12.6 Modified from Sandford et al. (2017) © American Meteorological Society. Used with permission.

Fig. 12.7 Reprinted from Mallinson and Lasher-Trapp (2019). © American Meteorological Society. Used with permission.

Fig. 12.8 Figure by S. Lasher-Trapp.

Fig. 12.9 Modified from Zuidema et al. (2017). Distributed under the Creative Commons Attribution 4.0 License (CC-BY).

Fig. 13.1 Figure by S. Lasher-Trapp.

Fig. 13.2 Photos left and top right by S. Lasher-Trapp; photo bottom right from WikimediaCommons, released by user CambridgeBayWeather into the public domain.

Fig. 13.3 Modified from Wang et al. (2008). © American Meteorological Society. Used with permission.

Fig. 13.4 Figure by G. Petty based on data from Burt (2005).

Fig. 13.5 Figure by G. Petty.

Fig. 13.6 Figure by G. Petty.

Fig. 13.7 Photo courtesy of The Community Collaborative Rain, Hail, and Snow Network via cocorahs.org.

Fig. 13.8 Figure by S. Lasher-Trapp.

Fig. 13.9 Courtesy of Dr. R. Jeffrey Trapp, Dept. of Climate, Meteorology, and Atmospheric Sciences, University of Illinois Urbana-Champaign.

Fig. 13.10 Figure by S. Lasher-Trapp.

Fig. 13.11 Modified from Doswell et al. (1996). Published 1996 by the American Meteorological Society.

Bibliography

Ayala, O., B. Rosa, L.-P. Wang, and W. W. Grabowski, 2008: Effects of turbulence on the geometric collision rate of sedimenting droplets. Part 1. Results from direct numerical simulation. *New Journal of Physics*, **10 (7)**, 075 015.

Baumgardner, D., 1989: Airborne measurements for cloud microphysics. URL http://n2t.net/ark:/85065/d71z45n6.

Baumgardner, D., and Coauthors, 2017: Cloud ice properties: In situ measurement challenges. *Meteorological Monographs*, **58**, 9.1–9.23.

Beard, K. V., 1976: Terminal velocity and shape of cloud and precipitation drops aloft. *Journal of the Atmospheric Sciences*, **33 (5)**, 851–864.

Beard, K. V., J. Q. Feng, and C. Chuang, 1989: A simple perturbation model for the electrostatic shape of falling drops. *Journal of the Atmospheric Sciences*, **46 (15)**, 2404–2418.

Beard, K. V., and H. T. Ochs, 1984: Collection and coalescence efficiencies for accretion. *Journal of Geophysical Research: Atmospheres*, **89 (D5)**, 7165–7169.

Bluestein, H. B., 1992: *Synoptic-dynamic Meteorology in Midlatitudes: Observations and Theory of Weather Systems*, Vol. 2. Taylor & Francis.

Blyth, A. M., W. A. Cooper, and J. B. Jensen, 1988: A study of the source of entrained air in Montana cumuli. *Journal of the Atmospheric Sciences*, **45 (24)**, 3944–3964.

Blyth, A. M., S. G. Lasher-Trapp, and W. A. Cooper, 2005: A study of thermals in cumulus clouds. *Quarterly Journal of the Royal Meteorological Society*, **131 (607)**, 1171–1190.

Böhm, H. P., 1989: A general equation for the terminal fall speed of solid hydrometeors. *Journal of the Atmospheric Sciences*, **46 (15)**, 2419–2427.

Brandes, E. A., K. Ikeda, G. Zhang, M. Schönhuber, and R. M. Rasmussen, 2007: A statistical and physical description of hydrometeor distributions in Colorado snowstorms using a video disdrometer. *Journal of Applied Meteorology and Climatology*, **46 (5)**, 634–650.

Browning, K. A., and G. Foote, 1976: Airflow and hail growth in supercell storms and some implications for hail suppression. *Quarterly Journal of the Royal Meteorological Society*, **102 (433)**, 499–533.

Burt, S., 2005: Cloudburst upon Hendraburnick down: the Boscastle storm of 16 August 2004. *Weather*, **60 (8)**, 219–227.

Chueh, C.-C., P. K. Wang, and T. Hashino, 2018: Numerical study of motion of falling conical graupel. *Atmospheric Research*, **199**, 82–92.

Connolly, P., C. Emersic, and P. Field, 2012: A laboratory investigation into the aggregation efficiency of small ice crystals. *Atmospheric Chemistry and Physics*, **12 (4)**, 2055–2076.

Cooper, W. A., S. G. Lasher-Trapp, and A. M. Blyth, 2013: The influence of entrainment and mixing on the initial formation of rain in a warm cumulus cloud. *Journal of the Atmospheric Sciences*, **70 (6)**, 1727–1743.

Curry, J. A., and P. J. Webster, 1998: *Thermodynamics of Atmospheres and Oceans*. Elsevier.

Cziczo, D. J., L. Ladino, Y. Boose, Z. A. Kanji, P. Kupiszewski, S. Lance, S. Mertes, and H. Wex, 2017: Measurements of ice nucleating particles and ice residuals. *Meteorological Monographs*, 8.1–8.13.

DeMott, P. J., and Coauthors, 2010: Predicting global atmospheric ice nuclei distributions and their impacts on climate. *Proceedings of the National Academy of Sciences*, **107 (25)**, 11 217–11 222.

Doswell, C. A., H. E. Brooks, and R. A. Maddox, 1996: Flash flood forecasting: An ingredients-based methodology. *Weather and Forecasting*, **11 (4)**, 560–581.

Field, P. R., and Coauthors, 2017: Secondary ice production: Current state of the science and recommendations for the future. *Meteorological Monographs*, **58**, 7.1–7.20.

Fletcher, N. H., 1962: *The Physics of Rainclouds*. Cambridge University Press.

Fukuta, N., and T. Takahashi, 1999: The growth of atmospheric ice crystals: A summary of findings in vertical supercooled cloud tunnel studies. *Journal of the Atmospheric Sciences*, **56 (12)**, 1963–1979.

GIllespie, D. T., 1975: Three models for the coalescence growth of drops. *Journal of the Atmospheric Sciences*, **32**, 600–607.

Harrington, J. Y., K. Sulia, and H. Morrison, 2013: A method for adaptive habit prediction in bulk microphysical models. part i: Theoretical development. *Journal of the Atmospheric Sciences*, **70 (2)**, 349–364.

Hobbs, P. V., S. Chang, and J. D. Locatelli, 1974: The dimensions and aggregation of ice crystals in natural clouds. *Journal of Geophysical Research*, **79 (15)**, 2199–2206.

Houze, R. A., Jr., 2014: *Cloud Dynamics*. 2nd ed., Academic Press.

Huang, J., 2018: A simple accurate formula for calculating saturation vapor pressure of water and ice. *Journal of Applied Meteorology and Climatology*, **57 (6)**, 1265–1272.

Hudson, J. G., and S. S. Yum, 2001: Maritime–continental drizzle contrasts in small cumuli. *Journal of the Atmospheric Sciences*, **58 (8)**, 915–926.

Hueholt, D. M., S. E. Yuter, and M. A. Miller, 2022: Revisiting diagrams of ice growth environments. *Bulletin of the American Meteorological Society*, **103 (11)**, E2584–E2603.

Jayaweera, K., and R. Cottis, 1969: Fall velocities of plate-like and columnar ice crystals. *Quarterly Journal of the Royal Meteorological Society*, **95 (406)**, 703–709.

Jiang, Z., J. Verlinde, E. E. Clothiaux, K. Aydin, and C. Schmitt, 2019: Shapes and fall orientations of ice particle aggregates. *Journal of the Atmospheric Sciences*, **76 (7)**, 1903–1916.

Kajikawa, M., 1972: Measurement of falling velocity of individual snow crystals. *Journal of the Meteorological Society of Japan. Ser. II*, **50 (6)**, 577–584.

Kanji, Z. A., L. A. Ladino, H. Wex, Y. Yvonne Boose, M. Burkert-Kohn, D. J. Cziczo, and M. Krämer, 2017: Overview of ice nucleating particles. *Meteorological Monographs*, **58**, 1.1–1.33.

Khain, A., M. Pinsky, M. Shapiro, and A. Pokrovsky, 2001: Collision rate of small graupel and water drops. *Journal of the Atmospheric Sciences*, **58 (17)**, 2571–2595.

Kidd, C., A. Becker, G. J. Huffman, C. L. Muller, P. Joe, G. Skofronick-Jackson, and D. B. Kirschbaum, 2017: So, how much of the Earth's surface is covered by rain gauges? *Bulletin of the American Meteorological Society*, **98 (1)**, 69–78.

Kikuchi, K., T. Kameda, K. Higuchi, A. Yamashita, and Coauthors, 2013: A global classification of snow crystals, ice crystals, and solid precipitation based on observations from middle latitudes to polar regions. *Atmospheric Research*, **132**, 460–472.

Klett, J. D., and M. Davis, 1973: Theoretical collision efficiencies of cloud droplets at small Reynolds numbers. *Journal of the Atmospheric Sciences*, **30 (1)**, 107–117.

Knight, C. A., and N. C. Knight, 1970: Hailstone embryos. *Journal of the Atmospheric Sciences*, **27 (4)**, 659–666.

Knight, C. A., and N. C. Knight, 2001: Hailstorms. *Severe Convective Storms, Meteorological Monographs*, **28 (50)**, 223–248.

Korolev, A., and T. Leisner, 2020: Review of experimental studies of secondary ice production. *Atmospheric Chemistry and Physics*, **20 (20)**, 11,767–11,797.

Kumjian, M. R., and Coauthors, 2020: Gargantuan hail in Argentina. *Bulletin of the American Meteorological Society*, **101 (8)**, E1241–E1258.

Lamb, D., 2001: Rain production in convective storms. *Severe Convective Storms, Meteorological Monographs*, **28 (50)**, 299–321.

Lamb, D., and J. Verlinde, 2011: *Physics and Chemistry of Clouds*. Cambridge University Press.

Lasher-Trapp, S., S. Anderson-Bereznicki, A. Shackelford, C. H. Twohy, and J. G. Hudson, 2008: An investigation of the influence of droplet number concentration and giant aerosol particles upon supercooled large drop formation in wintertime stratiform clouds. *Journal of Applied Meteorology and Climatology*, **47 (10)**, 2659–2678.

Lasher-Trapp, S., E. Jo, L. R. Allen, B. N. Engelsen, and R. J. Trapp, 2021: Entrainment in a simulated supercell thunderstorm. Part I: The evolution of different entrainment mechanisms and their dilutive effects. *Journal of the Atmospheric Sciences*, **78 (9)**, 2725–2740.

Lasher-Trapp, S., D. C. Leon, P. J. DeMott, C. M. Villanueva-Birriel, A. V. Johnson, D. H. Moser, C. S. Tully, and W. Wu, 2016: A multisensor investigation of rime splintering in tropical maritime cumuli. *Journal of the Atmospheric Sciences*, **73 (6)**, 2547–2564.

Lasher-Trapp, S., and J. P. Stachnik, 2007: Giant and ultragiant aerosol particle variability over the eastern great lakes region. *Journal of Applied Meteorology and Climatology*, **46 (5)**, 651–659.

Lasher-Trapp, S. G., C. A. Knight, and J. M. Straka, 2001: Early radar echoes from ultragiant aerosol in a cumulus congestus: Modeling and observations. *Journal of the Atmospheric Sciences*, **58 (23)**, 3545–3562.

Magono, C., and C. W. Lee, 1966: Meteorological classification of natural snow crystals. *Journal of the Faculty of Science, Hokkaido University. Series 7, Geophysics*, **2 (4)**, 321–335.

Mallinson, H. M., and S. G. Lasher-Trapp, 2019: An investigation of hydrometeor latent cooling upon convective cold pool formation, sustainment, and properties. *Monthly Weather Review*, **147 (9)**, 3205–3222.

Marshall, J. S., and W. M. K. Palmer, 1948: The distribution of raindrops with size. *Journal of Meteorology*, **5 (4)**, 165–166.

McDonald, J. E., 1958: The physics of cloud modification. *Advances in Geophysics*, Vol. 5, Elsevier, 223–303.

Murray, B., D. O'sullivan, J. Atkinson, and M. Webb, 2012: Ice nucleation by particles immersed in supercooled cloud droplets. *Chemical Society Reviews*, **41 (19)**, 6519–6554.

Ono, A., 1970: Growth mode of ice crystals in natural clouds. *Journal of the Atmospheric Sciences*, **27 (4)**, 649–658.

Petters, M., and S. Kreidenweis, 2007: A single parameter representation of hygroscopic growth and cloud condensation nucleus activity. *Atmospheric Chemistry and Physics*, **7 (8)**, 1961–1971.

Petty, G. W., 2008: *A First Course in Atmospheric Thermodynamics*. Sundog Publishing.

Praz, C., Y.-A. Roulet, and A. Berne, 2017: Solid hydrometeor classification and riming degree estimation from pictures collected with a multi-angle snowflake camera. *Atmospheric Measurement Techniques*, **10 (4)**, 1335–1357.

Pruppacher, H. R., and J. D. Klett, 2010: *Microphysics of Clouds and Precipitation*. 2nd ed., Academic Press.

Rasmussen, R., and H. Pruppacher, 1982: A wind tunnel and theoretical study of the melting behavior of atmospheric ice particles. I: A wind tunnel study of frozen drops of radius< 500 μm. *Journal of the Atmospheric Sciences*, **39 (1)**, 152–158.

Rasmussen, R., and Coauthors, 2012: How well are we measuring snow: The NOAA/FAA/NCAR winter precipitation test bed. *Bulletin of the American Meteorological Society*, **93 (6)**, 811–829.

Rauber, R. M., and S. W. Nesbitt, 2018: *Radar Meteorology. A First Course*. John Wiley & Sons.

Rauber, R. M., and Coauthors, 2007: Rain in shallow cumulus over the ocean: The RICO campaign. *Bulletin of the American Meteorological Society*, **88 (12)**, 1912–1928.

Raupach, T. H., and Coauthors, 2021: The effects of climate change on hailstorms. *Nature Reviews Earth & Environment*, **2 (3)**, 213–226.

Rinehart, R. E., 2010: *Radar for Meteorologists*. 5th ed., Rinehart Pub.

Rogers, R. R., and M. K. Yau, 1989: *A Short Course in Cloud Physics*. Elsevier.

Ryzhkov, A. V., S. E. Giangrande, and T. J. Schuur, 2005: Rainfall estimation with a polarimetric prototype of WSR-88D. *Journal of Applied Meteorology and Climatology*, **44 (4)**, 502–515.

Sandford, C., A. Illingworth, and R. Thompson, 2017: The potential use of the linear depolarization ratio to distinguish between convective and stratiform rainfall to improve radar rain-rate estimates. *Journal of Applied Meteorology and Climatology*, **56 (11)**, 2927–2940.

Sasyo, Y., and T. Matsuo, 1980: On the statistical investigation fall velocity of snowfiakes. *Papers in Meteorology and Geophysics*, **31 (2)**, 61–79.

Skeie, R. B., T. K. Berntsen, G. Myhre, K. Tanaka, M. Kvalevåg, and C. R. Hoyle, 2011: Anthropogenic radiative forcing time series from pre-industrial times until 2010. *Atmospheric Chemistry and Physics*, **11 (22)**, 11 827–11 857.

Stith, J. L., and Coauthors, 2018: 100 years of progress in atmospheric observing systems. *Meteorological Monographs*, **59**, 2.1–2.55.

Stout, J. R., C. D. Westbrook, T. H. Stein, and M. W. McCorquodale, 2024: Stable and unstable fall motions of plate-like ice crystal analogues. *EGUsphere*, **2024**, 1–37, doi: 10.5194/egusphere-2024-319.

Sulia, K. J., Z. J. Lebo, V. M. Przybylo, and C. G. Schmitt, 2021: A new method for ice–ice aggregation in the adaptive habit model. *Journal of the Atmospheric Sciences*, **78 (1)**, 133–154.

Twomey, S., 1959: The nuclei of natural cloud formation. Part I: The chemical diffusion method and its application to atmospheric nuclei. *Geofisica Pura e Applicata*, **43**, 227–242.

Wallace, J. M., and P. V. Hobbs, 2006: *Atmospheric Science: An Introductory Survey*, Vol. 92. Elsevier.

Wang, J., B. L. Fisher, and D. B. Wolff, 2008: Estimating rain rates from tipping-bucket rain gauge measurements. *Journal of Atmospheric and Oceanic Technology*, **25 (1)**, 43–56.

Wang, P. K., and W. Ji, 2000: Collision efficiencies of ice crystals at low–intermediate Reynolds numbers colliding with supercooled cloud droplets: A numerical study. *Journal of the Atmospheric Sciences*, **57 (8)**, 1001–1009.

Wang, W., and Coauthors, 2021: Measurement report: Comparison of wintertime individual particles at ground level and above the mixed layer in urban Beijing. *Atmospheric Chemistry and Physics*, **21 (7)**, 5301–5314.

Welti, A., and Coauthors, 2020: Ship-based measurements of ice nuclei concentrations over the Arctic, Atlantic, Pacific and Southern Ocean. *Atmospheric Chemistry and Physics Discussions*, **2020**, 1–22.

Young, K. C., 1975: The evolution of drop spectra due to condensation, coalescence and breakup. *Journal of the Atmospheric Sciences*, **32 (5)**, 965–973.

Young, K. C., 1977: A numerical examination of some hail suppression concepts. *Hail: A Review of Hail Science and Hail Suppression*, G. B. Foote, and C. A. Knight, Eds., American Meteorological Society, Boston, MA, 195–214, doi: 10.1007/978-1-935704-30-0_10.

Young, K. C., 1993: *Microphysical Processes in Clouds*. Oxford University Press.

Zuidema, P., G. Torri, C. Muller, and A. Chandra, 2017: A survey of precipitation-induced atmospheric cold pools over oceans and their interactions with the larger-scale environment. *Surveys in Geophysics*, **38 (6)**, 1283–1305.

Index

Note: Unitalicized page numbers refer to topics in the text, while italicized page numbers refer to figures or tables.

www.ingramcontent.com/pod-product-compliance
Lightning Source LLC
Chambersburg PA
CBHW041258210926
51847CB00009B/62
9781944441005